Bio-Nano Filtration in Industrial Effluent Treatment

The ever-increasing number of pollutants discharged into the environment drives the search for new treatment technologies or the modification of the existing ones. In this sense, innovation in bio-nano filtration systems seems very promising, and, therefore, a book on the current advances and innovations on this topic is highly appropriate.

Bio-nano filtration is a relatively new emerging technology applied to the treatment of wastewater and other toxic compounds. In the last two decades, this technology has begun to emerge as an economically viable process to treat the great variety of recalcitrant pollutants discharged into the environment. Thus, it is speculated that the US biofiltration market will reach over $100 million by 2020. This book aims to present how innovation in bio-nano filtration can provide effective solutions to overcome the serious problem of water pollution worldwide. The removal of contaminants will be the result of the combined effects of biological oxidation, adsorption, and filtration processes.

Features:

- Describes the microbial ecology of bio-nano filtration.
- Describes the modelling of bio-nano filtration.
- Describes the design of bio-nanofillers.

Wastewater Treatment and Research

Series Editor: Maulin P. Shah

Wastewater Treatment: Molecular Tools, Techniques, and Applications
Maulin P. Shah

Advanced Oxidation Processes for Wastewater Treatment: An Innovative Approach
Maulin P. Shah, Sweta Parimita Bera, Hiral Borasiya and Gunay Yildiz Tore

Emerging Technologies in Wastewater Treatment
Maulin P. Shah

Bio-Nano Filtration in Industrial Effluent Treatment: Advanced and Innovative Approaches
Maulin P. Shah

Membrane and Membrane-Based Processes for Wastewater Treatment
Maulin P. Shah

Phycoremediation Processes in Industrial Wastewater Treatment
Maulin P. Shah

For more information, please visit: www.routledge.com/Wastewater-Treatment-and-Research/book-series/WASTEWATER

Bio-Nano Filtration in Industrial Effluent Treatment
Advanced and Innovative Approaches

Edited by Maulin P. Shah

CRC Press
Taylor & Francis Group
Boca Raton London New York

CRC Press is an imprint of the
Taylor & Francis Group, an **informa** business

Cover image: © Shutterstock

First edition published 2023
by CRC Press
6000 Broken Sound Parkway NW, Suite 300, Boca Raton, FL 33487-2742

and by CRC Press
4 Park Square, Milton Park, Abingdon, Oxon, OX14 4RN

CRC Press is an imprint of Taylor & Francis Group, LLC

© 2023 selection and editorial matter, Maulin P. Shah; individual chapters, the contributors

Reasonable efforts have been made to publish reliable data and information, but the author and publisher cannot assume responsibility for the validity of all materials or the consequences of their use. The authors and publishers have attempted to trace the copyright holders of all material reproduced in this publication and apologize to copyright holders if permission to publish in this form has not been obtained. If any copyright material has not been acknowledged please write and let us know so we may rectify in any future reprint.

Except as permitted under U.S. Copyright Law, no part of this book may be reprinted, reproduced, transmitted, or utilized in any form by any electronic, mechanical, or other means, now known or hereafter invented, including photocopying, microfilming, and recording, or in any information storage or retrieval system, without written permission from the publishers.

For permission to photocopy or use material electronically from this work, access www.copyright.com or contact the Copyright Clearance Center, Inc. (CCC), 222 Rosewood Drive, Danvers, MA 01923, 978-750-8400. For works that are not available on CCC please contact mpkbookspermissions@tandf.co.uk

Trademark notice: Product or corporate names may be trademarks or registered trademarks and are used only for identification and explanation without intent to infringe.

Library of Congress Cataloging-in-Publication Data
Names: Shah, Maulin P., editor.
Title: Bio-nano filtration in industrial effluent treatments: advanced and innovative approaches /
    edited by Maulin P. Shah.
Description: First edition. | Boca Raton: CRC Press, 2023. | Series: Wastewater treatment and research |
    Includes bibliographical references.
Identifiers: LCCN 2022048418 (print) | LCCN 2022048419 (ebook)
Subjects: LCSH: Factory and trade waste—Purification—Biological treatment. | Nanofiltration.
Classification: LCC TD897.5 .B565 2023 (print) | LCC TD897.5 (ebook) | DDC 628.44—dc23/eng/20221212 LC
    record available at https://lccn.loc.gov/2022048418LC ebook record available at https://lccn.loc.gov/2022048419

ISBN: 978-0-367-76013-7 (hbk)
ISBN: 978-0-367-76019-9 (pbk)
ISBN: 978-1-003-16514-9 (ebk)

DOI: 10.1201/9781003165149

Typeset in Times New Roman
by Apex CoVantage, LLC

Contents

Editor Bio..vii

List of Contributors ix

Chapter 1   Biofiltration-Based Methods for the Removal of Volatile Organic
Compounds and Heavy Metals in Industrial Effluent Treatment Plants....................1

*Basil K. Munjanja, Nilesh S. Wagh, Philiswa N. Nomngongo,
Jaya Lakkakula, and Nomvano Mketo*

Chapter 2   Bio-Based Iron-Nanoparticles (NPs) Incorporated Polymeric
Nanofiltration Membranes (PNCMs) for Wastewater Treatment............................17

Piyal Mondal, Anweshan, and Mihir Kumar Purkait

Chapter 3   Nanofiltration: Unravelling the Potential of the Future.............................35

Komal Agrawal and Pradeep Verma

Chapter 4   Recent Advances in Biological Remediation of Volatile Organic
Compounds (VOCs) and Heavy Metals...49

*Khyati Arora, Vanshika Kumar, Shubham Jyoti Nayak, Aditya Surya,
and Shobana Sugumar*

Chapter 5   Tailor-Made Microbial Wastewater Treatment...69

Shaon Ray Chaudhuri

Chapter 6   Systematic Industrial Wastewater Treatment by Biomaterial Fabricated
Nanofiltration Membrane...85

Puja Ghosh, Supriya Ghule, Nilesh S. Wagh, and Jaya Lakkakula

Chapter 7   Biological-Based Methods for the Removal of VOCs and
Heavy Metals 105

Amrin Pathan and Anupama Shrivastav

Chapter 8   Bacterial Biofilters for Arsenic Removal..123

Rahul Nitnavare, Joorie Bhattacharya, and Sougata Ghosh

Chapter 9   Bio-nanoparticle: Synthesis and Application in Wastewater
Treatment 137

Swatilekha Pati, Somok Banerjee, and Shaon Ray Chaudhuri

vi Contents

Chapter 10 Applications of Nanofiltration for Wastewater Treatment...157

Charles Oluwaseun Adetunji, Olugbemi T. Olaniyan, Kshitij RB Singh, Ruth Ebunoluwa Bodunrinde, Abel Inobeme, John Tsado Mathew, Ogundolie Abimbola Frank, Olalekan Akinbo, Jay Singh, Vanya Nayak, and Ravindra Pratap Singh

Chapter 11 Bio-nano Filtration as an Abatement Technique Used in the Management and Treatment of Impurities in Industrial Wastewater...171

Osikemekha Anthony Anani, Maulin P. Shah, Paul Atagamen Aidonojie, and Alex Ajeh Enuneku

Chapter 12 Removal of VOC and Heavy Metals through Microbial Approach 183

Osikemekha Anthony Anani, Abel Inobeme, Maulin P. Shah, Kenneth Kennedy Adama, and Ikenna Benedict Onyeachu

Editor Bio

Maulin P. Shah is very interested in genetic adaptation processes in bacteria, the mechanisms by which they deal with toxic substances, how they react to pollution in general, and how we can apply microbial processes in a useful way (like bacterial bioreporters). One of his major interests is the study of how bacteria evolve and adapt to use organic pollutants as novel growth substrates. Bacteria with new degradation capabilities are often selected in polluted environments and have accumulated small (mutations) and large genetic changes (transpositions, recombination, and horizontally transferred elements). His work has been focused to assess the impact of industrial pollution on microbial diversity of wastewater following cultivation-dependent and cultivation-independent analysis. He has more than 280 research publications in highly reputed national and international journals. He is Editorial Board Member for *CLEAN-Soil, Air, Water* (Wiley); *Current Pollution Reports* (Springer Nature); *Environmental Technology & Innovation* (Elsevier), the *Journal of Biotechnology & Biotechnological Equipment* (Taylor & Francis); *Current Microbiology* (Springer Nature); *Eco Toxicology (Microbial Eco Toxicology)* (Springer Nature); *Geo Microbiology* (Taylor & Francis); *Applied Water Science* (Springer Nature); *Archives of Microbiology* (Springer); *Journal of Applied Microbiology* (Wiley); *Letters in Applied Microbiology* (Wiley); *Green Technology, Resilience and Sustainability* (Springer); *Biomass Conversion & Bio Refinery* (Springer,) the *Journal of Basic Microbiology* (Wile); *Energy Nexus* (Elsevier); *e Prime* (Elsevier), *IET Nano Biotechnology* (Wiley); *Cleaner and Circular Bioeconomy* (Elsevier); and *International Microbiology* (Springer). He has edited 75 books in wastewater microbiology and industrial wastewater treatment. He has edited 20 special issues on the theme of industrial wastewater treatment and has researched for high-impact factor journals with Elsevier, Springer, Wiley, and Taylor & Francis.

Contributors

Abel Inobeme
Edo University Iyamho, Edo State, Nigeria

Aditya Surya
SRM Institute of Science and Technology, Chennai, Tamil Nadu, India

Alex Ajeh Enuneku
University of Benin, Edo State, Nigeria

Amrin Pathan
Parul University, Gujarat, India

Anupama Shrivastav
Parul University, Gujarat, India

Anweshan
Indian Institute of Technology, Assam, India

Basil K. Munjanja
University of South Africa, Johannesburg, South Africa

Charles Oluwaseun Adetunji
Edo University Iyamho, Edo State, Nigeria

Ikenna Benedict Onyeachu
Edo State University, Uzairue, Nigeria

Jay Singh
Banaras Hindu University, Varanasi, India

Jaya Lakkakula
Amity University, Mumbai, Maharashtra, India

John Tsado Mathew
University Lapai, Niger State, Nigeria

Joorie Bhattacharya
International Crops Research Institute for the Semi-Arid Tropics, India and Osmania University, Telangana, India

Kenneth Kennedy Adama
Department of Chemical Engineering, Faculty of Engineering, Edo State University, Uzairue, Edo State, Nigeria

Khyati Arora
SRM Institute of Science and Technology, Chennai, Tamil Nadu, India

Komal Agrawal
Department of Microbiology, School of Bio Engineering and Biosciences, Lovely Professional University, Phagwara, India

Kshitij RB Singh
Autonomous College, Chhattisgarh, India

Maulin P. Shah
Division of Applied & Environmental Microbiology, Enviro Technology Limited, India

Mihir Kumar Purkait
Indian Institute of Technology, Assam, India

Nilesh S. Wagh
Amity University, Mumbai, Maharashtra, India

Nomvano Mketo
University of South Africa, Johannesburg, South Africa

Olugbemi T. Olaniyan
Edo University Iyamho, Edo State, Nigeria

Paul Atagamen Aidonojie
Edo State University Uzairue, Nigeria

Philiswa N. Nomngongo
University of Johannesburg, Johannesburg, South Africa

Piyal Mondal
Indian Institute of Technology, Assam, India

Pradeep Verma
Central University of Rajasthan, Rajasthan, India

Puja Ghosh
Pondicherry University, India

Rahul Nitnavare
University of Nottingham, United Kingdom,
Rothamsted Research, United Kingdom

Ravindra Pratap Singh
Indira Gandhi National Tribal University, MP,
India

Ruth Ebunoluwa Bodunrinde
University of Technology, Akure, Nigeria

Shaon Ray Chaudhuri
Tripura University, India

Shobana Sugumar
SRM Institute of Science and Technology,
Chennai, Tamil Nadu, India

Shubham Jyoti Nayak
SRM Institute of Science and Technology,
Chennai, Tamil Nadu, India

Somok Banerjee
Tripura University, India

Sougata Ghosh
RK University, Rajkot, India

Supriya Ghule
Amity University, Mumbai, India

Swatilekha Pati
Tripura University, India

Vanshika Kumar
SRM Institute of Science and Technology,
Chennai, Tamil Nadu, India

Vanya Nayak
Indira Gandhi National Tribal University, MP,
India

1 Biofiltration-Based Methods for the Removal of Volatile Organic Compounds and Heavy Metals in Industrial Effluent Treatment Plants

Basil K. Munjanja, Nilesh S. Wagh, Philiswa N. Nomngongo, Jaya Lakkakula, and Nomvano Mketo

CONTENTS

1.1 Introduction 1
1.2 Application of Biofiltration in Removal of VOCs in Industrial Effluent 2
    1.2.1 Biofilters 4
    1.2.2 Biotrickling Filters 5
    1.2.3 Bioscrubbers 6
    1.2.4 Membrane Bioreactor 6
    1.2.5 Hybrid Processes 7
1.3 Application of Biofiltration in Removal of Heavy Metals in Industrial Effluent ... 8
    1.3.1 Biofiltration with Genetically Modified Microorganisms 9
1.4 Conclusion and Prospects 12
Bibliography 12

1.1 INTRODUCTION

Environmental pollution has become a major challenge globally due to industrial, mining, and agricultural activities. For example, in Canada some 41 mega tonnes of pollutants were released into the atmosphere, representing about 1.3 tonnes per person (Delhoménie and Heitz, 2005). As a result, many governments have enacted laws and policies to reduce pollution. For instance, the Clean Air Act of 1990 in the United States of America has been drafted to regulate hazardous air pollutants (Gallastegui et al., 2011). Some of the hazardous air pollutants include volatile organic compounds (VOCs) and heavy metals. These pollutants are a concern because of the adverse effects that they have on human health, other living organisms, and the environment.

Volatile organic compounds (VOCs) are carbon-based organic compounds that include both human-made and naturally occurring compounds. Most of these VOCs are created from anthropogenic activities of manufacturing, petrochemical industries, and emissions from vehicles (Kauselya et al., 2015). These compounds have boiling points ranging from (50–1000 °C) to (240–2600 °C) and are usually alkanes, alkenes, aromatic, and cyclic hydrocarbons. Furthermore, VOCs can be classified as hydrophilic, moderately hydrophilic, and hydrophobic based on their solubility in water (Malakar et al., 2017). If VOCs are released into the environment, they present ecological and health hazards. For instance, methane is capable of causing global warming, while other hydrocarbons can lead to photochemical smog with nitrogen oxides and produce ozone near the ground

DOI: 10.1201/9781003165149-1

(West and Fiore, 2005). The removal of VOCs from industrial effluent has been mainly carried out using conventional methods such as adsorption and thermal and catalytic oxidation, all of which are characterised by high gas flow rates and low contaminant concentrations (Song et al., 2003). Recently, vapour phase bioreactors have emerged as alternatives in the removal of VOCs from industrial effluent from painting manufacturers and petroleum refineries because they are environmentally friendly and cheap (Moe and Qi, 2005). Treatment of VOCs in industrial effluent is mainly carried out using biofilters, bioscrubbers, or biotrickling filters. The basic pollutant removal mechanisms of all these bioreactors are almost the same although differences exist in the microorganisms used, packing media, and pollutant concentration levels (Mudliar et al., 2010). With regard to performance, conventional biofilters are superior in the removal of hydrophobic contaminants, while biotrickling filters offer enhanced removal of hydrophilic VOCs (Gospodarek et al., 2019).

Heavy metals are elements with atomic weights between 63.5 and 200.6, and (Rahman and Singh, 2020) most of the point sources of heavy metals are wastewater from mining, metal processing, tanneries, and pharmaceuticals, among others. These metals are transported by runoff water and contaminate water downstream from the industrial site (Srivastava and Majumder, 2008). If heavy metals enter water bodies, they may accumulate in the tissues of living organisms leading to various diseases such as cancer (Sall et al., 2020), and they can also destroy aquatic life (Meena et al., 2021). Additionally, high levels of As, Cd, Co, Cr, Cr, Fe, Hg, Ni, and Pb are known to reduce the growth rate of a plant as most affect the rate of photosynthesis and some kill the organelles (mitochondria) that are key in generating energy needed for growth (Edokpayi et al., 2017). For this reason, it is of paramount importance to monitor industrial effluent to ensure maximum removal of heavy metals from the environment. Conventional methods that have been used to treat heavy metals from industrial wastewater include ion exchange, chemical precipitation, and adsorption, among others. However, these techniques have limitations such as high costs, high energy consumption, and generation of secondary pollution due to the use of chemicals. Biofiltration is one of the most promising techniques for the treatment of heavy metals in contaminated industrial wastewater because it has low operating costs and has high removal efficiency at lower levels of contamination (Srivastava and Majumder, 2008). The principle of operation of these bioreactors, their advantages, and disadvantages are summarised in **Table 1.1**.

1.2 APPLICATION OF BIOFILTRATION IN REMOVAL OF VOCS IN INDUSTRIAL EFFLUENT

The major operational parameters for all types of bioreactors are packing material, temperature, pH, pressure drop, bed porosity, moisture content, and empty bed residence time (Kauselya et al., 2015). The commonly used materials in biofiltration beds are classified as either organic (peat, soil, compost, and wood chips) or inert (plastic, perlite, ceramics, zeolite, activated carbon; Mudliar et al., 2010, 2008; Ortiz et al., 2003; Fu et al., 2011). Organic materials have a high microbial load and nutrients, while inert materials require the addition of mineral nutrients and have uniform structure and size (Ortiz et al., 2003). Four differently packed biofilters were compared for the suitability of coconut fibre, digested sludge compost from a wastewater treatment plant, peat, and pine leaves as packing material for biofiltration of toluene (Maestre et al., 2007). The pH of the medium in the bioreactor affected the biodegradation efficiency, hence biofiltration required optimum pH to proceed. The ideal pH for bioreactors, especially the traditional biofilters and biotrickling filters, is between 7.5 and 8 (Kauselya et al., 2015). Temperature is another critical parameter in the operation of biofilters. The biofiltration process is divided into mesophilic and thermophilic based on temperature requirements, where the former describes reactors that grow best in moderate temperatures of between 20 and 45°C, while the latter describes reactors that thrive in high temperatures above 45°C (Fulazzaky et al., 2014). On the other hand, fungi are able to grow under both neutral and acidic conditions and are metabolically active over a wide pH range between 2 and 7 (Aydin et al., 2012). Regarding moisture, excess water leads to the development of

TABLE 1.1
Advantages and Disadvantages of the Different Bioreactors

Bioreactor	Principle of Operation	Advantage	Disadvantage	Reference
Biofilter	• Contaminated gases flow through porous material and oxidation occurs after diffusion into biofilm.	• High gas phase surface area. • Low capital and operation costs. • Low pressure drops. • Does not produce waste.	• Clogging of medium due to particulate medium. • Large area is required. • Problem in treating acidic gases. Packing has a limited life.	(Malakar et al., 2017)
Biotrickling Filter	• Gas flow through fixed bed that is continuously irrigated with aqueous containing nutrients. Pollutants absorbed by aqueous film that surrounds biofilm, and degradation takes place.	• More consistent due to better control of pressure drop, pH, and nutrient concentration. • Less operating and capital constraints. • Capable of degrading acid degradation products of VOCs.	• Low surface area for mass transfer. • Accumulation of excess biomass in filter bed. • Complex operation.	(López et al., 2013; Malakar et al., 2017; Delhoménie and Heitz, 2005)
Bioscrubber	• Consists of absorption tower and bioreactor. • Gaseous contaminants are converted to liquid phase in the absorption unit. • Separated liquid phase is pumped towards aerated bioreactor.	• Better control of reaction conditions. • Low pressure drops. • Hot polluted air is treatable.	• Difficult start up procedures. • Produces wastewater. • Can be complicated to operate.	(Malakar et al., 2017; Delhoménie and Heitz, 2005)
Membrane Bioreactor	• Purification is based on separation of polluted gas stream passing through membrane and contact with microorganisms on the other side of membrane.	• Presence of a discrete water phase allows optimal moistening of biomass and removal of degradation products. • Gas and liquid flow can be varied without problems of flooding, loading, or foaming.	• Biofouling.	(Kumar et al., 2008)
Hybrid Processes	• Consist of photo oxidation technique prior to biofiltration.	• Improved removal efficiency.	• Harmful products may be released during UV photo oxidation. • Higher operational costs of photo catalysts.	(Cheng et al., 2016b)

anaerobic zones within the bed and the reduction of the specific surface available for gas/biofilm exchanges, leading to bed compaction and an increased pressure drop (Aydin et al., 2012). Empty bed residence time (EBRT) is a critical parameter in the design and operation of biofilters, which is defined as the quotient of total filter bed volume divided by the air flow rate (Malakar et al., 2017). Shorter EBRT reduces the removal efficiency, the elimination capacity of the process due to limitations in mass transfer, and biological reactions (Jiang et al., 2012). Pressure drop is a key parameter in bioreactor design, and it depends on airflow rate, moisture content, type of media,

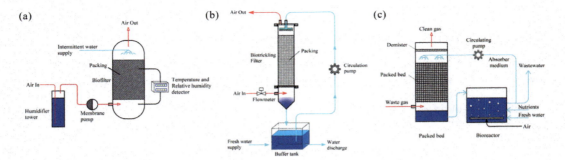

FIGURE 1.1 Schematic diagram of Biofilter, BioTrickling filter and Bioscruber (Figure 1.4 from Y.-C. Wang, M.-F. Han, T.-P. Jia, et al., Emissions, measurement, and control of odor in livestock farms: A review, Science of the Total Environment (2021), https://doi.org/10.1016/j.scitotenv.2021.145735)

and media pore size (Kauselya et al., 2015). Microorganisms are a critical component of the bioreactors because they degrade the compounds. The major advantage of fungal biofilters is their ability to degrade substrates under extreme environmental conditions of pH, low water content, and low nutrient concentration (Kennes and Veiga, 2004). In addition, fungal biofilters are capable of increasing availability of nutrients by colonising unoccupied space with their aerial hyphae and penetrating the solid support. When fungal and bacterial biofilters were compared under identical operating conditions, the former supported lower elimination capacities than the latter, which exhibited a final pressure drop of 60%, higher than that of the bacterial biofilter due to mycelial growth (Estrada et al., 2013). However, these biofilters also have their own disadvantages, which include clogging challenges as a result of high head losses being experienced in filamentous fungi (Kennes and Veiga, 2004). This problem was solved by adding mites to the filter bed to feed on the fungal mycelium and to use of chemical, air sparging, and backwashing methods (Mendoza et al., 2004). The schematic diagram of biofilter, biotrickling filter, and bioscrubber is shown in Figure 1.1. Furthermore, the application of bacteria and fungi in the removal of VOCs by various biofilters is illustrated in the following subsections.

1.2.1 Biofilters

Biofilters are reactors in which a humid polluted air stream is passed through a porous packed bed (generally a peat on compost mixture) and is degraded by microorganisms into carbon dioxide, water, and additional biomass (Malakar et al., 2017). An important feature of this type of bioreactor is that, prior to entering the biofilter, the gas stream should be humidified. The major designs are open design biofilters with ascending gas flows installed outside the VOC generation units. However, close design biofilters with either ascending or descending gas flows installed in closed rooms are also used.

Studies have been carried out to investigate the performance of biofilters under short-term shutdown conditions, long-term shut down, or daily discontinuous operation, and to challenge the microorganisms (Ralebitso-Senior et al., 2012). The potential of two biofilters for removing and degrading VOCs associated with reformulated paint even in continuous and discontinuous loading scenarios was evaluated (Moe and Qi, 2005). It was reported that both strategies achieved high contaminant removal efficiency (> 99%) at a target contaminant mass loading rate of 80.3 g m^{-3} h^{-1} and an empty bed residence time of 59 sec. However, the biofilter subjected to intermittent loading conditions at start up took considerably longer time (27 days) to reach high performance. In another study, the performance of a compost biofilter without external microbial inoculation was investigated for the removal of VOC mixtures from a polluted airstream under intermittent loading conditions (Zamir et al., 2012). The biofilter was operated for 10 h per day at different empty bed residence times, and inlet concentrations of hexane and toluene were used. Steady-state removal efficiency profiles reaching more than 90% for both pollutants were observed after 44 days of operation. In a related study, a polyurethane biofilter inoculated with *Rhodococcus* sp. EH831 was evaluated under different transient loading conditions such as shutdown, intermittent, and fluctuation in the removal of benzene and toluene (Lee et al., 2009). A comparison of the removal capacity under continuous and intermittent loading revealed that the constant and periodic loading (8 h on/16 h off per day) and a 2-day shutdown did not significantly influence the performance of the biofilter. However, the removal of benzene and toluene was relatively unstable and lower under intermittent loading during the initial week.

1.2.2 Biotrickling Filters

A biotrickling filter is another environmentally friendly and cost-effective air pollution control strategy. It operates by adding a liquid stream that trickles through the biological media to provide nutrients and buffering (Engleman, 2010). The microorganisms grow on the packing material of the bioreactor as biofilm, and the pollutant to be treated is initially absorbed by the aqueous film that surrounds the biofilm, and then the biodegradation takes place within the biofilm (Wu et al., 2018). The packing material is usually chemically inert material such as polyurethane foam, lava rock, pall rings, or coal material among others that can be arranged either randomly or in an orderly manner (Mathur et al., 2006). It has to be noted that water must be sprayed onto the packing material occasionally to maintain high moisture content, and the leachate can be recirculated to the biofilters to avoid wastewater generation (Wu et al., 2018). A distinct advantage of biotrickling filters over biofilters and bioscrubbers is their ability to treat recalcitrant VOCs and acidic or alkaline compounds.

Mass transfer occurs in biotrickling filters, and it consists of gas-liquid and liquid-biofilm transfer between the pollutants and the microorganisms. In addition, the critical factors affecting mass transfer include liquid flow and the cyclic frequency of the trickling liquid. Biotrickling filters are more adapted to remove water-soluble VOCs. However, these filters can also be used to remove hydrophobic compounds by adding surfactants that make these compounds more bioavailable by reducing mass transfer between the gas phase and liquid (or solid) phase (Avalos Ramirez et al., 2012). The anionic surfactant, sodium dodecyl sulphate (SDS), enhanced n-hexane removal in two biotrickling filters with negligible toxicity to microorganisms (Cheng et al., 2016a). The effect of a nonionic surfactant, Triton X-100 on styrene removal was examined by using two biotrickling filters (Song et al., 2012). It was reported that the average removal efficiency improved in the presence of the surfactant. In addition, the use of a surfactant controlled the excessive biomass accumulation.

Several publications have reported the use of the biotrickling filter to remove various VOCs from gaseous emissions of different industries. For example, the biotrickling filter was scaled up and operated in-situ for the treatment of gaseous emissions from a paint and varnish industrial plant (Bastos et al., 2003). A microbial culture capable of degrading the target compounds was enriched, and a laboratory vapour phase bioreactor was established. Removal efficiencies higher than 90% were obtained when exposed to organic loads of approximately 50 g $h^{-1} - m^{-3}$ of reactor. However, challenges were faced with mass production of microbial cultures. To investigate the performance of a one-stage biotrickling filter for the removal of methanol and alpha pinene, autotrophic bacteria, *Candida sp, Rhodococcus sp,* and *Ophiostoma sp* were used (López et al., 2013). The authors reported that the elimination capacity improved with operation time (EBRT 26 s), reaching 302 and 175 $gm^{-3}h^{-1}$, respectively. In addition, some of the inoculated bacteria were still detected in the biotrickling filter after long term operation. Recently, a biotrickling filter system packed with polyurethane foam was used to investigate treatment of polluted air containing cyclohexane (Salamanca et al., 2017). The obtained removal efficiencies and elimination capacities were 80–99% and 5.4 g – 38 g C m^{-3} h^{-1}, respectively. The influence of influent loading rate on removal efficiency and elimination capacity was investigated in another study by varying both the influent concentration,\ and the apparent air residence time (Zilli and Nicolella, 2012). It was reported that pollutant removal was completed at influent concentrations of up to 1.25, 0.75, and 0.20 g.m^{-3} and apparent air residence times of 60, 30, and 15 s. Lastly, the potential of a biotrickling filter inoculated with fungi to remove benzene was explored (Aly Hassan and Sorial, 2010). It is remarkable to note that the loading rate used was relatively high compared to most biofilter operations, and the maximum elimination capacity was superior to the one obtained from literature. However, when the pH was changed, there was a lack in the acclimation period (Aly Hassan and Sorial, 2010).

1.2.3 BIOSCRUBBERS

A bioscrubber consists of an absorption unit where input gaseous contaminants are transferred to the liquid phase and the bioreactor unit, where the microbial strains suspended in the aqueous phase in nutrient solution are found (Mudliar et al., 2010). In the absorption tower, diffusion of gas-phase contaminants into aqueous solution takes place by counter-current gas-liquid flow through inert packing, and the washed gas is emitted from the top, while the contaminated liquid is pumped through an aerated bioreactor (Berenjian et al., 2012). The microorganism or activated sludge in the bioreactor is suspended in nutrient-rich media and residence time for treatment varies according to concentration and type of VOCs in the feed from 20 to 40 days (Delhoménie and Heitz, 2005). After complete degradation of the contaminants, the medium is filtered, and biomass is left to the sediments with portions being recycled through the bioscrubbing process again.

The potential of a bioscrubber to remove methane from pig farm wastewater was explored by first optimising the influential filtering parameters (Liu et al., 2019). High removal efficiency of 25% was achieved with improved methanotrophic activity due to the extra nutrient support from the wastewater. In a study to compare the bioscrubber and the biotrickling filter in the removal of ethyl acetate, it was reported that the former was easier to operate and control than the latter, while the enhancement of oxygen mass transfer could potentially increase its performance by up to three times (Koutinas et al., 2005). Another study was carried out to explore the effects of different bioscrubbers on styrene waste gas removal in a bioscrubber (Liu et al., 2020). The highest removal rate of styrene was 88.52% which was obtained with hollow-sided spheres.

1.2.4 MEMBRANE BIOREACTOR

Membrane bioreactors (see Figure 1.2) are suitable for removal of hydrophobic contaminants from waste air because they provide a large gas-liquid interface and favourable mass transfer conditions (Barbusiński et al., 2020). Their operating principle is based on the diffusion of the pollutants across the gas-liquid interface through the membrane and to the film attached to the other side of the membrane where the pollutants are degraded. Membranes can be classified as dense phase, porous, or composite, depending on the material matrix (Barbusiński et al., 2020). In addition, their performance is determined by the membrane material, which can be either polydimethylsiloxane (PDMS), polypropylene (PP), polyethylene (PP), or polyvinylidene difluoride (PVDF) (Kumar et al., 2008).

A study was conducted to compare the membrane bioreactor, biofilter, and biotrickling filter in the removal of methyl-mercaptan, toluene, alpha-pinene, and hexane from wastewater treatment plants, and the results showed that the membrane bioreactor was very efficient as it completely removed methyl-mercaptan and toluene (Lebrero et al., 2014). However, when it came to alpha-pinene, the membrane bioreactor suffered from excessive bioaccumulation. The potential of a two-stage membrane bioreactor in the removal of VOCs has also been investigated in three industrial

Biofiltration-Based Methods

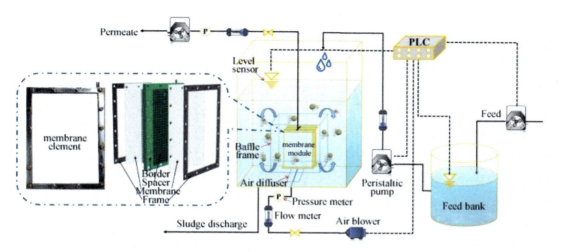

FIGURE 1.2 Schematic of a laboratory-scale MBR system (Figure 1.1 from J. Teng et al./Chemosphere 248 (2020) 126013).

plants, and the removal efficiencies were between 89 and 98% (Rolewicz-Kalinska et al., 2021). The advantage of using a membrane as the second stage of biofiltration was enhancement of the removal of insoluble compounds, especially at low concentration levels.

1.2.5 Hybrid Processes

One major shortcoming of biofiltration processes is that they may lack consistency and are slow to adapt to changes in VOC concentrations in the gas stream (Wang et al., 2009). For this reason, hybrid systems coupling advanced oxidation processes (photolytic and photocatalytic) and biofiltration processes have been introduced to treat hydrophobic and recalcitrant VOCs, as shown in Figure 1.3. Recently, photocatalytic oxidation, which utilises semiconductors like TiO_2, ZnO, or $FeTiO_3$ to carry out the photo-induced redox process, has emerged as the method of choice because of low energy consumption, moderate operating cost, and environmental safety (Babbitt et al., 2009). The potential of a hybrid system consisting of a UV/TiO_2-In photooxidation coupled with a biofiltration process was evaluated for the degradation of ethylbenzene vapours. The greatest degradation rate of 0.414 ng m^{-2} min^{-1} was obtained with TiO_2 – In1%/365 nm photocatalytic system (Hinojosa-Reyes et al., 2012). It is worth in noting that the hybrid system provided 36% additional removal compared to the individual degradation of the photocatalytic treatment and the control biofilter. In another study, photocatalytic activity of N-TiO_2/zeolite coupled to a biofilter was utilised to treat gases containing toluene, and high removal efficiencies (> 90%) were reported (Wei et al., 2010).

UV photooxidation is another technique that can be coupled to bioreactors. Several studies have been reported to combine UV and biofilters for the treatment of VOCs. The removal efficiencies obtained with a combined ultraviolet-biotrickling filter were much higher (57–98%) than those of a single biotrickling filter (25–76%) under the same filtration conditions (Cheng et al., 2011). In a related study, the performances of a biotrickling filter and a combined UV-biotrickling filter were compared for styrene removal (Runye et al., 2015). The presence of UV enhanced the performance of the biotrickling filter because styrene was converted into compounds that are more easily degraded for subsequent biological treatment. In addition, the combined system exhibited better resistance to shock loads and intermittent operations. A hollow fibre membrane reactor coupled to ozone used to remove xylene showed higher performance and stability and avoided the formation of additional biomass (Wang et al., 2013). However, the variation in outlet concentration compared with the hollow fibre membrane reactor was very small.

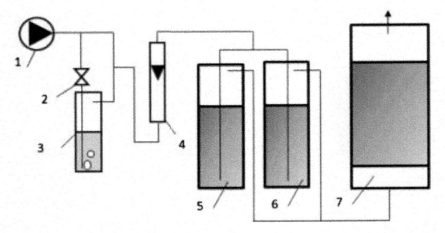

FIGURE 1.3 Schematic diagram of laboratory scale hybrid bioreactors and biofilter: (1 – air pump; 2 – air valve; 3 – tank with contaminants; 4 – rotameter; 5 and 6 – hybrid bioreactors; 7 – biofilter; Figure 1.1 from A. Tabernacka et al./Journal of Hazardous Materials 264 (2014) 363–369).

1.3 APPLICATION OF BIOFILTRATION IN REMOVAL OF HEAVY METALS IN INDUSTRIAL EFFLUENT

Heavy metals are removed by passing them through a moist, biologically active film containing biological waste material that can use exchange, surface adsorption, or complexation (Gallardo-Rodríguez et al., 2019). Microorganisms such as bacteria and fungi present in the active layer of the biofilter oxidise or reduce these heavy metals into less toxic species (Meena et al., 2021). These microorganisms are used because they can utilise their substrate rapidly, and their small size produces a high surface area to volume ratio for rapid pollutant uptake (Srivastava and Majumder, 2008). In addition, these microorganisms can withstand unconducive environments, hence their use.

Several studies have been performed for the removal of different metals from industrial wastewater. The removal of nickel (II) from mining and electroplating effluents was conducted by using an anaerobic, zeolite packed bioreactor inoculated with heterotrophic consortium as bioadsorbents (Parades-Aguilar et al., 2021). The removal rate of nickel (II) was 99% for 250–500 ppm tested effluents with an efficient alkalinity rate (0.5–0.7) and high production of biogas. Moreover, *Fervidobacterium* and *Geobacter* genus were absent at the end of the bioreactor treatment, suggesting that they play a key role in the beginning of nickel (II) removal during an anaerobic treatment. In another study, arsenic was removed from ground water using bacteria *Gallionella ferruginea* and *Leptothrix ochracea* (Katsoyiannis and Zouboulis, 2004). During the removal process, iron oxides were deposited into the filter medium, along with the microorganisms, which offered an optimum environment for arsenic removal. Excellent removal rates of up to 95% were obtained. The potential of manganese removal was explored using microbial cell immobilised biochar with *Streptomyces violarus* strain SBP1 (Youngwilai et al., 2020). Using autochthonous iron and manganese oxidising bacteria (IMOB), the feasibility of adopting an intermittent operation was investigated with biofilters in a ground water treatment plant (Zeng et al., 2019). The results showed that the intermittent operation had very little effect on the cultivation of the biofilter because dissolved oxygen would be gradually exhausted, making the filter layer anaerobic, thus inhibiting the growth and reproduction of the bacteria. The immobilised biochar contained a higher proportion of oxygen removing groups, leading to better manganese removal rates of up to 78%. Recently, a living biomass biofilter using natural fibres (*Furcraea andina*) as bacterial support was used to adsorb Pb in wastewaters (Gallardo-Rodríguez et al., 2019). A distinct advantage of the method compared to physicochemical methods was that the use of live biofilms simplified the process since the microorganisms were

Biofiltration-Based Methods

produced in situ. A lab scale biofilter packed with a mixture of coal and compost was used to remove copper (II) from aqueous solution (Majumder et al., 2015). The bacterial strain *Acinetobacter guillouiae* was used to obtain a removal efficiency of 97.5% for an inlet concentration of 20 mg. L^{-1}.

Hexavalent chromium is very toxic and therefore requires efficient treatment to remove it from industrial effluent. Several studies have been carried out on its removal using biofiltration as either batch reactors, fixed-film (Chirwa and Wang, 1997), or continuous-flow bioreactors (Shen and Wang, 1995). Another study was also performed by using *Arthrobacter Cr-47* to form Cr(VI), reducing biofilms in gravel packed bed reactors (Córdoba et al., 2008). It is worth noting that *Arthrobacter* sp. packed bed film reactors achieved Cr (VI) reduction rates comparable to other aerobic and anaerobic fixed film bioreactors previously reported. Another microorganism that has been used in the removal of Cr(VI) from wastewater is *Sphaerotilus natans* (Caravelli and Zaritzky, 2009). It was reported that the *Sphaerotilus natans* exhibited optimal growth in batch and continuous systems exposed to Cr (VI) concentrations as high as 80 mg. L^{-1}. For instance, the performance of a laboratory cultivated microalgae (*Chlorococcum sp.*) and a commercially available granular activated carbon for the removal of copper and chromium from wastewater was evaluated (Jacinto et al., 2009). The performance of the algae-based biofilter was better than the granular activated carbon.

The first study to utilise a trickling filter to biologically reduce Cr (VI) was operated as a sequencing batch reactor with recirculation, and it gave very high reduction rates of 530 g m^{-2} d^{-1} (Dermou et al., 2005). In addition, the trickling filters proved to be very promising devices that provided supporting material for consistent biofilm structure development, while minimising operating cost. It is also worth noting that the sequencing batch reactor operation with recirculation proved to be very effective, since it ensured even wetting of the filter and distribution of the precipitates all over the filter volume. The same research group reported higher reduction rates of 1117 g m^{-2} d^{-1} in the reduction of (VI) using mixed cultures originating from industrial sludge under continuous operation with recirculation in a trickling filter (Dermou and Vayenas, 2008). Hence, this system can be used for detoxification of chromate wastewater effluents. Lastly, a kinetic study of biological Cr (VI) reduction in trickling filters operated in sequencing batch reactor mode with recirculation using plastic and calcitic gravel media revealed that the former was more efficient than the latter because it was porous (Dermou and Vayenas, 2007). Moreover, the latter has higher specific surface area but lower working wastewater volume.

The performance of a membrane bioreactor has also been assessed for the removal of copper, lead, nickel, and zinc from wastewater (Katsou et al., 2011). The average removal efficiencies obtained were 80% for copper, 98% for lead, 77% for zinc, and 50% for nickel, and these were attributed to the addition of vermiculite. However, the addition of the vermiculite resulted in membrane fouling and mixed liquor suspended solids. Recently, a membrane bioreactor combined with granular activated carbon has been utilised for the removal of nickel, lead, arsenic, and zinc (Çalik et al., 2020). Very high removal efficiencies of over 99% were obtained for all heavy metals. In addition, coupling granular activated carbon to a membrane bioreactor enhanced the performance of the system.

1.3.1 Biofiltration with Genetically Modified Microorganisms

Biofilter efficiency is also improved by genetic modification of microorganisms so that they can have high affinity for the target analyte. This leads to an enhanced removal of contaminants at a lower cost and increased ability to withstand environmental conditions (Srivastava and Majumder, 2008).

A study was conducted using genetically engineered *Escherichia coli* JM109, which simultaneously expressed a nickel transport system and metallothionein to remove and recover Ni^{2+} from aqueous solution. A more than sixfold binding capacity was obtained by genetically engineered *E.coli* cells (9.89 mg/g) compared with original host *E.coli* cells (1.52 mg/g) (Deng et al., 2003). Furthermore, genetically engineered *E. coli* did not require an extra nutrient for Ni^{2+} bioaccumulation.

Table 1.2 summarises the various applications of biofiltration in the removal of heavy metals and VOCs in treatment of industrial effluent. It is evident that in most of the studies, high removal

TABLE 1.2
Application of Biofiltration in Removal of VOCs and Heavy Metals

Microorganism	Pollutant	Matrix	Bioreactor	Elimination Capacity	Removal Efficiency (%)	Removal Duration (Day)	Reduction rate	Reference
Gallionella ferruginea *Leptothrix ochracea*	Arsenic	Drinking water	NS	NS	≥ 95	NS	NS	(Katsoyiannis and Zouboulis, 2004)
NS	Arsenic, nickel, zinc, vanadium	Petrochemical effluent	Membrane bioreactor	NS	40–70	NS	NS	(Malamis et al., 2015)
Iron and manganese oxidising bacteria	Iron, manganese	Groundwater	Pilot biofilter	NS	95	30–50 in continuous mode	0.8–1.0 mg/L	(Zeng et al., 2019)
Pseudomonas bacteria	Lead		Living biomass biofilter	0.16–0.76 g m^{-3} h^{-1}	90	NS	48.75 mg/g	(Gallardo-Rodríguez et al., 2019)
Mycobacterium vaccae *Rhodococcus erythropolis* *Nocardia fluminae*	Ethylene	Artificially contaminated airstream	Zeolite packed biofilter	4.44 g m^{-3} h^{-1}	≥ 96	15	NS	(Fu et al., 2011)
NS	Ethylbenzene	NS	Laboratory scale bioreactor	120 g m^{-3} h^{-1}	> 95	NS	NS	(Álvarez-Hornos et al., 2008)
Pseudomonas veronii PT *Pseudomonas fluorescens* ZW	Alpha-pinene	NS	Ultraviolet-biotrickling filter	94.2 mg m^{-3} h^{-1}	> 90	13	NS	(Cheng et al., 2011)
Candida boidinii *Rhodococcus erythropolis* *Ophiostoma stenoceras*	Methanol Alpha pinene	Effluent from wood industry	One stage single reactor	175–302	> 99	150	NS	(López et al., 2013)
NS	NS	Waste gases from paint emissions	Biotrickling filter	12 g m^{-3} h^{-1}	> 90	18	NS	(Bastos et al., 2003)
Fungi and bacteria	Methyl ethyl ketone Toluene Ethylbenzene	Waste gases from paint manufacture	Polyurethane packed biofilter	NS	> 99	32	NS	(Moe and Qi, 2005)
Paecilomyces varioitt	Propanal Toluene hexanol	Wastewater	Vermiculite packed biofilter	27.7–45.6 g C m^{-3} h^{-1}	> 90	11	NS	(Estrada et al., 2013)

Pseudomonas veronii PT *Pseudomonas fluorescens* ZW	Alpha pinene	NS	Lab scale anaerobic biofilter	50 g m^{-3} h^{-1}	94	10	0.053–0.081 mg/hr	(Jiang et al., 2012)
Acidovorax sp. CHX100	Cyclohexane	NS	Biotrickling filter	5.4–38 g C m^{-3} h^{-1}	80–99	18	NS	(Salamanca et al., 2017)
NS	Nickel, lead, arsenic, zinc	Wastewater	Membrane bioreactor	NS	> 99	NS	NS	(Çalik et al., 2020)
Arthrobacter viscosus	Chromium (VI)	NS	Packed bioreactor with GAC	NS	> 99	30	NS	(Quintelas et al., 2009)
Acinetobacter guillouiae	Copper (II)		Biofilter with compost and coal	0.57–0.61 mg L^{-1} min^{-1}	> 97.5	25	NS	(Majumder et al., 2015)
Arthrobacter CR47	Chromate (VI)	ND	Biofilter packed with gravel	0.46–0.79 mg. L^{-1} h^{-1}	100	1	NS	(Córdoba et al., 2008)
Klebsiella sp. 3S1	Lead (II)	wastewater	Fixed bed column of porous ceramic Raschig rings	NS	>99	28	67–153 mg/g	(Muñoz et al., 2016)
Sphaerotilus natans	Cr (VI)	wastewaters	Batch and continuous reactors	NS	50	1–2	NS	(Caravelli and Zaritzky, 2009)

HRT – hydraulic retention time, EBRT – empty bed residence time, GAC – granular activated carbon, NS – not specified.

efficiencies are obtained. Hence biofiltration is proving to be a suitable alternative in treatment of industrial effluent.

1.4 CONCLUSION AND PROSPECTS

The use of microorganisms, plants, and fungi in the removal of pollutants like volatile organic compounds and heavy metals can largely solve the problems of air and water pollution, respectively, since they are readily available. Additionally, the problem that comes with secondary waste because of the chemicals use in wastewater treatment is eliminated. Several studies have been conducted using different types of biofilters, and remarkable values of removal efficiency have been obtained for both heavy metals and VOCs. However, more studies have been carried out on applications of biofiltration in removal of VOCs than heavy metals. Hence more investigations need to be performed on the application of biofiltration in removal of heavy metals. In addition, more research still needs to be carried out on newer types of bioreactors that have not been explored and that possess superior properties to the traditional ones.

Furthermore, there is a lot of research that need to be conducted on the different species of plants, fungus, and bacteria since the adsorption capacity has been seen to differ from one species to another. The species that have high metal binding capacities need to be protected and frequently used in the cleaning up of wastewater. Additionally, more investigations need to be conducted on the genetic engineering of these organisms to improve the extraction efficiencies for all species and make these species efficiently adsorb to different compounds.

BIBLIOGRAPHY

Álvarez-Hornos, F. J., Gabaldón, C., Martínez-Soria, V., Martín, M., Marzal, P. & Penya-Roja, J. M. 2008. Biofiltration of ethylbenzene vapours: Influence of the packing material. *Bioresource Technology*, 99, 269–276.

Aly Hassan, A. & Sorial, G. A. 2010. Removal of benzene under acidic conditions in a controlled Trickle Bed Air Biofilter. *Journal of Hazardous Materials*, 184, 345–349.

Avalos Ramirez, A., Jones, J. P. & Heitz, M. 2012. Methane treatment in biotrickling filters packed with inert materials in presence of a non-ionic surfactant. *Journal of Chemical Technology & Biotechnology*, 87, 848–853.

Aydin, B., Natalie, C. & Hoda Jafarizadeh, M. 2012. Volatile organic compounds removal methods: A review. *American Journal of Biochemistry and Biotechnology*, 8, 220–229.

Babbitt, C. W., Stokke, J. M., Mazyck, D. W. & Lindner, A. S. 2009. Design-based life cycle assessment of hazardous air pollutant control options at pulp and paper mills: A comparison of thermal oxidation to photocatalytic oxidation and biofiltration. *Journal of Chemical Technology & Biotechnology*, 84, 725–737.

Barbusiński, K., Urbaniec, K., Kasperczyk, D. & Thomas, M. 2020. Biofilters versus bioscrubbers and biotrickling filters: State-of-the-art biological air treatment. In: Soreanu, G. & Dumont, É. (eds.) *From Biofiltration to Promising Options in Gaseous Fluxes Biotreatment*. Elsevier.

Bastos, F. S. C., Castro, P. M. L. & Ferreira Jorge, R. 2003. Biological treatment of a contaminated gaseous emission from a paint and varnish plant—from laboratory studies to pilot-scale operation. *Journal of Chemical Technology & Biotechnology*, 78, 1201–1207.

Berenjian, A., Chan, N. & Malmiri, H. J. 2012. Volatile organic compounds removal methods: A review. *American Journal of Biochemistry and Biotechnology*, 8, 220–229.

Çalik, S., Sözüdoğru, O., Massara, T. M., Yilmaz, A. E., Bakirdere, S., Katsou, E. & Komesli, O. T. 2020. Removal of heavy metals by a membrane bioreactor combined with activated carbon. *Analytical Letters*, 1–11.

Caravelli, A. H. & Zaritzky, N. E. 2009. About the performance of *Sphaerotilus natans* to reduce hexavalent chromium in batch and continuous reactors. *Journal of Hazardous Materials*, 168, 1346–1358.

Cheng, Y., He, H., Yang, C., Yan, Z., Zeng, G. & Qian, H. 2016a. Effects of anionic surfactant on n-hexane removal in biofilters. *Chemosphere*, 150, 248–253.

Biofiltration-Based Methods

Cheng, Z., Zhang, L., Chen, J., Yu, J., Gao, Z. & Jiang, Y. 2011. Treatment of gaseous alpha-pinene by a combined system containing photo oxidation and aerobic biotrickling filtration. *Journal of Hazardous Materials*, 192, 1650–1658.

Chirwa, E. M. N. & Wang, Y. 1997. Hexavalent chromium reduction by Bacillus sp. in a packed-bed bioreactor. *Environmental Science & Technology*, 31, 1446–1451.

Córdoba, A., Vargas, P. & Dussan, J. 2008. Chromate reduction by *Arthrobacter* CR47 in biofilm packed bed reactors. *Journal of Hazardous Materials*, 151, 274–279.

Delhoménie, M. & Heitz, M. 2005. Biofiltration of air: A review. *Critical Reviews in Biotechnology*, 25, 53–72.

Deng, X., Li, Q. B., Lu, Y. H., Sun, D. H., Huang, Y. L. & Chen, X. R. 2003. Bioaccumulation of nickel from aqueous solutions by genetically engineered Escherichia coli. *Water Research*, 37, 2505–2511.

Dermou, E. & Vayenas, D. V. 2007. A kinetic study of biological Cr(VI) reduction in trickling filters with different filter media types. *Journal of Hazardous Materials*, 145, 256–262.

Dermou, E. & Vayenas, D. V. 2008. Biological Cr(VI) reduction in a trickling filter under continuous operation with recirculation. *Journal of Chemical Technology & Biotechnology*, 83, 871–877.

Dermou, E., Velissariou, A., Xenos, D. & Vayenas, D. V. 2005. Biological chromium(VI) reduction using a trickling filter. *Journal of Hazardous Materials*, 126, 78–85.

Edokpayi, J. N., Odiyo, J. O. & Durowoju, O. S. 2017. Impact of wastewater on surface water quality in developing countries: A case study of South Africa. In: Tutu, H. (ed.) *Water Quality*. Intechopen.

Engleman, V. S. 2010. Updates on choices of appropriate technology for control of VOC emissions. *Metal Finishing*, 108, 305–317.

Estrada, J. M., Hernández, S., Muñoz, R. & Revah, S. 2013. A comparative study of fungal and bacterial biofiltration treating a VOC mixture. *Journal of Hazardous Materials*, 250–251, 190–197.

Fu, Y., Shao, L., Tong, L. & Liu, H. 2011. Ethylene removal efficiency and bacterial community diversity of a natural zeolite biofilter. *Bioresource Technology*, 102, 576–584.

Fulazzaky, M. A., Talaiekhozani, A., Ponraj, M., Abd Majid, M. Z., Hadibarata, T. & Goli, A. 2014. Biofiltration process as an ideal approach to remove pollutants from polluted air. *Desalination and Water Treatment*, 52, 3600–3615.

Gallardo-Rodríguez, J. J., Rios-Rivera, A. C. & Von Bennevitz, M. R. 2019. Living biomass supported on a natural-fiber biofilter for lead removal. *Journal of Environmental Management*, 231, 825–832.

Gallastegui, G., Ávalos Ramirez, A., Elías, A., Jones, J. P. & Heitz, M. 2011. Performance and macrokinetic analysis of biofiltration of toluene and p-xylene mixtures in a conventional biofilter packed with inert material. *Bioresource Technology*, 102, 7657–7665.

Gospodarek, M., Rybarczyk, P., Szulczyński, B. & Gębicki, J. 2019. Comparative evaluation of selected biological methods for the removal of hydrophilic and hydrophobic odorous VOCs from air. *Processes*, 7, 187.

Hinojosa-Reyes, M., Rodríguez-González, V. & Arriaga, S. 2012. Enhancing ethylbenzene vapors degradation in a hybrid system based on photocatalytic oxidation UV/TiO2–In and a biofiltration process. *Journal of Hazardous Materials*, 209–210, 365–371.

Jacinto, M. L. J. A. J., David, C. P. C., Perez, T. R. & De Jesus, B. R. 2009. Comparative efficiency of algal biofilters in the removal of chromium and copper from wastewater. *Ecological Engineering*, 35, 856–860.

Jiang, Y., Li, S., Cheng, Z., Zhu, R. & Chen, J. 2012. Removal characteristics and kinetic analysis of an aerobic vapor-phase bioreactor for hydrophobic alpha-pinene. *Journal of Environmental Sciences*, 24, 1439–1448.

Katsou, E., Malamis, S. & Loizidou, M. 2011. Performance of a membrane bioreactor used for the treatment of wastewater contaminated with heavy metals. *Bioresource Technology*, 102, 4325–4332.

Katsoyiannis, I. A. & Zouboulis, A. I. 2004. Application of biological processes for the removal of arsenic from groundwaters. *Water Research*, 38, 17–26.

Kauselya, K., Narendiran, R. & Ravi, R. 2015. Biofiltration emerging technology for removal of Volatile Organic Compounds (VOC's)—A review. *International Journal of Environment and Bioenergy*, 10, 1–8.

Kennes, C. & Veiga, M. C. 2004. Fungal biocatalysts in the biofiltration of VOC-polluted air. *Journal of Biotechnology*, 113, 305–319.

Koutinas, M., Peeva, L. G. & Livingston, A. G. 2005. An attempt to compare the performance of bioscrubbers and biotrickling filters for degradation of ethyl acetate in gas streams. *Journal of Chemical Technology & Biotechnology*, 80, 1252–1260.

Kumar, A., Dewulf, J. & Van Langenhove, H. 2008. Membrane-based biological waste gas treatment. *Chemical Engineering Journal*, 136, 82–91.

Lebrero, R., Gondim, A. C., Pérez, R., García-Encina, P. A. & Muñoz, R. 2014. Comparative assessment of a biofilter, a biotrickling filter and a hollow fiber membrane bioreactor for odor treatment in wastewater treatment plants. *Water Research*, 49, 339–350.

Lee, E., Ryu, H. & Wcho, K. 2009. Removal of benzene and toluene in polyurethane biofilter immobilized with Rhodococcus sp. EH831 under transient loading. *Bioresource Technology*, 100, 5656–5663.

Liu, F., Fiencke, C., Guo, J., Lyu, T., Dong, R. & Pfeiffer, E. 2019. Optimisation of bioscrubber systems to simultaneously remove methane and purify wastewater from intensive pig farms. *Environmental Science and Pollution Research*, 26, 15847–15856.

Liu, S., Ren, A., Chen, Z., Guo, B., Hou, X. & Gu, D. 2020. Study and optimization of styrene waste gas purification process in bioscrubber. *IOP Conference Series: Earth and Environmental Science*, 474, 052059.

López, M. E., Rene, E. R., Malhautier, L., Rocher, J., Bayle, S., Veiga, M. C. & Kennes, C. 2013. One-stage biotrickling filter for the removal of a mixture of volatile pollutants from air. Performance and microbial community analysis. *Bioresource Technolog*, 138, 245–252.

Maestre, J. P., Gamisans, X., Gabriel, D. & Lafuente, J. 2007. Fungal biofilters for toluene biofiltration: Evaluation of the performance with four packing materials under different operating conditions. *Chemosphere*, 67, 684–692.

Majumder, S., Gangadhar, G., Raghuvanshi, S. & Gupta, S. 2015. Biofilter column for removal of divalent copper from aqueous solutions: Performance evaluation and kinetic modeling. *Journal of Water Process Engineering*, 6, 136–143.

Malakar, S., Das Saha, P., Baskaran, D. & Rajamanickam, R. 2017. Comparative study of biofiltration process for treatment of VOCs emission from petroleum refinery wastewater—A review. *Environmental Technology & Innovation*, 8, 441–461.

Malamis, S., Katsou, E., Di Fabio, S., Frison, N., Cecchi, F. & Fatone, F. 2015. Treatment of petrochemical wastewater by employing membrane bioreactors: A case study of effluents discharged to a sensitive water recipient. *Desalination and Water Treatment*, 53, 3397–3406.

Mathur, A. K., Sundaramurthy, J. & Balomajumder, C. 2006. Kinetics of the removal of mono-chlorobenzene vapour from waste gases using a trickle bed air biofilter. *Journal of Hazardous Materials*, 137, 1560–1568.

Meena, M., Sonigra, P. & Yadav, G. 2021. Biological-based methods for the removal of volatile organic compounds (VOCs) and heavy metals. *Environmental Science and Pollution Research*, 28, 2485–2508.

Mendoza, J. A., Prado, Ó. J., Veiga, M. C. & Kennes, C. 2004. Hydrodynamic behaviour and comparison of technologies for the removal of excess biomass in gas-phase biofilters. *Water Research*, 38, 404–413.

Moe, W. M. & Qi, B. 2005. Biofilter treatment of volatile organic compound emissions from reformulated paint: Complex mixtures, intermittent operation, and startup. *Journal of the Air & Waste Management Association*, 55, 950–960.

Mudliar, S., Giri, B., Padoley, K., Satpute, D., Dixit, R., Bhatt, P., Pandey, R., Juwarkar, A. & Vaidya, A. 2010. Bioreactors for treatment of VOCs and odours—A review. *Journal of Environmental Management*, 91, 1039–1054.

Mudliar, S. N., Padoley, K. V., Bhatt, P., Kumar, S. M., Lokhande, S. K., Pandey, R. A. & Vaidya, A. N. 2008. Pyridine biodegradation in a novel rotating rope bioreactor. *Bioresource Technology*, 99, 1044–1051.

Muñoz, A. J., Espinola, F. & Ruiz, E. 2016. Removal of Pb (II) in a packed bed column by a Klebsiella sp. 3S1 biofilm supported on porous ceramic Raschig rings. *Journal of Industrial and Engineering Chemistry*, 40, 118–127.

Ortiz, I., Revah, S. & Auria, R. 2003. Effects of packing material on the biofiltration of benzene, toluene and xylene vapours. *Environmental Technology*, 24, 265–275.

Parades-Aguilar, J., Reyes-Martínez, V., Bustamante, G., Almendáriz-Tapia, F. J., Martínez-Meza, G., Vílchez-Vargas, R., Link, A., Certucha-Barragán, M. T. & Calderón, K. 2021. Removal of nickel(II) from wastewater using a zeolite-packed anaerobic bioreactor: Bacterial diversity and community structure shifts. *Journal of Environmental Management*, 279, 111558.

Quintelas, C., Fonseca, B., Silva, B., Figueiredo, H. & Tavares, T. 2009. Treatment of chromium(VI) solutions in a pilot-scale bioreactor through a biofilm of *Arthrobacter viscosus* supported on GAC. *Bioresource Technology*, 100, 220–226.

Rahman, Z. & Singh, V. P. 2020. Bioremediation of toxic heavy metals (THMs) contaminated sites: Concepts, applications and challenges. *Environmental Science and Pollution Research*, 27, 27563–27581.

Ralebitso-Senior, T. K., Senior, E., Di Felice, R. & Jarvis, K. 2012. Waste gas biofiltration: Advances and limitations of current approaches in microbiology. *Environmental Science & Technology*, 46, 8542–8573.

Rolewicz-Kalinska, A., Lelicinska-Serafin, K. & Manczarski, P. 2021. Volatile organic compounds, ammonia, and hydrogen sulphide removal using a two-stage membrane biofiltration process. *Chemical Engineering Research and Design*, 165, 69–80.

Runye, Z., Christian, K., Zhuowei, C., Lichao, L., Jianming, Y. & Jianmeng, C. 2015. Styrene removal in a biotrickling filter and a combined UV–biotrickling filter: Steady- and transient-state performance and microbial analysis. *Chemical Engineering Journal*, 275, 168–178.

Salamanca, D., Dobslaw, D. & Engesser, K.-H. 2017. Removal of cyclohexane gaseous emissions using a biotrickling filter system. *Chemosphere*, 176, 97–107.

Sall, M. L., Diaw, A. K. D., Gningue-Sall, D. & Aaron, J. 2020. Toxic heavy metals: Impact on the environment and human health, and treatment with conducting organic polymers, a review. *Environmental Science and Pollution Research*, 27, 29927–29942.

Shen, H. & Wang, T. 1995. Hexavalent chromium removal in two-stage bioreactor system. *Journal of Environmental Engineering*, 121, 798–804.

Song, J., Kimney, K. A. & John, P. 2003. Influence of nitrogen supply and substrate interactions on the removal of paint VOC mixtures in a hybrid bioreactor. *Environmental Progress*, 22, 137–144.

Song, T., Yang, C., Zeng, G., Yu, G. & Xu, C. 2012. Effect of surfactant on styrene removal from waste gas streams in biotrickling filters. *Journal of Chemical Technology & Biotechnology*, 87, 785–790.

Srivastava, N. K. & Majumder, C. B. 2008. Novel biofiltration methods for the treatment of heavy metals from industrial wastewater. *Journal of Hazardous Materials*, 151, 1–8.

Wang, C., Xi, J., Hu, H. & Yao, Y. 2009. Effects of UV pretreatment on microbial community structure and metabolic characteristics in a subsequent biofilter treating gaseous chlorobenzene. *Bioresource Technology*, 100, 5581–5587.

Wang, Z., Xiu, G., Qiao, T., Zhao, K. & Zhang, D. 2013. Coupling ozone and hollow fibers membrane bioreactor for enhanced treatment of gaseous xylene mixture. *Bioresource Technology*, 130, 52–58.

Wei, Z., Sun, J., Xie, Z., Liang, M. & Chen, S. 2010. Removal of gaseous toluene by the combination of photocatalytic oxidation under complex light irradiation of UV and visible light and biological process. *Journal of Hazardous Materials*, 177, 814–821.

West, J. J. & Fiore, A. M. 2005. Management of tropospheric ozone by reducing methane emissions. *Environmental Science & Technology*, 39, 4685–4691.

Wu, H., Yan, H., Quan, Y., Zhao, H., Jiang, N. & Yin, C. 2018. Recent progress and perspectives in biotrickling filters for VOCs and odorous gases treatment. *Journal of Environmental Management*, 222, 409–419.

Youngwilai, A., Kidkhunthod, P., Jearanaikoon, N., Chaiprapa, J., Supanchaiyamat, N., Hunt, A. J., Ngernyen, Y., Ratpukdi, T., Khan, E. & Siripattanakul-Ratpukdi, S. 2020. Simultaneous manganese adsorption and biotransformation by Streptomyces violarus strain SBP1 cell-immobilized biochar. *Science of The Total Environment*, 713, 136708.

Zamir, M., Halladj, R., Sadraei, M. & Nasernejad, B. 2012. Biofiltration of gas-phase hexane and toluene mixture under intermittent loading conditions. *Process Safety and Environmental Protection*, 90, 326–332.

Zeng, H., Yin, C., Zhang, J. & Li, D. 2019. Start-up of a biofilter in a full-scale groundwater treatment plant for iron and manganese removal. *International Journal of Environmental Research and Public Health*, 16, 698.

Zilli, M. & Nicolella, C. 2012. Removal of monochlorobenzene from air in a trickling biofilter at high loading rates. *Journal of Chemical Technology & Biotechnology*, 87, 1141–1149.

2 Bio-Based Iron-Nanoparticles (NPs) Incorporated Polymeric Nanofiltration Membranes (PNCMs) for Wastewater Treatment

Piyal Mondal, Anweshan, and Mihir Kumar Purkait

CONTENTS

2.1 Introduction 17
2.2 Biosynthesis of Iron Nanoparticle (NPs) 18
    2.2.1 Plant-Mediated Synthesis 18
    2.2.2 Microbial Assisted Synthesis 19
    2.2.3 Plant Wastes and Other Biological Substances 21
    2.2.4 Microwave and Hydrothermal Synthesis 22
2.3 Different Routes for Preparing Iron NPs-Based Nanofiltration Membrane 23
    2.3.1 Phase Inversion Technique 23
        2.3.1.1 Immersion Precipitation 23
        2.3.1.2 Precipitation via Regulated Evaporation 23
        2.3.1.3 Precipitation from the Vapor Phase 23
        2.3.1.4 Thermally Induced Phase Separation 24
    2.3.2 Interfacial Polymerisation Method 24
    2.3.3 Electrospinning Method 24
2.4 Applications of Iron and Iron-Based Nanomaterial Incorporated Nanofiltration
    Membrane 24
    2.4.1 Heavy Metal Removal from Wastewater 25
    2.4.2 Dye Removal from Textile Industry Wastewater 25
    2.4.3 Pollutant Removal from Wastewater 25
    2.4.4 Catalytic Membrane for Degrading Contaminants 26
2.5 Advancements of Bio-Based Nanomaterial Impregnated Nanofiltration Membranes 26
    2.5.1 Performance Optimisation 26
    2.5.2 Membrane Stability 28
2.6 Environmental Hazards and Toxicity Study of Iron NPs 29
2.7 Future Scope of Research Work 29
2.8 Conclusion 30
References 30

2.1 INTRODUCTION

The severity of the potable water scarcity is intensifying each year; fuelled by the insatiable demand of commercialisation and globalisation processes, it is engulfing both the developed and

DOI: 10.1201/9781003165149-2

the developing countries in its clasps (Agboola et al., 2020). The conceptualisation of polymer materials is crucial for synthesising nanofiltration (NF) membranes used globally for water purification in treatment processes.

NF exhibit properties intermediate to reverse osmosis (RO) and ultrafiltration (UF), hence providing a unique set of advantages such as low osmotic pressure differential, as well as low capital, operational, and maintenance costs as compared to RO but higher permeate flux and retention of ions and molecules heavier than 300 Da (Hilal et al., 2004). Many investigations have evaluated the NF membranes' performance based on permeate flux, salt rejection, and COD retention (Sharma et al., 2017; Sinha et al., 2015). Also, several studies on membrane attributes and the effects of the various operating parameters for wastewater treatment exist. The vast literature on NF membranes conclusively proves their suitability as a separation media to remediate effluents with good rejections (Lau and Ismail, 2009).

The energy efficiency and separation efficacy of the polymeric membranes are well understood. However, phenomena such as swelling, fouling, chemical and thermal degradation, flux reductions, and densification inevitably decrease the membrane's separation efficacy in due course of filtration operations (Mondal et al., 2017b). All such drawbacks have motivated researchers to concentrate their efforts on combining polymers with nanoparticle incorporation to construct nanocomposite membranes. Unique polymeric membranes with exceptional chemical and electrochemical attributes are achieved by blending polymers with diverse nanoparticles (Mondal and Purkait, 2019; Mondal et al., 2019). The evolution of polymeric membranes with integrated nanoparticles has resulted in membranes with high thermal and chemical stability and vertically aligned pores with a uniform pore size distribution, revolutionising the membrane fabrication processes. The popularity and fame of the nanocomposite membrane are due to the concept of amalgamation of the polymer's physicochemical features to the electrochemical and surface attributes of the nanoparticle for gaining superior separation efficacy.

Mixed matrix membranes currently provide a platform for effective water treatment via enhanced chemical, mechanical, and thermal stability along with high selectivity and functionality (Mondal et al., 2020a, 2020b). In order to reap meaningful benefits without compromising the membrane's virtue, the interaction between reinforcing properties and impairment in membrane formulation must be balanced. NF membranes with high porosity and pore interconnectivity are preferred for the majority of water purification processes.

2.2 BIOSYNTHESIS OF IRON NANOPARTICLE (NPS)

Green chemistry prescribes the preferable use of renewables with minimal energy consumption and waste discharge for an efficient and environmentally sustainable approach for synthesising nanoparticles (Anastas and Williamson, 1996). So, utilisation of bio-polymers, microbes, plant parts, and waste materials with benign solvents and minimal energy requisite is necessary for the green synthesis process (Bolade et al., 2020). Several studies have reported water as a solvent with polyphenols as the bioactive component isolated from plants for reduction, capping, and stabilisation of NPs. Polyphenolic compounds can be extracted from any part of a plant and even waste biomass (Yew et al., 2020). Green synthesis techniques have significantly resolved the issue of agglomeration of NPs that plagued the chemical synthesis route. The polyphenols and other bioactive components surround the NPs, arresting agglomeration. This phenomenon is most evident in microbial and plant-mediated synthesis.

2.2.1 PLANT-MEDIATED SYNTHESIS

Plant-mediated synthesis of NPs has a significantly fast production rate compared to the microbial method of preparation, forming well-dispersed metal NPs (Dhillon et al., 2012). Plant extracts have been well known for being the best choice for the rapid synthesis of high-yielding metallic NPs with good stability. Also, the process is quite facile, inexpensive, and repeatable (Iravani,

Bio-Based Iron-Nanoparticles

2011). Plant-mediated synthesis of nanomaterials enjoys significant preference over other methods because a plethora of robust biomolecules, isolated from plant extracts, can be used to gain specific functionalities. A high concentration of phytochemicals having antioxidant properties can be extracted from the stem, leaf, fruit, and seed of a multitude of plants and herbs. The utilisation of phytochemcials of botanical origins in the preparation of nanomaterials significantly enhances the environmental sustainability and economic feasibility of the process, even setting benchmarks for upcoming green technologies (Zambre et al., 2012). The production of nanomaterials from processes utilising botanical phytochemicals significantly reduces environmental pollution, thereby setting benchmarks in economically feasible and ultra-sustainable clean and green technologies (Zambre et al., 2012).

Huang et al. (2015) looked into the factors that influence the efficiency of iron nanoparticles (FeNPs) made from green tea leaf extracts in removing malachite green dye. The study examined how various tea extract concentrations, temperatures, and pHs affected dye removal and observed a negative effect on increasing extract volume and pH in the solution. Increased temperature, however, increased nanoparticle reactivity. Dye degradation of up to 90.56% was achieved with the prepared FeNPs (Huang et al., 2015). Wang et al. (2014a, 2014b) used a dried eucalyptus leaf extract to make spheroidal 20–80 nm FeNPs to treat eutrophic wastewater. The study reported a removal efficiency of about 71.7% for total nitrogen and 84.5% for chemical oxygen demand (COD; Wang et al., 2014a, 2014b).

Machado et al. (2013a) used leaf extracts from 26 different tree species to determine each extract's efficacy in reducing trivalent iron to form zero-valent iron nanoparticles (ZVI NPs) with diameters of 10–20 nm. The investigation also reported that the antioxidant content of dried leaf extracts was higher than that of fresh leaf extracts, and the best extraction conditions differed for each leaf. At 80 °C, the leaf extracts from oak, pomegranate, and green tea performed the best of all (Machado et al., 2013a). The use of the ZVI NPs, prepared via green synthesis, to remove ibuprofen, one of the most frequently used anti-inflammatory drugs, from sandy soils was further explored by Machado et al. (2013b). In the aqueous phase, ZVI NPs reportedly achieved a removal efficacy of approximately 54–66%, while approximately 95% degradation was accomplished when the ZVI NPs were used as the catalyst for the ibuprofen removal process in Fenton (Machado et al., 2013b).

Fazlzadeh et al. (2017) have prepared ZVI NPs with irregular morphology having a mean particle size of 100 nm from extracts derived from three plants, namely Rosa damascena and *Thymus vulgaris*, and *Urtica dioica,* for the removal of hexavalent chromium from water (Fazlzadeh et al., 2017).

Manquián-Cerda et al. (2017) used blueberries as a source for phytochemical extraction to synthesise ZVI NPs. The fruit extract's characterisation highlighted its high total phenolic content, antioxidant activity, and ferric ion reducing potency. Highly stable ZVI NPs were obtained with an average particle diameter of 52.4 nm, having a surface area of 70.7 m^2g^{-1}. The resultant ZVI NPs could remove As (V), with 52.23 ± 6.06 mg g^{-1} sorption capacity (Manquián-Cerda et al., 2017). **Table 2.1** summarises the variety of iron nanoparticles synthesised from plant sources.

2.2.2 MICROBIAL ASSISTED SYNTHESIS

Due to its advantages over traditional chemical methods, the use of microorganisms to synthesise nanoparticles has become increasingly popular. The advantages include ambient temperature synthesis, high energy efficiency, limited release of adverse by-products, renewable and safe precursors, and robust and facile process scale-up (Park et al., 2016). Nanoparticles can be synthesised by intracellular or extracellular mechanisms in microorganisms like fungi, bacteria, and yeasts. These processes include the enzymatic reduction of metal ions resulting in well-dispersed nanoparticles with a smaller overall particle size distribution. As capping agents, natural proteins, tannins, and peptides are introduced onto the nanoparticle surface. By minimising agglomeration, such a

TABLE 2.1

Green Synthesised Iron-Based Nanoparticles from Plant Sources

Part	Name of Source	Average Size	Morphology
Plant	Soya bean sprouts	~8 nm	Spherical
	Aloe vera	93–227 nm	Spherical
	Aloe vera	~6–30 nm	Agglomerated irregular
Marine plant	*Sargassum muticum*	18 ± 4 nm	Cubic
	Kappaphycus alvarezii	14.7 ± 1.8 nm	Spherical
	Padina pavonica	10–19.5 nm	Spherical
Seed	Grape seed proanthocyanidin	~30 nm	Irregular shape
	Syzygium cumini	9–20 nm	Agglomerated spherical
	Carom and clove	0.088–3.95	Spherical, irregular
Leaf	Carob	4–8 nm	Well monodisperse
	Tridax procumbens	< 100 nm	Irregular shape
	Artemisia annua	3–10 nm	Spherical
	Green tea	5.7 ± 4.1 nm	Spherical
	Zea mays L.	< 100 nm	Aggregated spherical
Fruit peel	Plantain peel	30–50 nm	Spherical
	Punica Granatum	D: 40 nm, L: > 200 nm	Rod shaped
	Plantain peel	< 50 nm	Spherical
	Banana peel	10–25 nm	Agglomerated
Fruit	*Passiflora tripartita*	18.2–24.7 nm	Spherical
	Averrhoa carambola	1.9–3.1 nm	Spherical
	Lemon	14–17 nm	Spherical
Root & Shoot	*Mimosa pudica* (root)	60–80 nm	Agglomerated rough spherical
	Vaccinium corymbosum (shoot)	52.4 nm	Irregular shape, non-agglomerated
Stolon	Potato	40 ± 2.2 nm	Cubic
Waste	Tea waste	5–25 nm	Cuboid/pyramid
	Rice straw	9.9 ± 2.4 nm	Aggregated spherical
	Coffee waste hydrochar	10–40 nm	Spherical

(Reproduced with permission from Mondal et al. (2020a) © Elsevier)

surface coating improves nanoparticle stability and dispersion (Singh et al., 2016). The intracellular process involves metal ions diffusing into the membrane, which are then reduced by enzymes to form nanoparticles. The extracellular system entails the electrostatic attraction of metal ions to the cell wall and the enzymatic reduction of metal ions thereafter.

Bharde et al. (2008), in their study, reported the preparation of spherical iron oxide (Fe_3O_4) nanoparticles under aerobic conditions using *Actinobacter* sp. (Bharde et al., 2008). To make the FeNPs, *Actinobacter* sp. activates an extracellular process involving the iron reductase enzyme. Moon et al. (2010) used *Thermoanaerobacter* sp., an extremophile, to synthesise mono-dispersed, sphere-shaped 13.1 nm magnetite Fe_3O_4 NPs from FeOOH under anaerobic conditions (Moon et al., 2010). The estimated cost of production of commercial nanomagnetite of size 25 to 50 nm via chemical synthesis is around $500 kg^{-1}, which is many times higher than the production cost of nanomagnetites (5–90 nm) made using the microbial route.

Bharde et al. (2006) documented the synthesis of ferrimagnetic nanoparticles and the insignificant spontaneous magnetisation through the use of fungi at low temperatures (Bharde et al., 2006). Magnetic NPs of various sizes at room temperature can be synthesised using fungi like

Verticillium sp. and *Fusarium oxysporum* via extracellular mechanisms. Fungi such as *A. wentii, A. fumigates, C. globosum, C. lunata,* and *P. chlamydosporium* were used by Kaul et al. (2012) to produce FeNPs (Kaul et al., 2012). Pavani and Kumar (2013) used *Aspergillus* sp. segregated from a soil sample taken near a metal-plating plant in Hyderabad, India, to synthesise FeNPs and expunge a high concentration of iron from wastewater (Pavani and Kumar, 2013). Mohamed et al. (2015) employed a fungus, identified as *Alternaria alternate*, in a pitch-dark room to produce cube-shaped FeNPs within the size range of 6–12 nm with a broad-spectrum antibacterial effect (Mohamed et al., 2015). Another study by Sarkar et al. (2017) reported the synthesis of magnetic iron oxide NPs, using the same fungus. They documented that the pathway of NP formation was through an extracellular mechanism (Sarkar et al., 2017).

Algae is a complex genus of polyphyletic photosynthetic eukaryotic species. They consist of both single and multicellular species. Mahdavi et al. (2013) prepared superparamagnetic FeNPs via green synthesis using aqueous extracts of *Sargassum muticum*, a sea-dwelling macroalgae. The NPs thus formed were crystalline and cube-shaped with sizes ranging from 14–22 nm. The study group also reported high functional bioactivity (Mahdavi et al., 2013). A study carried out by Subramaniyam et al. (2015) documented 20–50 nm-sized spherical FeNPs using *Chlorococcum* sp., a soil microalgae, and an aqueous solution of ferric chloride as a precursor. The study proposed that nanoparticle formation was caused by the amine and carbonyl functional groups of the polysaccharides and glycoproteins present in the algal cell and confirmed by FTIR spectrogram (Subramaniyam et al., 2015). **Table 2.2** summarises different forms of Iron nanoparticles synthesised various microorganisms.

2.2.3 PLANT WASTES AND OTHER BIOLOGICAL SUBSTANCES

Several anthropogenic activities generate enormous quantities of vegetative solid wastes. The utilisation of such materials to produce value-added products enhances the sustainability of an economic enterprise. Researchers are actively working on reducing solid waste generation from

TABLE 2.2
Microorganism Mediated Synthesis of Iron-Based Nanoparticles

Microorganisms	Species Name	Nanoparticle	Average Size	Morphology
Bacteria	*Actinobacter sp.*	Fe_3O_4	10–40 nm	Cubical
	Actinobacter sp.	γ-Fe_2O_3	< 50 nm	Spherical
	Thermoanaerobacter sp.	Fe_3O_4	~13 nm	Spherical
	Bacillus subtilis	Fe_3O_4	60–80 nm	Spherical
	Thiobacillus thioparus	Fe_3O_4	-	-
	Microbac-terium marinilacus	Magnetic Iron oxide	2–10 nm	Spherical
Fungi	*Fusarium oxysporum and Verticillium sp.*	Fe_3O_4	20–50 nm	Spherical
	P. chlamydosporium, A. fumigates, A. wentii, C. lunata and C. globosum	Maghemite (Fe_2O_3)	~12–50 nm	Spherical
	Aspergillus sp.	Fe_3O_4	50–200 nm	Spherical
	Alternaria alternate	Fe Nanoparticles	~9nm	Cubical
Algae	*Sargassum muticum*	Magnetic Fe_3O_4 nanoparticles	18 ± 4 nm	Cubical
	Chlorococcum sp.	Iron nanoparticles	20–50 nm	Spherical

(Reproduced with permission from Mondal et al. (2020a) © Elsevier)

various processes by re-utilising it as raw materials for producing valuable by-products. The synthesis of metal NPs is one such endeavour. Herrera-Becerra et al. (2007) reported the biosynthesis of magnetite (Fe_3O_4) nanoclusters along with wuestite-like ($Fe_{0.902}O$) nanostructures using biomass of *Medicago sativa* (alfalfa) as raw material. The iron oxide nanoparticles obtained were less than 5 nm in average diameter. In this study, aqueous ferrous ammonium sulphate was used as a precursor. Further, the group reported the pH of the precursor mix as the size-constraint factor during the synthesis. Experimental observations indicated that NPs with an average diameter of 3.6 nm could be achieved at an optimum pH condition viz., pH 10. A detailed characterisation of the nanoparticles was documented, which reports the use of techniques such as Electron Energy Loss Spectroscopy (EELS), High-Resolution Transmission Electron Microscopy (HRTEM), and Fast Fourier Transform (FFT) spectra analysis. The analysis confirmed the coexistence of magnetite and wustite-like clusters in the nanostructures (Herrera-Becerra et al., 2007).

Njagi et al. (2011), in their study, explored the impact of the copious amount of phenolic compounds present in sorghum bran on FeNPs synthesis. Aqueous extract of sorghum bran was mixed with 0.1 M ferric chloride solution in 2:1 volume ratio and kept unperturbed for 1 h after 1 min of vigorous shaking. The research group reported the formation of iron oxide NPs with spherical morphology and amorphous nature, having an average particle size in the range of 40 to 50 nm. The resultant NPs exhibited promising prospects in the catalytic degradation of dye. In the presence of hydrogen peroxide, the 0.66 mM of iron oxide NPs could degenerate 90% of bromomethyl blue at 500 ppm concentration within 30 min (Njagi et al., 2011).

The applications of iron oxide NPs can vary based on the substrate used for synthesis. Due to their large amounts of polyphenolic material, natural waste in everyday life, including rice hay, fruit peels, coffee and tea discards, and other waste remnants, have essential ingredients to be employed in the green production of nanoparticles. Tea residue was employed for producing iron oxide NPs by Lunge et al. (2014), and then the latter was explored as an adsorbent for arsenic removal. The resultant NPs were tetrahedral in shape and showed a significant adsorptive capability for arsenic. The maximum adsorption capacity reported were 188.69 mg g^{-1} for arsenic (III) and 153.8 mg g^{-1} for arsenic (V; Lunge et al., 2014). Edathil et al. (2018) reported a successful synthesis of a low-cost magnetic adsorbent from coffee discards by dispersing iron oxide NPs on its surface via the precipitation process. The researchers then assessed its capability as an adsorbent. The NPs were able to comprehensively remove lead (II) from an aqueous solution at pH 7.3 in 30 min, with a maximum adsorption capacity of 41.15 mg g^{-1} (Edathil et al., 2018). Such experiments demonstrate that natural residues have excellent stabilising attributes and can be used for various environmental purposes.

2.2.4 MICROWAVE AND HYDROTHERMAL SYNTHESIS

Microwave-aided synthesis makes use of electromagnetic radiation, in which the precursor solvents and reducing agents are activated in the reaction through ionic and molecular conduction. Łuczak et al. (2016) employed microwaves to directly achieve the required temperature within the sample to initiate the reaction. Microwave irradiation achieves the target temperature faster than conventional heating, thereby increasing the energy efficiency and paving a greener route for nanoparticle synthesis (Amores et al., 2016). Microwave enhances nucleation and ensures dispersion uniformity of the NPs. The microwave-aided synthesis process produces NPs within minutes, with precise size regulation, uniform dispersion, and high crystallinity (Blanco-Andujar et al., 2015).

Kombaiah et al. (2017) provided evidence that microwave-aided heating is both economical and efficient than conventional thermal methods, thus reducing the cost of NPs synthesis further (Kombaiah et al., 2017). Schneider et al. (2017) have identified a rapid synthesis of magnetic nanoparticles covered with alendronic acid, a biphosphate with high reactivity, by a green microwave-aided method. The study reported the use of triethylene glycol (TEG) and iron acetylacetonate as precursors. The synthesis of NPs was carried out at 200°C in an inert atmosphere.

Due to its small isoelectric point, lack of toxicity, high boiling point, and high viscosity, TEG is considered a green component (Schneider et al., 2017).

In hydrothermal synthesis, desired precursors, i.e., the green extract and the aqueous solution of the metal ion of specific molarity, are taken in a Teflon-lined autoclave, and the reaction is carried out under predetermined conditions of temperature and pressure for sufficient time. Such a method is helpful as crystalline materials from the precursor need a lower temperature than calcination.

In a study by Ahmmad et al. (2013), extract from the leaf of the green tea (*Camellia sinensis*) plant was used to prepare α-Fe_2O_3 (hematite) nanoparticles via hydrothermal process. The group reported the NPs to be highly porous, having spherical morphology with a mean size of 60 nm. The prepared NPs also exhibited enhanced photocatalytic capability compared to the commercial hematite NPs, with a greater surface area of 22.5 m^2g^{-1} (Ahmmad et al., 2013).

Phumying et al. (2013), in their study, documented the preparation of Fe_3O_4 nanoparticles by the hydrothermal pathway utilising extracts from the aloe vera herb. As for the salt precursor, ferric acetylacetonate ($Fe(C_5H_8O_2)_3$) solution was employed. Images from a transmission electron microscope (TEM) show that the nanoparticles have a spherical shape with sizes ranging from 6–30 nm. The resulting iron oxide NPs had inverse cubic spinel morphology with no phase impurities, confirmed through HR-TEM and XRD analysis. The coercivity experiment validated the superparamagnetism attribute exhibited by the NPs (Phumying et al., 2013). The study showed that increased reaction temperature and time led to enhanced crystallinity and magnetic saturation of the iron oxide nanoparticles.

2.3 DIFFERENT ROUTES FOR PREPARING IRON NPS-BASED NANOFILTRATION MEMBRANE

2.3.1 PHASE INVERSION TECHNIQUE

The controlled transition of a polymer from a liquid phase to a solid phase is known as phase inversion. Phase inversion membranes can be fabricated fundamentally via four procedures, namely: (a) immersion precipitation, (b) precipitation via regulated evaporation, (c) precipitation from the vapor phase, and (d) temperature-induced phase transition. Among these processes, polymeric membranes are most popularly fabricated via the immersion precipitation method.

2.3.1.1 Immersion Precipitation

The most popular method for membrane preparation via phase inversion is the immersion precipitation technique. A blend of polymer and solvent is cast on a support media and then immersed in a bath of a non-solvent moiety. The polymer precipitates out as the solvent exchanges with the non-solvent. The membrane composition is affected by the combination of phase separation and mass transfer rates.

2.3.1.2 Precipitation via Regulated Evaporation

A concoction of solvent and non-solvent is prepared to dissolve the polymer, as such that the solvent is relatively highly volatile compared to the non-solvent. Thus, the solvent gets evaporated rapidly, leaving the mixture skewed with higher non-solvent and polymer composition that ultimately leads to precipitation of the polymer. A thin film membrane can be obtained through this process.

2.3.1.3 Precipitation from the Vapor Phase

A polymer dissolved in a solvent is at first cast into a film and then placed in an atmosphere of vapor containing a non-solvent component saturated with invariably the same solvent. The high saturation of the solvent in the vapor phase prevents the further transition of the solvent present in the liquid phase to the vapor phase. However, diffusion of the non-solvent into the cast film occurs, causing membrane formation. The membranes fabricated using this process are porous with no top layer.

2.3.1.4 Thermally Induced Phase Separation

A specific or blended solvent is employed to dissolve the polymer. The solution is then allowed to cool to facilitate phase separation. The evaporation of the solvent enables skinned membrane formation. Microfiltration membranes are often fabricated using this technique.

2.3.2 INTERFACIAL POLYMERISATION METHOD

The three most common ways of interface polymerisation are liquid-solid interfaces, liquid-liquid interfaces, and liquid-in-liquid emulsion interfaces (Song et al., 2017). Monomers can be found at the liquid-liquid and liquid-in-liquid emulsion interfaces in one or both fluid phases (Wittbecker and Morgan, 1959; Song et al., 2017). Other kinds of interfaces, including liquid-gas, solid-gas, and solid-solid, are seldom used (Song et al., 2017).

Polymerisation occurs at the interface of a liquid-solid interface, which ends in a polymer bound to the solid phase's surface. Polymerisation occurs on just one side of a liquid-liquid interface with monomer dissolved in one form, while polymerisation occurs on both sides of a liquid-liquid interface with monomer dissolved in both phases (Raaijmakers and Benes, 2016). A stirred or unstirred interfacial polymerisation reaction is possible. The two phases are mixed with intense agitation in a stirred reaction, resulting in a higher interfacial surface area and a higher polymer yield (Wittbecker and Morgan, 1959; Raaijmakers and Benes, 2016). The capsule size is determined directly by the stirring rate of the emulsion in capsule synthesis (Raaijmakers and Benes, 2016).

2.3.3 ELECTROSPINNING METHOD

Electrospinning is a technique for drawing out threads from polymer melts or solutions to produce nanofibres under the influence of an electric field. The driving force in this process is the electrostatic force regulated through voltage, and electro-hydrodynamics influence the fibre attributes. A typical configuration for this technique essentially consists of a storage tank for the polymer solution with a blunt needle shaped as an outlet (basically a syringe), a high voltage electric-power source, a pump for fluid transport, and a collector.

When the solution is injected at a steady flow rate and a specific voltage is applied to generate an electric field between the needle tip and the collector, the spinning phase begins. At the liquid's surface, an electrostatic potential difference builds up due to charge accumulation. The liquid meniscus gets deformed into a conically shaped configuration known as the Taylor cone when electrostatic repulsion exceeds the surface tension of the liquid.

Only after the formation of the Taylor cone is the charged fluid jet flung out to the collector. A collector can be of any shape; rotating drums, disks, mandrels, and immobile flat plates are most common. Phase change occurs to form fibres when the solvent evaporates from the whipping motion during its flight time from the Taylor cone to the collector, depending on the solution viscosity. A non-woven fibre mat is collected on the collector as a result.

2.4 APPLICATIONS OF IRON AND IRON-BASED NANOMATERIAL INCORPORATED NANOFILTRATION MEMBRANE

Immobilisation of nanoparticles in ceramic, polymeric, and zeolitic support structures overcomes limitations such as reduction in adsorptive and catalytic efficacies caused due to agglomeration. Nanoparticle embedded membranes enjoy considerable advantages over the conventional fundamental membranes as they can be customised with relative ease to suit the separation conditions

Bio-Based Iron-Nanoparticles

with enhanced attributes and functionalities. The application of such membranes in various fields is discussed in this section.

2.4.1 HEAVY METAL REMOVAL FROM WASTEWATER

Liu et al. (2019) applied the principles of green synthesis first to prepare spherical FeNPs using extracts from green tea leaves and then embedded the metal NPs in the calcium alginate (CaAlg) hydrogel membrane. The TEM and SEM images of the membrane established the well-dispersed immobilisation of 4.96 ± 2.03 nm FeNPs over the surface of the membrane. The study also reports 99.5% removal of 1 ppm hexavalent chromium (Cr(VI)) within 10 min with only a 0.6g membrane. A synergistic effect between FeNPs and CaAlg was observed that improved the removal efficiency. The CaAlg membrane shielded the FeNPs from oxidation that boosted the shelf life, ion selectivity, and reusability of the composite (Liu et al., 2019).

Gholami et al. (2014) prepared iron oxide NPs immobilised in polyvinyl chloride (PVC) blended cellulose acetate (CA) nanocomposite membrane using the phase-inversion technique for removing lead from wastewaters. The study compared the attributes and performance of the nanocomposite membranes of different blends to the membranes with pristine composition. The composite membrane having 10% (w/w) CA and 0.1% (w/w) iron oxide NPs was reported to have performed exceptionally well in lead removal (Gholami et al., 2014).

Tandon et al. (2013) documented the fabrication of ZVI NPs impregnated on montmorillonite K10 support for arsenic removal. In the study, ZVI NPs having a size in the range 59.08 ± 7.81 nm was prepared using tea liquor. The removal efficiency for trivalent arsenic (As(III)) was reported to be 90% within 30 min at both low and high pH levels. A synergetic enhancement in the composite's performance was observed here as well (Tandon et al., 2013). In another study Prasad et al. (2014) produced chitosan composite impregnated with FeNPs, synthesised using *Mentha spicata* L. leaf extract. The mean particle size of the green synthesised FeNPs was reported in the range of 20 to 45 nm. A core-shell structure gives the best description of the FeNPs mentioned here, with many functional groups covering the outer surface. The composite displayed a remarkable affinity towards the removal of both arsenite (As(III)) and arsenate (As(V)) from water samples. The results showed that eliminating 100 mg L^{-1} As(III) and As(V) using a 2 g L^{-1} Fe NP-chitosan composite was effective at 98.79% and 99.65%, respectively, after 60 mins (Prasad et al., 2014).

2.4.2 DYE REMOVAL FROM TEXTILE INDUSTRY WASTEWATER

In their investigation, Saranya et al. (2015) studied the impact of the ZVI NPs in the morphology and pore structure alteration of the CA/ZVI mixed matrix membrane (MMM). ZVI NPs prepared using green tea leaf extracts significantly improved the thermal stability and surface morphology of the CA/ZVI MMM. The study reported removal efficiencies of 96.7%, 90.2%, and 55.5% for BOD, COD, and sulfate ions present in textile industry effluent. The study highlighted the use of adsorptive nanoparticles to improve rejection performance, which would significantly impact the relevance of nanostructural alterations of polymeric membranes (Saranya et al., 2015).

2.4.3 POLLUTANT REMOVAL FROM WASTEWATER

Lakhotia et al. (2018) reported the fabrication of thin-film nanocomposite membrane by initiating prior-seeding interfacial polymerisation technique using cerium oxide NPs. Furthermore, a thorough characterisation study was conducted for the novel membrane. The study reports more than 90% rejection of polyethylene glycol (MW 1,500 Da) by the membrane. In addition, significant improvement in antibacterial property, flux, hydrophilicity, salt rejection, and the surface charge was also documented. The membrane's performance was also assessed using seawater for fouling investigation, and it was reported that a cerium oxide NPs embedded boosted the rejection of hydrophobic pollutants and significantly decreased fouling (Lakhotia et al., 2018).

Mondal and Purkait (2018) fabricated a composite membrane by incorporating iron oxide NPs in a pH-responsive polymeric membrane to remove fluoride and catalytically convert nitrobenzene to aniline concurrently. Iron oxide NPs were synthesised using clove extracts having an average particle size of 13.5 nm with a core-shell type structure, wherein iron oxide constitutes the core. The study reported a fluoride rejection efficiency of 72% for a 20 mg L^{-1} fluoride stock solution along with a removal efficacy of 86.7% for nitrobenzene in 39 min at pH 3 (Mondal and Purkait, 2018).

Another study by Chaudhary et al. (2020) developed a Fe–Al–Mn@chitosan-loaded cellulose acetate-based MMM to remove fluoride from groundwater. The membrane was synthesised applying the phase inversion technique, and a detailed characterisation study of the composite membrane has been documented. The MMM can treat 2,000 L m^{-2} of groundwater at optimised conditions with a high sustenance time of 275 h continuously along with substantially high antimicrobial attributes. In addition, the study reported that the fluoride is first adsorbed on the membrane surface, and then separation occurs due to the accumulation of repulsive charges on the surface (Chaudhary and Maiti, 2020).

2.4.4 CATALYTIC MEMBRANE FOR DEGRADING CONTAMINANTS

Polyvinylidene fluoride (PVDF) membrane was utilised as support to incorporate green synthesised Fe and bimetallic Fe/Pd nanoparticles (Smuleac et al., 2011). Green tea extract was used a source of reducing agent for preparing Fe and Fe/Pd nanoparticles. Modification of the membrane were done by polyacrylic acid (PAA), utilising polymerisation reaction. Average particle diameter of 20–30 nm was found to be incorporated within the PVDF/PAA modified membrane. The toxic organic pollutant trichloroethylene (TCE) was utilised to study its degradation process utilising the prepared catalytic membrane. With increase in nanoparticle weight percentage catalytic reactivity of the membrane was enhanced, and it efficiently dechlorinated TCE compound. In comparison to Fe NPs, Fe/Pd incorporation was found to enhance the catalytic process much more.

Saikia et al. (2017) reported a cheap green route for synthesising iron oxide nanoparticles and utilised it as catalyst through silica support. The catalyst was utilised for preparing phenol from boronic acids without using hydrogen peroxide (H_2O_2); instead ipso-hydroxylation reaction was performed. About 98% of yield in reaction was obtained utilising 4 mg of nanocatalyst at 50 °C for 2h.

Synthesis of iron nanoparticles was performed by Mondal and Purkait (2017a) employing cardamom extract as reducing agent. Obtained nanoparticles were in the range of 30–32 nm diameter as shown in Figure 2.1. Catalytic membrane was formed by incorporating Fe nanoparticles within pH-responsive poly (vinylidene fluoride-co-hexafluro propylene; PVDF-co-HFP) membrane. Degradation of nitrobenzene (NB) through reduction process was studied utilising the catalytic membrane. About 57% of NB degradation was obtained from 100 ppm nitrobenzene solution with aniline yield of 9.65 ppm at an optimum pH of 11.9, iron content of 0.01 wt%, and operating time of 50 min. Figure 2.2 shows the surface and cross-sectional morphology of the synthesised Fe NPs impregnated polymeric membrane.

2.5 ADVANCEMENTS OF BIO-BASED NANOMATERIAL IMPREGNATED NANOFILTRATION MEMBRANES

2.5.1 PERFORMANCE OPTIMISATION

In order to fabricate optimised nanoparticle impregnated membranes with defect-free property, the main challenge lies in the compatibility issue between nanomaterials with a polymer matrix. Dispersity of functional nanomaterials into membrane layer also plays a vital role

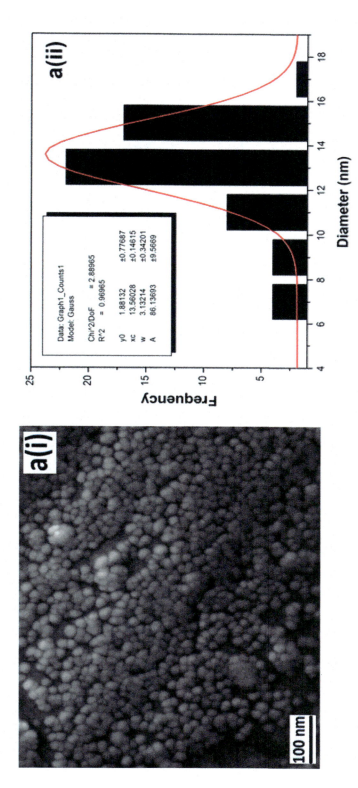

FIGURE 2.1 a(i) FESEM images of prepared Fe NPs, a(ii) average particle size analysis of Fe NPs.

Source: Reproduced with permission from Mondal et al. (2018) © Elsevier

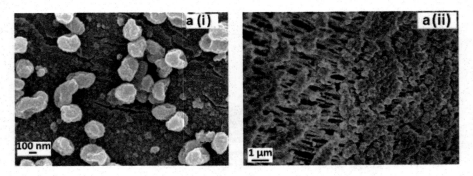

FIGURE 2.2 a(i) Surface morphology of Fe NPs impregnated membrane (ii) Cross-section of Fe NPs impregnated membrane.
Source: Reproduced with permission from Mondal et al. (2018) © Elsevier

(Zhang et al., 2019). Moreover, studies reveal that nanomaterials synthesised through green routes are organic in nature and biocompatible in nature, having better dispersity in polymer matrix in comparison to inorganic nanoparticles. Hence, introducing such organic nanomaterials is an advantage for minimising interfacial defects. However, due to their higher surface energy, the tendency for agglomeration persists in polymeric layer, giving rise to defects in polymeric thin film membranes.

Homogenous incorporation of sufficient nanomaterials within polymeric film depends on two effective methods: (a) surface modification of nanomaterials and (b) improved method for nanomaterial incorporation. In order to improve the dispersity of nanomaterials in the aqueous phase, various water-soluble polymers such as PEG are utilised for modifying the nanomaterial surface. Moreover, to increase the dispersity in organic phase, higher molecular weight alkyl chains are functionalised on the nanomaterial surface. Green synthesised nanomaterials, due to the sufficient hydroxyl containing functional groups on their surface inherited from plant extracts, are found to be sufficiently well dispersed in the aqueous phase. The surface modification could also enhance the compatibility between green synthesised nanomaterials and a polymer matrix and could result in forming selective microstructures (Yang et al., 2017). Moreover, for preparing defect-free high-performance polymeric membranes, efficient new methodologies have been studied for loading sufficient nanomaterials in polymeric membrane layers. High-performance thin-film polymeric membranes with low nanomaterial consumption were prepared via bio-inspired layer assisted vacuum filtration, evaporation-controlled nanofiller positioning (EFP), dip-coating, layer-by-layer assembly, etc. In the future, preparing efficient functionalised nanomaterials along with their better dispersing strategies within a membrane layer could further optimise the performance of nanomaterial-embedded thin-film nanocomposite membranes.

2.5.2 Membrane Stability

Researchers are extensively working towards developing thin-film nanocomposite polymeric membranes for higher scalability along with better stability performance. Utilisation of organic nanomaterials as fillers within polymeric membranes could result in low stability and might be degraded under real operation conditions. Such demerits could affect the membrane performance and lower its efficiency. Further research must be devoted to studying long-term membrane stability, which includes pH tolerance, temperature sensitivity, solvent resistance, chlorine and fouling resistance, and antimicrobial activity. Such parameters are found to be common in almost every filtration-related application in our daily life.

Bio-Based Iron-Nanoparticles

Moreover, though green synthesised nanomaterials possess a high affinity towards a polymeric matrix, still it is possible for such materials to leach out into the permeate streams during operation. Such activities may be denoted due to the nanosizes or due to the lack of covalent bonding of nanomaterials with polymeric matrix. Such a leaching out process of metallic nanomaterials, which are generally toxic in nature, imposes environmental concerns. Presence of such nanomaterials in higher amounts violates the safety regulation of drinking water. The prospective thing about green synthesised nanomaterials is that they are bioinspired and thus biocompatible in nature, hence less toxicity is associated with them. In this regard, chemically synthesised nanomaterials possess a higher threat, and researchers should find a better route for strongly binding the nanomaterials within a polymeric membrane by chemical bonding. A better option for such nanomaterial-embedded membranes is to search for bio-inspired nanoparticles, which are biocompatible and environmentally friendly in nature and thus are a better alternative.

2.6 ENVIRONMENTAL HAZARDS AND TOXICITY STUDY OF IRON NPS

Iron-based nanomaterials, despite their tremendous environmental applications, possess a high risk for environmental concern. Improper management of wastewater and other trash from industries could lead to specific environmental hazards in groundwater and soil, since synthesised iron nanomaterial for specific remediation purposes can be transported from one medium to another. Zero-valent iron nanoparticles (nZVI) are considered to be the most reactive, among other iron nanoparticles. During the treatment of groundwater contaminants with zero-valent iron nanoparticles (nZVI), it was observed that it is transported through the permeable soil and undergoes transformation on interacting with contaminants as well with the exposed environment.

The environment is directly or indirectly affected due to the toxic impacts generated by iron NPs on microorganisms and soil fauna. It was reported that the induced stress of nanosized iron exerts a toxic effect on soil microorganisms and causes an alteration in microbial biomass (Sacca et al., 2014). The nanosized iron particles interact strongly with enzymes of living bodies and metals in the environmental arena due to its unique morphological features, particle size, surface energy, self-assembly, and synthesis process. However, the biosynthesised nanoparticles being deployed from harmful chemicals like hydrazine as capping agents make it a mild toxic agent for the environment.

Auffan et al. (2008), through experimental studies, have drawn a direct relationship between the cytotoxicity and oxidation state of iron nanoparticles. The cytotoxic impacts of different iron oxides such as nZVI, magnetite, and maghemite were tested towards gram-negative bacteria *E. Coli* and were compared. From the tests it was found that nZVI had a higher toxicity effect as compared to other iron oxide NPs. Hence, the study suggested the oxidation state of iron nanoparticles played a vital role in toxicity. Such a phenomenon arises due to the generation of oxidative stress from reactive oxygen species (ROS). ROS denotes highly unstable radicals such as superoxide radicals and hydroxyl radicals, which disrupt the functioning of cells upon being adsorbed over the cell membrane.

2.7 FUTURE SCOPE OF RESEARCH WORK

From previous reports it was obtained that when preparing iron nanoparticle embedded NF membranes the following conditions should be satisfied: well-dispersed nanomaterials in the polymeric material, better interaction and stability of the nanomaterial with polymer, and finally nanomaterial's compatibility with the polymer. Despite better material properties, most of the nanomaterial embedded polymeric NF membranes have limitations regarding processing and solubility, which makes it less effective for commercialisation. Safety and health concerns of the iron nanomaterials leaching out of the polymeric membranes into the permeate has been of great concern nowadays. But, the use of green synthesised nanomaterials such as iron reduces the risk of environmental impact. Future scope of research should focus on the stability of such biobased nanomaterial incorporated NF membranes and their cost of preparation for economic and realistic application.

Further, keeping in mind the adverse effects of fouling, which plays a vital role towards overall performance, advance antifouling techniques and strategies should be taken into account. Biobased nanomaterials and their mechanisms of interaction towards fouling within the membrane should be analysed deeply for controlling the persistent membrane fouling issues. Research should focus on the optimisation of biobased nanomaterial content, stability, and cleaning performance of the nanocomposite NF membranes in order to restore the permeate flux along with maintaining the water quality, membrane replacement, and overall membrane operation cost. Process feasibility and sustainability of the membrane process depends on the advanced techniques utilised in preparing novel NF membranes with enhanced fouling control and cleaning efficiency properties. Such novel advancements would lead to significant reduction in total operation cost for the membrane process and finally would ease the process of providing clean and safe water. The main problems associated with NF membranes are high cost of operation and volume reduction of permeate along with its disposal. Though NF membranes are marked with high selectivity, permeability, and antifouling properties, still the cost of membrane fabrication and energy usage surpasses the membrane properties and increases the total cost.

2.8 CONCLUSION

Membrane technology has been one of the vital and versatile separation technologies that plays an immense role in today's world to meet the water demand and provide water supply to society. This review provides a deep insight into the green synthesis of iron nanoparticles through various routes and discusses its advantages over chemical methods. The study shows the morphological variations and applications of various green synthesised iron NPs. Moreover, the nanofiltration membrane incorporating such biobased iron NPs has been extensively discussed along with various techniques of preparation and fabrication methods. The utilisation of such biobased iron NPs and doped NF membranes for various environmental remediation applications has been highlighted and discussed in detail. This study also focuses on the various hazards related to the leaching of such nanomaterials in the environment and discusses the future scope of research in order to tackle such problems along with searching for various techniques and approaches to minimise fouling along with the operating cost of NF membranes. Such review articles that are specific towards fabrication of NF membranes embedded with biobased iron NPs and their various environmental remediation applications are scant and have been extensively discussed in this study. Moreover, the future scope of research could possibly help the research community to render more advanced techniques to deal with fabrication of such nanomaterial modified NF membranes with better antifouling and stability analysis.

REFERENCES

Agboola, O., Sunday Isaac Fayomi, O., Sadiku, R., Popoola, P., Adeniyi Alaba, P., Adegbola, A.T., 2020. Polymers blends for the improvement of nanofiltration membranes in wastewater treatment: A short review. *Mater. Today:. Proc.* https://doi.org/10.1016/j.matpr.2020.05.387

Ahmmad, B., Leonard, K., Shariful Islam, M., Kurawaki, J., Muruganandham, M., Ohkubo, T., Kuroda, Y., 2013. Green synthesis of mesoporous hematite (α-Fe2O3) nanoparticles and their photocatalytic activity. *Adv. Powder Technol.* 24, 160–167. https://doi.org/10.1016/j.apt.2012.04.005

Amores, M., Ashton, T.E., Baker, P.J., Cussen, E.J., Corr, S.A., 2016. Fast microwave-assisted synthesis of Li-stuffed garnets and insights into Li diffusion from muon spin spectroscopy. *J. Mater. Chem. A* 4, 1729–1736. https://doi.org/10.1039/C5TA08107F

Anastas, P.T., Williamson, T.C. (Eds.). 1996. *Green Chemistry: Designing Chemistry for the Environment*, ACS Symposium Series. Washington, DC: American Chemical Society. https://doi.org/10.1021/bk-1996-0626

Auffan, M., Achouak, W., Rose, J., Roncato, M.-A., Chanéac, C., Waite, D.T., Masion, A., Woicik, J.C., Wiesner, M.R., Bottero, J.-Y., 2008. Relation between the redox state of iron-based nanoparticles and their cytotoxicity toward Escherichia coli Environ. *Sci. Technol.*, 42, pp. 6730–6735

Bharde, A.A., Parikh, R.Y., Baidakova, M., Jouen, S., Hannoyer, B., Enoki, T., Prasad, B.L.V., Shouche, Y.S., Ogale, S., Sastry, M., 2008. Bacteria-mediated precursor-dependent biosynthesis of superparamagnetic iron oxide and iron sulfide nanoparticles. *Langmuir* 24, 5787–5794. https://doi.org/10.1021/la704019p

Bharde, A.A., Rautaray, D., Bansal, V., Ahmad, A., Sarkar, I., Yusuf, S.M., Sanyal, M., Sastry, M., 2006. Extracellular biosynthesis of magnetite using fungi. *Small* 2, 135–141. https://doi.org/10.1002/smll.200500180

Blanco-Andujar, C., Ortega, D., Southern, P., Pankhurst, Q.A., Thanh, N.T.K., 2015. High performance multi-core iron oxide nanoparticles for magnetic hyperthermia: Microwave synthesis, and the role of core-to-core interactions. *Nanoscale* 7, 1768–1775. https://doi.org/10.1039/C4NR06239F

Bolade, O.P., Williams, A.B., Benson, N.U., 2020. Green synthesis of iron-based nanomaterials for environmental remediation: A review. *Environ. Nanotechnol. Monit. Manage.* 13, 100279. https://doi.org/10.1016/j.enmm.2019.100279

Chaudhary, M., Maiti, A., 2020. Fe–Al–Mn@chitosan based metal oxides blended cellulose acetate mixed matrix membrane for fluoride decontamination from water: Removal mechanisms and antibacterial behavior. *J. Membr. Sci.* 611, 118372. https://doi.org/10.1016/j.memsci.2020.118372

Dhillon, G.S., Brar, S.K., Kaur, S., Verma, M., 2012. Green approach for nanoparticle biosynthesis by fungi: current trends and applications. *Crit. Rev. Biotechnol.* 32, 49–73. https://doi.org/10.3109/07388551.2010.550568

Edathil, A.A., Shittu, I., Hisham Zain, J., Banat, F., Haija, M.A., 2018. Novel magnetic coffee waste nanocomposite as effective bioadsorbent for Pb(II) removal from aqueous solutions. *J. Environ. Chem. Eng.* 6, 2390–2400. https://doi.org/10.1016/j.jece.2018.03.041

Fazlzadeh, M., Rahmani, K., Zarei, A., Abdoallahzadeh, H., Nasiri, F., Khosravi, R., 2017. A novel green synthesis of zero valent iron nanoparticles (NZVI) using three plant extracts and their efficient application for removal of Cr(VI) from aqueous solutions. *Adv. Powder Technol.* 28, 122–130. https://doi.org/10.1016/j.apt.2016.09.003

Gholami, A., Moghadassi, A.R., Hosseini, S.M., Shabani, S., Gholami, F., 2014. Preparation and characterization of polyvinyl chloride based nanocomposite nanofiltration-membrane modified by iron oxide nanoparticles for lead removal from water. *J. Ind. Eng. Chem.* 20, 1517–1522. https://doi.org/10.1016/j.jiec.2013.07.041

Herrera-Becerra, R., Zorrilla, C., Ascencio, J.A., 2007. Production of iron oxide nanoparticles by a biosynthesis method: An environmentally friendly route. *J. Phys. Chem. C* 111, 16147–16153. https://doi.org/10.1021/jp072259a

Hilal, N., Al-Zoubi, H., Darwish, N.A., Mohamma, A.W., Abu Arabi, M., 2004. A comprehensive review of nanofiltration membranes: Treatment, pretreatment, modelling, and atomic force microscopy. *Desalination* 170, 281–308. https://doi.org/10.1016/j.desal.2004.01.007

Huang, L., Luo, F., Chen, Z., Megharaj, M., Naidu, R., 2015. Green synthesized conditions impacting on the reactivity of Fe NPs for the degradation of malachite green. *Spectrochim. Acta, Part A* 137, 154–159. https://doi.org/10.1016/j.saa.2014.08.116

Iravani, S., 2011. Green synthesis of metal nanoparticles using plants. *Green Chem.* 13, 2638–2650. https://doi.org/10.1039/C1GC15386B

Kaul, R.K., Kumar, P., Burman, U., Joshi, P., Agrawal, A., Raliya, R., Tarafdar, J.C., 2012. Magnesium and iron nanoparticles production using microorganisms and various salts. *Mater. Sci-Pol.* 30, 254–258. https://doi.org/10.2478/s13536-012-0028-x

Kombaiah, K., Vijaya, J.J., Kennedy, L.J., Bououdina, M., 2017. Optical, magnetic and structural properties of ZnFe2O4 nanoparticles synthesized by conventional and microwave assisted combustion method: A comparative investigation. *Optik* 129, 57–68. https://doi.org/10.1016/j.ijleo.2016.10.058

Lakhotia, S.R., Mukhopadhyay, M., Kumari, P., Mukhopadhyay, M., Kumari, P., 2018. Cerium oxide nanoparticles embedded thin-film nanocomposite nanofiltration membrane for water treatment. *Sci. Rep.* 8, 4976. https://doi.org/10.1038/s41598-018-23188-7

Lau, W.-J., Ismail, A.F., 2009. Polymeric nanofiltration membranes for textile dye wastewater treatment: Preparation, performance evaluation, transport modelling, and fouling control—a review. *Desalination, Engineering with Membranes 2008* 245, 321–348. https://doi.org/10.1016/j.desal.2007.12.058

Liu, H., Sun, Y., Yu, T., Zhang, J., Zhang, X., Zhang, H., Zhao, K., Wei, J., 2019. Plant-mediated biosynthesis of iron nanoparticles-calcium alginate hydrogel membrane and its eminent performance in removal of Cr(VI). *Chem. Eng. J.* 378, 122120. https://doi.org/10.1016/j.cej.2019.122120

Łuczak, M. Paszkiewicz, A. Krukowska, A. Malankowska, A. Zaleska-Medynska Ionic liquids for nano- and microstructures preparation. Part 2: application in synthesis Adv. Colloid Interface Sci., 227 (2016), pp. 1–52

Lunge, S., Singh, S., Sinha, A., 2014. Magnetic iron oxide (Fe3O4) nanoparticles from tea waste for arsenic removal. *J. Magn. Magn. Mater.* 356, 21–31. https://doi.org/10.1016/j.jmmm.2013.12.008

Machado, S., Pinto, S.L., Grosso, J.P., Nouws, H.P.A., Albergaria, J.T., Delerue-Matos, C., 2013a. Green production of zero-valent iron nanoparticles using tree leaf extracts. *Sci. Total Environ.* 445–446, 1–8. https://doi.org/10.1016/j.scitotenv.2012.12.033

Machado, S., Stawiński, W., Slonina, P., Pinto, A.R., Grosso, J.P., Nouws, H.P.A., Albergaria, J.T., Delerue-Matos, C., 2013b. Application of green zero-valent iron nanoparticles to the remediation of soils contaminated with ibuprofen. *Sci. Total Environ.* 461–462, 323–329. https://doi.org/10.1016/j.scitotenv.2013.05.016

Mahdavi, M., Namvar, F., Ahmad, M.B., Mohamad, R., 2013. Green biosynthesis and characterization of magnetic iron oxide (Fe3O4) nanoparticles using seaweed (Sargassum muticum) aqueous extract. *Molecules* 18, 5954–5964. https://doi.org/10.3390/molecules18055954

Manquián-Cerda, K., Cruces, E., Angélica Rubio, M., Reyes, C., Arancibia-Miranda, N., 2017. Preparation of nanoscale iron (oxide, oxyhydroxides and zero-valent) particles derived from blueberries: Reactivity, characterization and removal mechanism of arsenate. *Ecotoxicol. Environ. Saf.* 145, 69–77. https://doi.org/10.1016/j.ecoenv.2017.07.004

Mohamed, Y.M., Azzam, A.M., Amin, B.H., Safwat, N.A., 2015. Mycosynthesis of iron nanoparticles by Alternaria alternata and its antibacterial activity. *Afr. J. Biotechnol.* 14, 1234–1241. https://doi.org/10.4314/ajb.v14i14

Mondal, P., Anweshan, A., Purkait, M. K., 2020a. Green synthesis and environmental application of Iron-based nano-materials and nanocomposite: A review. *Chemosphere* 259, 127509.

Mondal, P., Purkait, M.K., 2017a. Green synthesized iron nanoparticle-embedded pH-responsive PVDF-co-HFP membranes: Optimization study for NPs preparation and nitrobenzene reduction. *Sep. Sci. Technol.* 52, 2338–2355.

Mondal, P., Purkait, M.K., 2017b. Effect of Polyethylene glycol methyl ether blend Humic acid on poly (vinylidene fluoride-co-hexafluropropylene) PVDF-HFP membranes: pH responsiveness and antifouling behavior with optimization approach. *Pol. Test.* 61, 162–176.

Mondal, P., Purkait, M.K., 2018. Green synthesized iron nanoparticles supported on pH responsive polymeric membrane for nitrobenzene reduction and fluoride rejection study: Optimization approach. *J. Clean. Prod.* 170, 1111–1123.

Mondal, P., Purkait, M.K., 2019. Preparation and characterization of novel green synthesized iron–aluminum nanocomposite and studying its efficiency in fluoride removal. *Chemosphere* 235, 391–402.

Mondal, P., Samanta, N.S., Kumar, A., Purkait, M.K., 2020b. Recovery of H_2SO_4 from wastewater in presence of NaCl and $KHCO_3$ through pH responsive polysulfone membrane: Optimization approach. *Pol. Test.* 86, 106463.

Mondal, P., Samanta, N.S., Meghnani, V., Purkait, M.K., 2019. Selective glucose permeability in presence of various salts through tunable pore size of pH-responsive PVDF-co-HFP membrane. *Sep. Purif. Technol.* 221, 249–260.

Moon, J.-W., Rawn, C.J., Rondinone, A.J., Love, L.J., Roh, Y., Everett, S.M., Lauf, R.J., Phelps, T.J., 2010. Large-scale production of magnetic nanoparticles using bacterial fermentation. *J. Ind. Microbiol. Biotechnol.* 37, 1023–1031. https://doi.org/10.1007/s10295-010-0749-y

Njagi, E.C., Huang, H., Stafford, L., Genuino, H., Galindo, H.M., Collins, J.B., Hoag, G.E., Suib, S.L., 2011. Biosynthesis of iron and silver nanoparticles at room temperature using aqueous sorghum bran extracts. *Langmuir* 27, 264–271. https://doi.org/10.1021/la103190n

Park, T.J., Lee, K.G., Lee, S.Y., 2016. Advances in microbial biosynthesis of metal nanoparticles. *Appl. Microbiol. Biotechnol.* 100, 521–534. https://doi.org/10.1007/s00253-015-6904-7

Pavani, K.V., Kumar, N.S., 2013. Adsorption of iron and synthesis of iron nanoparticles by aspergillus species Kvp 12. *Am. J. Nanomater.* 1, 24–26. https://doi.org/10.12691/ajn-1-2-2

Phumying, S., Labuayai, S., Thomas, C., Amornkitbamrung, V., Swatsitang, E., Maensiri, S., 2013. Aloe vera plant-extracted solution hydrothermal synthesis and magnetic properties of magnetite (Fe3O4) nanoparticles. *Appl. Phys. A* 111, 1187–1193. https://doi.org/10.1007/s00339-012-7340-5

Prasad, K.S., Gandhi, P., Selvaraj, K., 2014. Synthesis of green nano iron particles (GnIP) and their application in adsorptive removal of As(III) and As(V) from aqueous solution. *Appl. Surf. Sci.* 317, 1052–1059. https://doi.org/10.1016/j.apsusc.2014.09.042

Raaijmakers, M.J.T., Benes, N.E., 2016. Current trends in interfacial polymerization chemistry. *Prog. Polym. Sci.* 63, 86–142. https://doi.org/10.1016/j.progpolymsci.2016.06.004

Sacca, M.L., Fajardo, C., Costa, G., Lobo, C., Nande, M., Martin M., 2014. Integrating classical and molecular approaches to evaluate the impact of nanosized zero-valent iron (nZVI) on soil organisms Chemosphere, 104, pp. 184–189

Saikia, I., Hazarika, M., Hussian, N., Das, M.R., Tamuly, C., 2017. Biogenic synthesis of $Fe_2O_3@SiO_2$ nanoparticles for ipso-hydroxylation of boronic acid in water. *Tetrahedron Lett.* 58(45), 4255–4259.

Saranya, R., Arthanareeswaran, G., Ismail, A.F., Dionysiou, D.D., Paul, D., 2015. Zero-valent iron impregnated cellulose acetate mixed matrix membranes for the treatment of textile industry effluent. *RSC Adv.* 5, 62486–62497. https://doi.org/10.1039/C5RA06948C

Sarkar, J., Mollick, Md.M.R., Chattopadhyay, D., Acharya, K., 2017. An eco-friendly route of γ-Fe2O3 nanoparticles formation and investigation of the mechanical properties of the HPMC-γ-Fe2O3 nanocomposites. *Bioproc. Biosyst. Eng.* 40, 351–359. https://doi.org/10.1007/s00449-016-1702-x

Schneider, T., Löwa, A., Karagiozov, S., Sprenger, L., Gutiérrez, L., Esposito, T., Marten, G., Saatchi, K., Häfeli, U.O., 2017. Facile microwave synthesis of uniform magnetic nanoparticles with minimal sample processing. *J. Magn. Magn. Mater.* 421, 283–291. https://doi.org/10.1016/j.jmmm.2016.07.063

Sharma, N., Purkait, M. K., 2017. Impact of synthesized amino alcohol plasticizer on the morphology and hydrophilicity of polysulfone ultrafiltration membrane. *J. Membr. Sci.* 522, 202–215.

Singh, P., Kim, Y.-J., Zhang, D., Yang, D.-C., 2016. Biological synthesis of nanoparticles from plants and microorganisms. *Trends Biotechnol.* 34, 588–599. https://doi.org/10.1016/j.tibtech.2016.02.006

Sinha, M.K., Purkait, M.K., 2015. Preparation of fouling resistant PSF flat sheet UF membrane using amphiphilic polyurethane macromolecules. *Desalination.* 355, 155–168.

Smuleac, V., Varma, R., Sikdar, S., Bhattacharyya, D., 2011. Green synthesis of Fe and Fe/Pd bimetallic nanoparticles in membranes for reductive degradation of chlorinated organics. *J. Membr. Sci.* 379, 131–137.

Song, Y., Fan, J.-B., Wang, S., 2017. Recent progress in interfacial polymerization. *Mater. Chem. Front.* 1, 1028–1040. https://doi.org/10.1039/C6QM00325G

Subramaniyam, V., Subashchandrabose, S.R., Thavamani, P., Megharaj, M., Chen, Z., Naidu, R., 2015. Chlorococcum sp. MM11—a novel phyco-nanofactory for the synthesis of iron nanoparticles. *J. Appl. Phycol.* 27, 1861–1869. https://doi.org/10.1007/s10811-014-0492-2

Tandon, P.K., Shukla, R.C., Singh, S.B., 2013. Removal of arsenic(III) from water with clay-supported zerovalent iron nanoparticles synthesized with the help of tea liquor. *Ind. Eng. Chem. Res.* 52, 10052–10058. https://doi.org/10.1021/ie400702k

Wang, T., Jin, X., Chen, Z., Megharaj, M., Naidu, R., 2014a. Green synthesis of Fe nanoparticles using eucalyptus leaf extracts for treatment of eutrophic wastewater. *Sci. Total Environ.* 466–467, 210–213. https://doi.org/10.1016/j.scitotenv.2013.07.022

Wang, T., Lin, J., Chen, Z., Megharaj, M., Naidu, R., 2014b. Green synthesized iron nanoparticles by green tea and eucalyptus leaves extracts used for removal of nitrate in aqueous solution. *J. Cleaner Prod.* 83, 413–419. https://doi.org/10.1016/j.jclepro.2014.07.006

Wittbecker, E.L., Morgan, P.W., 1959. Interfacial polycondensation. *I. J. Polym Sci.* 40, 289–297. https://doi.org/10.1002/pol.1959.1204013701

Yang, Z., Wu, Y., Guo, H., Ma, X.H., Lin, C.E., Zhou, Y., Cao, B., Zhu, B.K., Shih, K., Tang, C.Y., 2017. A novel thin-film nano-templated composite membrane with in situ silver nanoparticles loading: Separation performance enhancement and implications. *J. Membr. Sci.* 544, 351–358.

Yew, Y.P., Shameli, K., Miyake, M., Ahmad Khairudin, N.B.B., Mohamad, S.E.B., Naiki, T., Lee, K.X., 2020. Green biosynthesis of superparamagnetic magnetite Fe3O4 nanoparticles and biomedical applications in targeted anticancer drug delivery system: A review. *Arabian J. Chem.* 13, 2287–2308. https://doi.org/10.1016/j.arabjc.2018.04.013

Zambre, A., Upendran, A., Shukla, R., Chanda, N., Katti, K.K., Cutler, C., Kannan, R., Katti, K.V., 2012. Chapter 6: green nanotechnology—a sustainable approach in the nanorevolution. In: *Sustainable Preparation of Metal Nanoparticles*, pp. 144–156. https://doi.org/10.1039/9781849735469-00144

Zhang, L., Zhang, M., Lu, J., Tang, A., Zhu, L., 2019. Highly permeable thin-film nanocomposite membranes embedded with PDA/PEG nanocapsules as water transport channels. *J. Membr. Sci.* 586, 115–121.

3 Nanofiltration
Unravelling the Potential of the Future

Komal Agrawal and Pradeep Verma*

CONTENTS

3.1 Introduction 35
3.2 Basics of Nanofiltration..35
3.3 Fabrication of Nanofiltration Membrane..37
3.4 Strategies Employed for the Modification and Enhancement of Nanofiltration
Membrane..38
    3.4.1 Pore Size of Membrane 38
    3.4.2 Charge Distribution...38
    3.4.3 Antifouling 39
3.5 Application of Nanofiltration Membrane 39
    3.5.1 Bioremediation 39
    3.5.2 Desalination..39
    3.5.3 Pharmaceutical and Miscellaneous Applications 40
3.6 Limitations..41
3.7 Conclusion 42
Conflict of Interest 42
References...42

3.1 INTRODUCTION

Nanofiltration is a technology that lies in between ultrafiltration and reverse osmosis. Due to the integration and its flexibility, it has received significant attention from the scientific community. In case of the nanofiltration its membrane, due to its pore size (1 nm), has contributed to its diversity as an effective separation tool. In addition, due to the adsorption of the charged solutes or the dissociation of surface functional groups, the nanofiltration membrane possesses charge at moderate level. In the case of nanofiltration the separation process is highly competitive as it operates at no phase charge and exhibits rejections of multivalent inorganic salts and small organic molecules when modest pressures are applied, making the system more effective and economically viable compared to the conventional separation technologies. Due to the previously mentioned advantages nanofiltration has extended applications in various biotechnological, wastewater treatment, and pharmaceutical sectors. The past two decades have witnessed an upsurge in the research aspect of the nanofiltration (Figure 3.1).

Thus, the present chapter consists of basic of nanofiltration, the fabrication of nanofiltration membrane, various strategies employed for the modification and enhancement of the nanofiltration membrane along with its application in various sectors, limitations, and its possible solution.

3.2 BASICS OF NANOFILTRATION

The separation in nanofiltration relies on microhydrodynamics and the interfacial events that occur at the surface of the membrane and in the nanopores of the membrane. The integrated factors such

DOI: 10.1201/9781003165149-3

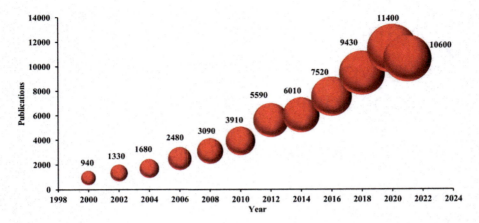

FIGURE 3.1 A graph of the number of publications on term 'nanofiltration' when searched on Google Scholar on 12-09-2021.

as the steric, Donnan (the equilibria and the potential membrane reactions in between the charged species and interface of the charged membrane), dielectric, and transport effects are responsible for the rejections from the nanofiltration membranes (Deen, 1987; Donnan, 1995). Also, it has to be noted that the charge in the membrane is due to the dissociation of the ionisable groups at the surface of the membrane and in the membrane pore structure (Ernst et al., 2000; Hagmeyer and Gimbel, 1998; Hall et al., 1997). The charge of the group will depend on the material used for the fabrication process. As the material is amphoteric in nature, the dissociation of the surface groups depends on the pH of the solution in contact. The nanofiltration membrane may exhibit an isoelectric point at a specific pH (Childress and Elimelech, 1996). Also, in addition to the previously mentioned causes of charge, the liquid in contact may be absorbed in the nanofiltration membrane, resulting in a slight change in charge (Afonso et al., 2001; Schaep and Vandecasteele, 2001). In the case of the membrane, the electrostatic repulsion/attraction depends on ion valence, and the localised ionic environment may also result in the variation (Yaroshchuk, 1998).

The nanofiltration membrane has porous layers present in most of the membranes (Oatley et al., 2012). Thus, efficient characterisation can be done on the basis of pore size and its distribution. Also, the understanding of the steric partitioning can allow better characterisation of the membrane. The various methods adapted to characterise the membrane included gas adsorption (enables measurement of pore size distribution using the Brunauer-Emmett-Teller (BET); Fang et al., 2014; Qian et al., 2013; Wang et al., 2013), atomic force microscopy (enables pore size and distribution, surface roughness, and the force interactions that exist between the colloid and the membrane; Carvalho et al., 2011; Johnson et al., 2012; Misdan et al., 2014; Stawikowska and Livingston, 2013), the neutral solute rejection model (indirect pore size measurement; García-Martín et al., 2014; Kiso et al., 2011; Oatley et al., 2013), and reverse surface impregnation along with TEM (pore size and distribution; Stawikowska and Livingston, 2012). The previously mentioned techniques' efficiency increases if they are used in combination. In addition, the charge of the nanoparticle is a very crucial factor and can be measured using streaming potential (enables measurement of zeta potential that is a pseudo measurement of Donnan potential; Cheng et al., 2011; Déon et al., 2012; Bauman et al., 2013; Rice et al., 2011), streaming current (analogous to streaming potential; Lee et al., 2011; Luxbacher, 2006; Xie et al., 2011), electro-osmosis (zeta potential; Teixeira et al., 2005), and a charged solution rejection study (measurement of charge density; Cheng et al., 2011; Kotrappanavar et al., 2011; Kumar et al., 2013; Gang et al., 2011). The other techniques such as SEM spectroscopy can be used for further characterisation of the membrane (Espinasse et al., 2012; Martínez et al., 2013; Panda and De, 2014; Baek et al., 2012;

Nanofiltration

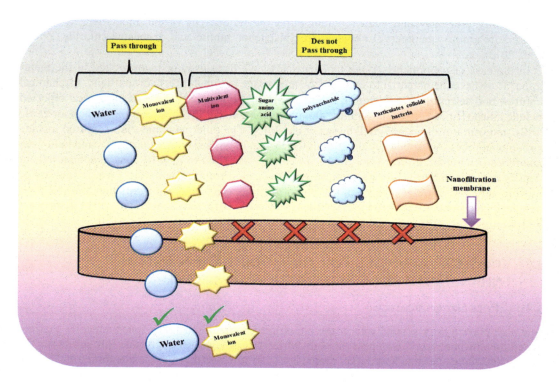

FIGURE 3.2 Diagrammatic representation of the effect of various components on nanofiltration membrane.

Tang et al., 2012; Hurwitz et al., 2010; Do et al., 2012; Lamsal et al., 2012; Montalvillo et al., 2014; Nanda et al., 2011).

Last, the prediction of the functioning of the membrane is critical and over a period of time the focus has shifted from the black box model based on irreversible thermodynamics to the Nernst–Planck equation (Levenstein et al., 1996; Tsuru et al., 1991). For the nanofiltration modelling to give effective results, the pore radius and the membrane charge should be inferred at near atomic scales. Due to the lack of the technologies the characterisation of the nanofiltration membranes has not been feasible. However, with the technological advancements, the nanofiltration modelling is also advancing. Thus, better technological advancements will lead to a better analysis of the nanofiltration membrane (Mohammad et al., 2015; Figure 3.2).

3.3 FABRICATION OF NANOFILTRATION MEMBRANE

The various techniques that have been used for the improvisation of the nanofiltration membrane are nanoparticles, interfacial polymerisation, UV treatment, etc. These methods enable development of higher selectivity and prevent fouling issues and rejection tendency. The nanofiltration system, despite the various technological advancements, has various limitations as well, such as membrane fouling, low separation and rejection efficiency, lifetime, and chemical resistance (Van der Bruggen et al., 2008). The various techniques that have been developed are interfacial polymerisation, nanomaterials, and grafting polymerisation.

Interfacial polymerisation enables the development of an active thin-film layer for nanofiltration and reverse osmosis. The utilisation of interfacial polymerisation for the development of thin-film composite membrane has enabled the development of a membrane with better characteristic features, e.g., fouling resistance and selectivity. This technique is easy to apply and can determine the permeability, efficiency, and solute retention of the membrane. The various monomers reported to be used are amine monomers (Li et al., 2014), bisphenol, tannic acid, trimesoyl chloride (Zhang et al., 2013), etc.

In the case of nanomaterials, the nanoparticles have been used or incorporated in the membrane. They have the potential to increase the membrane permeability, hydrophilicity, mechanical features, and selectivity (restricted to a few reports). The nanoparticles that have been reported include silica (Hu et al., 2012), titanium dioxide (Zhang et al., 2017), boehmite (Vatanpour et al., 2012), and osmium dioxide (Stawikowska et al., 2013). The other nanomaterials reported include carbon nanotubes (Vatanpour et al., 2014), electrospun nanofiber (Bui et al., 2011), and halloysite nanotubes (Zhu et al., 2014). Also, further research is required to identify new nanomaterials to identify new nanomaterials for its application in the nanofiltration membrane. Last, in the case of grafting polymerisation, it consists of UV/photo-grafting, electron beam irradiation, plasma treatment, and layer-by-layer surface modification (Deng et al., 2011; Linggawati et al., 2009; Kim et al., 2011; Lajimi et al., 2011).

3.4 STRATEGIES EMPLOYED FOR THE MODIFICATION AND ENHANCEMENT OF NANOFILTRATION MEMBRANE

3.4.1 PORE SIZE OF MEMBRANE

Pore size plays a very crucial role in the nanofiltration membrane. The ions/molecules that are larger than the pore size do not pass through, while the smaller ones easily pass through. Also, the certain ions/molecules that have larger dimensions can still pass through the membrane and reduce its efficiency due to unevenness in size or deterioration of the nanofiltration membrane (Chein and Liao, 2005). Thus, synthesising membranes with even pores (highly uniform) should be and is the prime focus when developing the nanofiltration membrane. Pore size has been controlled by creating nanopores on polymers by using different techniques such as the track-etching method (Wen et al., 2016; Wang et al., 2018). However, a membrane synthesised by the track-etching method due to its low membrane permeance because of low porosity has resulted in its low utility. Except for the track-etching method, the two methods that have gained utility in the commercial sector are phase inversion and interfacial polymerisation (Zhang et al., 2011). In the case of modifying interlayer spacing of the two-dimensional (2D) materials such as molybdenum disulfide, graphene-based materials and graphite phase carbo nitride have been used. The 2D membranes can have utility in the commercial sector, however, its limitation of mechanical strength has to be overcome prior to its application, as was attempted in the study by Yang et al. (2019). In the study by Yang et al. (2019), a 2D hybrid membrane was designed via the integration of single-walled carbon nanotubes in single-layer graphene nanomesh to enhance the overall mechanical strength. Over a period of time several attempts have be made to tune the interlayer spacing of 2D stacked membrane materials. The major issue is how to limit the spacing while regulating high water permeance, and this has resulted in limiting its utility in ion separation (Sun et al., 2019). Also, the selectivity of the membrane can be improved by adding nanofibers into the polymer matrix using metal organic frameworks and covalent organic frameworks (Wang et al., 2015; Zhu et al., 2017).

3.4.2 CHARGE DISTRIBUTION

The rejection performance of the nanofiltration membrane is affected due to the solute-size-related steric hindrance and electrostatic interaction resulting in Donnan effect. If, suppose, a membrane has a particular charge, it will repel ions of the same charge and attract the opposite charge. These types of interactions result in the production of Coulombs forces counter to the monovalent ions (though here the magnitude is smaller over the multivalent ions). The motility of the monovalent ions is also dependant on the pore size of the membrane, also the Donnan exclusion effect is

Nanofiltration

augmented in the case of the multivalent ions, where it allows the monovalent ions but rejects the multivalent ions (Epsztein et al., 2018). Thus, to overcome the limitations, the modifications of the membrane have been implemented using an amine that allows the separation of heavy metals ions (Zhang et al., 2020).

3.4.3 ANTIFOULING

The fouling of the membrane is one of the major issues of the nanofiltration membrane and cannot be avoided, which is the major limitation preventing/restricting its use in the commercial sector. The membrane fouling results in decline in flux and selectivity of the membrane, and efficiency, collectively. Thus, numerous efforts have been put forth to attain a better anti-fouling membrane with enhanced performance. Hydrophilic modification is an approach and has been used for improving antifouling of the membrane. In the case of hydrophilic modification, a tight hydration layer is formed on the hydrophilic surface that is responsible for weakening non-specific interactions that happen in between the pollutant and the membrane (Yang et al., 2018). It is also based on an approach of fighting fouling on the surface and the development of a low surface energy self-cleaning surface that is followed by the release of the foulants on the surface (Li et al., 2015).

3.5 APPLICATION OF NANOFILTRATION MEMBRANE

The nanofiltration membrane due to its effectiveness has tremendous applications and has been discussed as follows:

3.5.1 BIOREMEDIATION

Nanofiltration has been used for the treatment/remediation of polluted water to high-quality water after the treatment. Nanofiltration has been used for the treatment of ground water; though its practical implementation is yet to be achieved it has been studied at a laboratory and pilot scale using synthetically generated wastewater contaminated with arsenic (Chang et al., 2014; Yu et al., 2013; Harisha et al., 2010; Ahmed et al., 2010; Figoli et al., 2010; Saitua et al., 2011). In the case of surface water, is primarily focused on the softening of water and removal of persistent organic pollutants (Fang et al., 2013, 2014). As compared to the groundwater and surface water, the technique of nanofiltration has been used for the treatment of wastewater. It has been used for the removal of pharmaceutical pollutants and personal care product pollutants that are not easily removed from the wastewater (Zaviska et al., 2013; Lin et al., 2014) (Table 3.1).

3.5.2 DESALINATION

The scarcity of water can be resolved using membrane-based desalination technology. However, there are two major issues/limitations that need to be resolved prior to implementation at commercial scale i.e., membrane fouling and consumption of energy. The membrane fouling as stated earlier reduces the efficiency of the system, and high energy consumption results in the increase of the cost associated with the process, making it economically unviable as it accounts for approximately 44% of the cost of the water that is produced (Wilf, 2004). The nanofiltration is useful in the treatment of seawater desalination along with the integration of other technologies that will enhance the treatment process e.g., ultrafiltration (Song et al., 2012). Here the ultrafiltration-nanofiltration effectively removed 96.3 % of total organic carbon, though the issue of membrane fouling was observed despite chemical cleaning. Thus, later an integrated system of ultrafiltration-nanofiltration seawater reverse osmosis desalination was investigated. It was reported that the salt rejection was only 10%, whereas for SO_4^{2-} it was > 95%, and the calcium sulphate scale is more prone to form on the nanofiltration membrane surface than calcium carbonate at that nanofiltration

TABLE 3.1

The Application of Nanofiltration in Bioremediation

Sl. No	Nanofiltration	Application	Reference
1.	Nanofiltration using membranes NF70, NF45, UTC-20 and UTC-60	Removal of pesticides, nitrate, and hardness from ground water	Van der Bruggen et al. (2001)
2.	Pilot-scale hybrid nanofiltration and reverse osmosis membrane system	Brackish ground water	Srivastava et al. (2021)
3.	Thin-film composite nanofiltration membrane	Arsenic removal from drinking water	Harisha et al. (2010)
4.	Nanofiltration and ultrafiltration membrane	Endocrine disrupting compounds, pharmaceuticals, and personal care products consisting of 52 compounds with different physico-chemical properties	Yoon et al. (2006)
5.	Dually charged nanofiltration membrane by pH-responsive polydopamine	Removal of pharmaceuticals and personal care products	Ouyang et al. (2019)
6.	Positively charged hollow fibre nanofiltration membrane	Removal of pharmaceuticals and personal care products and environmental estrogens	Wei et al. (2021)
7.	Nanofiltration with in situ radical graft polymerisation	Refining the organic and biological fouling resistance and removal of pharmaceutical and personal care products	Lin et al. (2018)
8.	Nanofiltration and ultrafiltration membranes	Removal of endocrine disrupting compounds and pharmaceuticals	Yoon et al. (2007)
9.	Nanofiltration	Surface water treatment	Orecki et al. (2004)
10.	Coagulation-adsorption-nanofiltration	Pesticide contaminated surface water	Sarkar et al. (2007)

permeate recovery. Thus, it was concluded that the system was different from the conventional seawater reverse osmosis desalination processes (Song et al., 2013). A comparative study was done using nanofiltration-seawater reverse osmosis and only reverse osmosis. It was observed that an integrated system was more effective with high recovery rates and low energy consumption (Criscuoli and Drioli, 1999; Mehdizadeh, 2006). Thus recently the replacement to reverse osmosis in a seawater/brackish water desalination system has been considered as an alternative (Mohammad et al., 2015). Also, along with the traditional studies, other unconventional studies have also been reported such as the integration of nanofiltration with ion exchange and forward osmosis for desalination (Sarkar and SenGupta, 2008). Thus, from the previously reported studies it can be stated that due to the efficiency of each process the shortcoming can be overcome with an efficient integration of the system in desalination with better and enhanced implementation of the system.

3.5.3 PHARMACEUTICAL AND MISCELLANEOUS APPLICATIONS

Nanofiltration has been reported for its efficacy in removing active pharmaceutical ingredients that belong to common class of genotoxic impurities (Székely et al., 2012; Siew et al., 2013; Kim et al., 2014; Székely et al., 2011; Martínez et al., 2012). In the study by Omidvar et al. (2015), asymmetric polyethersulfone nanofiltration membranes were prepared using the phase inversion technique. In the study it was also observed that the addition of Brij 58 to the casting solution resulted in the formation of the membranes with superior pure water flux and higher rejection of amoxicillin and ceftriaxone as compared to pure asymmetric polyethersulfone membrane.

Nanofiltration

TABLE 3.2

The Application of Nanofiltration in Various Contaminants

Sl. No	Nanofiltration	Application	Reference
1.	Nanofiltration and reverse osmosis	Removal of five non-steroidal anti-inflammatory drugs, analgesics, and anti-pyretic: acetaminophen, ibuprofen, dipyrone, diclofenac, and caffeine	Licona et al. (2018)
2.	Desal-5 DK nanofiltration membrane	Treatment of complex pharmaceutical wastewater	Wei et al. (2010)
3.	Organic solvent nanofiltration membranes	Separation of oil compounds from vegetable oil and solvent recovery	Shi et al. (2019)
4.	Low-pressure nanofiltration membranes	Recovery and reuse of dairy industry effluents	Koyuncu et al. (2000)
5.	Nanofiltration and reverse osmosis membranes (a comparative study)	Treatment of dairy effluent	Balannec et al. (2005)
6.	Ultra- and nanofiltration	Whey protein and lactose utilisation	Atra et al. (2005)
7.	Nanofiltration membranes	Purification of solvent in oil industry	Darvishmanesh et al. (2011)
8.	Utilisation of two loose nanofiltration membranes and one dense ultrafiltration membrane	Fractionation of commercial oligosaccharide	Goulas et al. (2003)
9.	Nanofiltration	Concentration of flavonoids and phenolic compounds	Mello et al. (2010)
10.	Nanofiltration and tight ultrafiltration membranes	Recovery of polyphenols from agro-food by-products	Cassano et al. (2018)

Also, recently the membrane technology has been used in the food industry for the separation of trace amounts of compounds, i.e., antioxidants from vegetable oil (Sereewatthanawut et al., 2011). In the study by Sotoft et al. (2012), blackcurrant juice concentrate was produced using the integrated membrane process over the traditional multi-step evaporators, and it was reported that the annual production was 17,283 t of 66°Brix out of single-strength juice; also, the operational cost was reduced by 43%. Thus, the potential of the membrane system can contribute significantly to the food industry (Table 3.2).

3.6 LIMITATIONS

The technology of nanofiltration was developed in 1980 and, as with any new technique, the transitions process is always very rigorous and the problems encountered are unanticipated start-up issues, the believers vs non-believers, continuous detection of problem areas leading to rigorous troubleshooting with fruitful results, and technical improvisation with technological advancement. Nanofiltration is a technique in between reverse osmosis and ultrafiltration and rejects molecules/ions on the order of nanometres and was introduced for integrating softening with organic compound removal (Eriksson, 1988). Since its development its application has increased tremendously in various sectors. It has limitations such as membrane-fouling ash that restricted its use, and to overcome the issue numerous approaches have been developed as mentioned in the previous section. Further, the implementation of nanofiltration at an industrial scale suffers drawbacks such as the interaction between the membrane and the feedstock, scaling, fouling, low yield, high energy requirements, etc. Thus, to overcome the drawbacks, various integrated approaches ha e to be devised that would increase the efficiency as well as make it economically feasible to use at larger scale. In the case of the pharmaceutical sectors, the implementation membrane technology has limitations such as in the case of solvents where the issue is membrane stability and lack of basic understanding of the process that can be either modelled or simulated using bioinformatical tools (Gevers et al., 2006; Silva et al., 2005; Robinson et al., 2004; Dijkstra et al., 2006), though its use can result in beneficial output with less energy consumption (White, 2006; Boam and Nozari, 2006). Further, in the case of the food industry, the adaptation of new technologies is very fast and the dairy industry was the first to use the technique of nanofiltration (Räsänen et al., 2002). However, in the case of the food industry, the challenge is very high as the food, the end product, should be low in fat and calories and should fit the diet that requires intensive and enhanced separation. Despite the limitations, the advantages of the nanofiltration technique cannot be undermined, and regardless of the drawback it is widely used in various industries. A clearer understanding of the materials and process would allow better understanding and implementation of the process. The development/identification of novel membrane materials that are efficient and cost effective will allow extension of the application of nanofiltration in various sectors. The identified problem requires practical solutions and a number of attempts have been made in this direction and in the coming decade and will allow the development and commercialisation of an efficient system (Van der Bruggen et al., 2008).

3.7 CONCLUSION

Nanofiltration is a system with high efficiency, and it has tremendous applications in various sectors. With further research its application has extended to various unidentified/unexplored sectors. The technique of nanofiltration can allow efficient treatment and separation of pollutants/contaminants and the integration of various techniques will enhance the specificity of the system and extend its application and utility further. Thus, with intensive research in the future, the application and applicability of the nanofiltration technique can open new windows for a sustainable ecosystem.

CONFLICT OF INTEREST

Authors declare no conflict of interest.

REFERENCES

Afonso, M.D., Hagmeyer, G. and Gimbel, R., 2001. Streaming potential measurements to assess the variation of nanofiltration membranes surface charge with the concentration of salt solutions. *Separation and Purification Technology*, 22, pp. 529–541.

Ahmed, S., Rasul, M.G., Hasib, M.A. and Watanabe, Y., 2010. Performance of nanofiltration membrane in a vibrating module (VSEP-NF) for arsenic removal. *Desalination*, 252(1–3), pp. 127–134.

Atra, R., Vatai, G., Bekassy-Molnar, E. and Balint, A., 2005. Investigation of ultra-and nanofiltration for utilization of whey protein and lactose. *Journal of Food Engineering*, 67(3), pp. 325–332.

Baek, Y., Kang, J., Theato, P. and Yoon, J., 2012. Measuring hydrophilicity of RO membranes by contact angles via sessile drop and captive bubble method: A comparative study. *Desalination*, 303, pp. 23–28.

Balannec, B., Vourch, M., Rabiller-Baudry, M. and Chaufer, B., 2005. Comparative study of different nanofiltration and reverse osmosis membranes for dairy effluent treatment by dead-end filtration. *Separation and Purification Technology*, 42(2), pp. 195–200.

Bauman, M., Košak, A., Lobnik, A., Petrinić, I. and Luxbacher, T., 2013. Nanofiltration membranes modified with alkoxysilanes: Surface characterization using zeta-potential. *Colloids and Surfaces A: Physicochemical and Engineering Aspects*, 422, pp. 110–117.

Boam, A. and Nozari, A., 2006. Fine chemical: OSN–a lower energy alternative. *Filtration & Separation*, 43(3), pp. 46–48.

Bui, N.N., Lind, M.L., Hoek, E.M. and McCutcheon, J.R., 2011. Electrospun nanofiber supported thin film composite membranes for engineered osmosis. *Journal of Membrane Science*, 385, pp. 10–19.

Carvalho, A.L., Maugeri, F., Silva, V., Hernández, A., Palacio, L. and Pradanos, P., 2011. AFM analysis of the surface of nanoporous membranes: Application to the nanofiltration of potassium clavulanate. *Journal of Materials Science*, 46(10), pp. 3356–3369.

Cassano, A., Conidi, C., Ruby-Figueroa, R. and Castro-Muñoz, R., 2018. Nanofiltration and tight ultrafiltration membranes for the recovery of polyphenols from agro-food by-products. *International Journal of Molecular Sciences*, 19(2), p. 351.

Chang, F.F., Liu, W.J. and Wang, X.M., 2014. Comparison of polyamide nanofiltration and low-pressure reverse osmosis membranes on As (III) rejection under various operational conditions. *Desalination*, 334(1), pp. 10–16.

Chein, R. and Liao, W., 2005. Analysis of particle–wall interactions during particle free fall. *Journal of Colloid and Interface Science*, 288(1), pp. 104–113.

Cheng, S., Oatley, D.L., Williams, P.M. and Wright, C.J., 2011. Positively charged nanofiltration membranes: Review of current fabrication methods and introduction of a novel approach. *Advances in Colloid and Interface Science*, 164(1–2), pp. 12–20.

Childress, A.E. and Elimelech, M., 1996. Effect of solution chemistry on the surface charge of polymeric reverse osmosis and nanofiltration membranes. *Journal of Membrane Science*, 119(2), pp. 253–268.

Criscuoli, A. and Drioli, E., 1999. Energetic and exergetic analysis of an integrated membrane desalination system. *Desalination*, 124(1–3), pp. 243–249.

Darvishmanesh, S., Robberecht, T., Luis, P., Degrève, J. and Van der Bruggen, B., 2011. Performance of nanofiltration membranes for solvent purification in the oil industry. *Journal of the American Oil Chemists' Society*, 88(8), pp. 1255–1261.

Deen, W.M., 1987. Hindered transport of large molecules in liquid-filled pores. *AICHE Journal*, 33(9), pp. 1409–1425.

Deng, H., Xu, Y., Chen, Q., Wei, X. and Zhu, B., 2011. High flux positively charged nanofiltration membranes prepared by UV-initiated graft polymerization of methacrylatoethyl trimethyl ammonium chloride (DMC) onto polysulfone membranes. *Journal of Membrane Science*, 366(1–2), pp. 363–372.

Déon, S., Fievet, P. and Doubad, C.O., 2012. Tangential streaming potential/current measurements for the characterization of composite membranes. *Journal of Membrane Science*, 423, pp. 413–421.

Dijkstra, M.F.J., Bach, S. and Ebert, K., 2006. A transport model for organophilic nanofiltration. *Journal of Membrane Science*, 286(1–2), pp. 60–68.

Do, V.T., Tang, C.Y., Reinhard, M. and Leckie, J.O., 2012. Degradation of polyamide nanofiltration and reverse osmosis membranes by hypochlorite. *Environmental Science & Technology*, 46(2), pp. 852–859.

Donnan, F.G., 1995. Theory of membrane equilibria and membrane potentials in the presence of non-dialysing electrolytes. A contribution to physical-chemical physiology. *Journal of Membrane Science*, 100(1), pp. 45–55.

Epsztein, R., Shaulsky, E., Dizge, N., Warsinger, D.M. and Elimelech, M., 2018. Role of ionic charge density in donnan exclusion of monovalent anions by nanofiltration. *Environmental Science & Technology*, 52(7), pp. 4108–4116.

Eriksson, P., 1988. Nanofiltration extends the range of membrane filtration. *Environmental Progress*, 7(1), pp. 58–62.

Ernst, M., Bismarck, A., Springer, J. and Jekel, M., 2000. Zeta-potential and rejection rates of a polyethersulfone nanofiltration membrane in single salt solutions. *Journal of Membrane Science*, 165(2), pp. 251–259.

Espinasse, B.P., Chae, S.R., Marconnet, C., Coulombel, C., Mizutani, C., Djafer, M., Heim, V. and Wiesner, M.R., 2012. Comparison of chemical cleaning reagents and characterization of foulants of nanofiltration membranes used in surface water treatment. *Desalination*, 296, pp. 1–6.

Fang, W., Shi, L. and Wang, R., 2013. Interfacially polymerized composite nanofiltration hollow fiber membranes for low-pressure water softening. *Journal of Membrane Science*, 430, pp. 129–139.

Fang, W., Shi, L. and Wang, R., 2014. Mixed polyamide-based composite nanofiltration hollow fiber membranes with improved low-pressure water softening capability. *Journal of Membrane Science*, 468, pp. 52–61.

Fang, Y., Bian, L., Bi, Q., Li, Q. and Wang, X., 2014. Evaluation of the pore size distribution of a forward osmosis membrane in three different ways. *Journal of Membrane Science*, 454, pp. 390–397.

Figoli, A., Cassano, A., Criscuoli, A., Mozumder, M.S.I., Uddin, M.T., Islam, M.A. and Drioli, E., 2010. Influence of operating parameters on the arsenic removal by nanofiltration. *Water Research*, 44(1), pp. 97–104.

Gang, Y., Hong, S., Wenqiang, L., Weihong, X.I.N.G. and Nanping, X.U., 2011. Investigation of Mg2+/Li+ separation by nanofiltration. *Chinese Journal of Chemical Engineering*, 19(4), pp. 586–591.

García-Martín, N., Silva, V., Carmona, F.J., Palacio, L., Hernández, A. and Prádanos, P., 2014. Pore size analysis from retention of neutral solutes through nanofiltration membranes. The contribution of concentration–polarization. *Desalination*, 344, pp. 1–11.

Gevers, L.E., Meyen, G., De Smet, K., Van De Velde, P., Du Prez, F., Vankelecom, I.F. and Jacobs, P.A., 2006. Physico-chemical interpretation of the SRNF transport mechanism for solutes through dense silicone membranes. *Journal of Membrane Science*, 274(1–2), pp. 173–182.

Goulas, A.K., Grandison, A.S. and Rastall, R.A., 2003. Fractionation of oligosaccharides by nanofiltration. *Journal of the Science of Food and Agriculture*, 83(7), pp. 675–680.

Hagmeyer, G. and Gimbel, R., 1998. Modelling the salt rejection of nanofiltration membranes for ternary ion mixtures and for single salts at different pH values. *Desalination*, 117(1–3), pp. 247–256.

Hall, M.S., Lloyd, D.R. and Starov, V.M., 1997. Reverse osmosis of multicomponent electrolyte solutions Part II. Experimental verification. *Journal of Membrane Science*, 128(1), pp. 39–53.

Harisha, R.S., Hosamani, K.M., Keri, R.S., Nataraj, S.K. and Aminabhavi, T.M., 2010. Arsenic removal from drinking water using thin film composite nanofiltration membrane. *Desalination*, 252(1–3), pp. 75–80.

Hu, D., Xu, Z.L. and Chen, C., 2012. Polypiperazine-amide nanofiltration membrane containing silica nanoparticles prepared by interfacial polymerization. *Desalination*, 301, pp. 75–81.

Hurwitz, G., Guillen, G.R. and Hoek, E.M., 2010. Probing polyamide membrane surface charge, zeta potential, wettability, and hydrophilicity with contact angle measurements. *Journal of Membrane Science*, 349(1–2), pp. 349–357.

Johnson, D.J., Al Malek, S.A., Al-Rashdi, B.A.M. and Hilal, N., 2012. Atomic force microscopy of nanofiltration membranes: Effect of imaging mode and environment. *Journal of Membrane Science*, 389, pp. 486–498.

Kim, E.S., Yu, Q. and Deng, B., 2011. Plasma surface modification of nanofiltration (NF) thin-film composite (TFC) membranes to improve anti organic fouling. *Applied Surface Science*, 257(23), pp. 9863–9871.

Kim, J.F., Székely, G., Valtcheva, I.B. and Livingston, A.G., 2014. Increasing the sustainability of membrane processes through cascade approach and solvent recovery—pharmaceutical purification case study. *Green Chemistry*, 16(1), pp. 133–145.

Kiso, Y., Muroshige, K., Oguchi, T., Hirose, M., Ohara, T. and Shintani, T., 2011. Pore radius estimation based on organic solute molecular shape and effects of pressure on pore radius for a reverse osmosis membrane. *Journal of Membrane Science*, 369(1–2), pp. 290–298.

Kotrappanavar, N.S., Hussain, A.A., Abashar, M.E.E., Al-Mutaz, I.S., Aminabhavi, T.M. and Nadagouda, M.N., 2011. Prediction of physical properties of nanofiltration membranes for neutral and charged solutes. *Desalination*, 280(1–3), pp. 174–182.

Koyuncu, I., Turan, M., Topacik, D. and Ates, A., 2000. Application of low pressure nanofiltration membranes for the recovery and reuse of dairy industry effluents. *Water Science and Technology*, 41(1), pp. 213–221.

Kumar, V.S., Hariharan, K.S., Mayya, K.S. and Han, S., 2013. Volume averaged reduced order Donnan Steric Pore Model for nanofiltration membranes. *Desalination*, 322, pp. 21–28.

Lajimi, R.H., Ferjani, E., Roudesli, M.S. and Deratani, A., 2011. Effect of LbL surface modification on characteristics and performances of cellulose acetate nanofiltration membranes. *Desalination*, 266(1–3), pp. 78–86.

Lamsal, R., Harroun, S.G., Brosseau, C.L. and Gagnon, G.A., 2012. Use of surface enhanced Raman spectroscopy for studying fouling on nanofiltration membrane. *Separation and Purification Technology*, 96, pp. 7–11.

Lee, S., Lee, E., Elimelech, M. and Hong, S., 2011. Membrane characterization by dynamic hysteresis: Measurements, mechanisms, and implications for membrane fouling. *Journal of Membrane Science*, 366(1–2), pp. 17–24.

Levenstein, R., Hasson, D. and Semiat, R., 1996. Utilization of the Donnan effect for improving electrolyte separation with nanofiltration membranes. *Journal of Membrane Science*, 116(1), pp. 77–92.

Li, Y., Su, Y., Dong, Y., Zhao, X., Jiang, Z., Zhang, R. and Zhao, J., 2014. Separation performance of thin-film composite nanofiltration membrane through interfacial polymerization using different amine monomers. *Desalination*, 333(1), pp. 59–65.

Li, Y., Su, Y., Zhao, X., Zhang, R., Liu, Y., Fan, X., Zhu, J., Ma, Y., Liu, Y. and Jiang, Z., 2015. Preparation of antifouling nanofiltration membrane via interfacial polymerization of fluorinated polyamine and trimesoyl chloride. *Industrial & Engineering Chemistry Research*, 54(33), pp. 8302–8310.

Licona, K.P.M., Geaquinto, L.D.O., Nicolini, J.V., Figueiredo, N.G., Chiapetta, S.C., Habert, A.C. and Yokoyama, L., 2018. Assessing potential of nanofiltration and reverse osmosis for removal of toxic pharmaceuticals from water. *Journal of Water Process Engineering*, 25, pp. 195–204.

Lin, Y.L., Chiou, J.H. and Lee, C.H., 2014. Effect of silica fouling on the removal of pharmaceuticals and personal care products by nanofiltration and reverse osmosis membranes. *Journal of Hazardous Materials*, 277, pp. 102–109.

Lin, Y.L., Tsai, C.C. and Zheng, N.Y., 2018. Improving the organic and biological fouling resistance and removal of pharmaceutical and personal care products through nanofiltration by using in situ radical graft polymerization. *Science of the Total Environment*, 635, pp. 543–550.

Linggawati, A., Mohammad, A.W. and Ghazali, Z., 2009. Effect of electron beam irradiation on morphology and sieving characteristics of nylon-66 membranes. *European Polymer Journal*, 45(10), pp. 2797–2804.

Luxbacher, T., 2006. Electrokinetic characterization of flat sheet membranes by streaming current measurement. *Desalination (Amsterdam)*, 199(1–3), pp. 376–377.

Martínez, M.B., Jullok, N., Negrin, Z.R., Van der Bruggen, B. and Luis, P., 2013. Effect of impurities in the recovery of 1-(5-bromo-fur-2-il)-2-bromo-2-nitroethane using nanofiltration. *Chemical Engineering and Processing: Process Intensification*, 70, pp. 241–249.

Martínez, M.B., Van der Bruggen, B., Negrin, Z.R. and Alconero, P.L., 2012. Separation of a high-value pharmaceutical compound from waste ethanol by nanofiltration. *Journal of Industrial and Engineering Chemistry*, 18(5), pp. 1635–1641.

Mehdizadeh, H., 2006. Membrane desalination plants from an energy–exergy viewpoint. *Desalination*, 191(1–3), pp. 200–209.

Mello, B.C., Petrus, J.C.C. and Hubinger, M.D., 2010. Concentration of flavonoids and phenolic compounds in aqueous and ethanolic propolis extracts through nanofiltration. *Journal of Food Engineering*, 96(4), pp. 533–539.

Misdan, N., Lau, W.J., Ismail, A.F., Matsuura, T. and Rana, D., 2014. Study on the thin film composite poly (piperazine-amide) nanofiltration membrane: Impacts of physicochemical properties of substrate on interfacial polymerization formation. *Desalination*, 344, pp. 198–205.

Mohammad, A.W., Teow, Y.H., Ang, W.L., Chung, Y.T., Oatley-Radcliffe, D.L. and Hilal, N., 2015. Nanofiltration membranes review: Recent advances and future prospects. *Desalination*, 356, pp. 226–254.

Montalvillo, M., Silva, V., Palacio, L., Calvo, J.I., Carmona, F.J., Hernández, A. and Prádanos, P., 2014. Charge and dielectric characterization of nanofiltration membranes by impedance spectroscopy. *Journal of Membrane Science*, 454, pp. 163–173.

Nanda, D., Tung, K.L., Hung, W.S., Lo, C.H., Jean, Y.C., Lee, K.R., Hu, C.C. and Lai, J.Y., 2011. Characterization of fouled nanofiltration membranes using positron annihilation spectroscopy. *Journal of Membrane Science*, 382(1–2), pp. 124–134.

Oatley, D.L., Llenas, L., Aljohani, N.H., Williams, P.M., Martínez-Lladó, X., Rovira, M. and de Pablo, J., 2013. Investigation of the dielectric properties of nanofiltration membranes. *Desalination*, 315, pp. 100–106.

Oatley, D.L., Llenas, L., Pérez, R., Williams, P.M., Martínez-Lladó, X. and Rovira, M., 2012. Review of the dielectric properties of nanofiltration membranes and verification of the single oriented layer approximation. *Advances in Colloid and Interface Science*, 173, pp. 1–11.

Omidvar, M., Soltanieh, M., Mousavi, S.M., Saljoughi, E., Moarefian, A. and Saffaran, H., 2015. Preparation of hydrophilic nanofiltration membranes for removal of pharmaceuticals from water. *Journal of Environmental Health Science and Engineering*, 13(1), pp. 1–9.

Orecki, A., Tomaszewska, M., Karakulski, K. and Morawski, A.W., 2004. Surface water treatment by the nanofiltration method. *Desalination*, 162, pp. 47–54.

Ouyang, Z., Huang, Z., Tang, X., Xiong, C., Tang, M. and Lu, Y., 2019. A dually charged nanofiltration membrane by pH-responsive polydopamine for pharmaceuticals and personal care products removal. *Separation and Purification Technology*, 211, pp. 90–97.

Panda, S.R. and De, S., 2014. Preparation, characterization and performance of ZnCl2 incorporated polysulfone (PSF)/polyethylene glycol (PEG) blend low pressure nanofiltration membranes. *Desalination*, 347, pp. 52–65.

Qian, H., Zheng, J. and Zhang, S., 2013. Preparation of microporous polyamide networks for carbon dioxide capture and nanofiltration. *Polymer*, 54(2), pp. 557–564.

Räsänen, E., Nyström, M., Sahlstein, J. and Tossavainen, O., 2002. Comparison of commercial membranes in nanofiltration of sweet whey. *Le Lait*, 82(3), pp. 343–356.

Rice, G., Barber, A.R., O'Connor, A.J., Pihlajamaki, A., Nystrom, M., Stevens, G.W. and Kentish, S.E., 2011. The influence of dairy salts on nanofiltration membrane charge. *Journal of Food Engineering*, 107(2), pp. 164–172.

Robinson, J.P., Tarleton, E.S., Millington, C.R. and Nijmeijer, A., 2004. Solvent flux through dense polymeric nanofiltration membranes. *Journal of Membrane Science*, 230(1–2), pp. 29–37.

Saitua, H., Gil, R. and Padilla, A.P., 2011. Experimental investigation on arsenic removal with a nanofiltration pilot plant from naturally contaminated groundwater. *Desalination*, 274(1–3), pp. 1–6.

Sarkar, B., Venkateswralu, N., Rao, R.N., Bhattacharjee, C. and Kale, V., 2007. Treatment of pesticide contaminated surface water for production of potable water by a coagulation-adsorption-nanofiltration approach. *Desalination*, 212(1–3), pp. 129–140.

Sarkar, S. and SenGupta, A.K., 2008. A new hybrid ion exchange-nanofiltration (HIX-NF) separation process for energy-efficient desalination: Process concept and laboratory evaluation. *Journal of Membrane Science*, 324(1–2), pp. 76–84.

Schaep, J. and Vandecasteele, C., 2001. Evaluating the charge of nanofiltration membranes. *Journal of Membrane Science*, 188(1), pp. 129–136.

Sereewatthanawut, I., Baptista, I.I.R., Boam, A.T., Hodgson, A. and Livingston, A.G., 2011. Nanofiltration process for the nutritional enrichment and refining of rice bran oil. *Journal of Food Engineering*, 102(1), pp. 16–24.

Shi, G.M., Farahani, M.H.D.A., Liu, J.Y. and Chung, T.S., 2019. Separation of vegetable oil compounds and solvent recovery using commercial organic solvent nanofiltration membranes. *Journal of Membrane Science*, 588, p. 117202.

Siew, W.E., Livingston, A.G., Ates, C. and Merschaert, A., 2013. Molecular separation with an organic solvent nanofiltration cascade–augmenting membrane selectivity with process engineering. *Chemical Engineering Science*, 90, pp. 299–310.

Silva, P., Han, S. and Livingston, A.G., 2005. Solvent transport in organic solvent nanofiltration membranes. *Journal of Membrane Science*, 262(1–2), pp. 49–59.

Song, Y., Gao, X. and Gao, C., 2013. Evaluation of scaling potential in a pilot-scale NF–SWRO integrated seawater desalination system. *Journal of Membrane Science*, 443, pp. 201–209.

Song, Y., Su, B., Gao, X. and Gao, C., 2012. The performance of polyamide nanofiltration membrane for long-term operation in an integrated membrane seawater pretreatment system. *Desalination*, 296, pp. 30–36.

Sotoft, L.F., Christensen, K.V., Andrésen, R. and Norddahl, B., 2012. Full scale plant with membrane based concentration of blackcurrant juice on the basis of laboratory and pilot scale tests. *Chemical Engineering and Processing: Process Intensification*, 54, pp. 12–21.

Srivastava, A., Aghilesh, K., Nair, A., Ram, S., Agarwal, S., Ali, J., Singh, R. and Garg, M.C., 2021. Response surface methodology and artificial neural network modelling for the performance evaluation of pilot-scale hybrid nanofiltration (NF) & reverse osmosis (RO) membrane system for the treatment of brackish ground water. *Journal of Environmental Management*, 278, p. 111497.

Stawikowska, J., Jimenez-Solomon, M.F., Bhole, Y. and Livingston, A.G., 2013. Nanoparticle contrast agents to elucidate the structure of thin film composite nanofiltration membranes. *Journal of Membrane Science*, 442, pp. 107–118.

Stawikowska, J. and Livingston, A.G., 2012. Nanoprobe imaging molecular scale pores in polymeric membranes. *Journal of Membrane Science*, 413, pp. 1–16.

Stawikowska, J. and Livingston, A.G., 2013. Assessment of atomic force microscopy for characterisation of nanofiltration membranes. *Journal of Membrane Science*, 425, pp. 58–70.

Sun, Z., Wu, Q., Ye, C., Wang, W., Zheng, L., Dong, F., Yi, Z., Xue, L. and Gao, C., 2019. Nanovoid membranes embedded with hollow zwitterionic nanocapsules for a superior desalination performance. *Nano Letters*, 19(5), pp. 2953–2959.

Székely, G., Bandarra, J., Heggie, W., Sellergren, B. and Ferreira, F.C., 2011. Organic solvent nanofiltration: A platform for removal of genotoxins from active pharmaceutical ingredients. *Journal of Membrane Science*, 381(1–2), pp. 21–33.

Székely, G., Bandarra, J., Heggie, W., Sellergren, B. and Ferreira, F.C., 2012. A hybrid approach to reach stringent low genotoxic impurity contents in active pharmaceutical ingredients: Combining molecularly imprinted polymers and organic solvent nanofiltration for removal of 1, 3-diisopropylurea. *Separation and Purification Technology*, 86, pp. 79–87.

Nanofiltration

Tang, C.Y., Reinhard, M. and Leckie, J.O., 2012. Effects of hypochlorous acid exposure on the rejection of salt, polyethylene glycols, boron and arsenic (V) by nanofiltration and reverse osmosis membranes. *Water Research*, 46(16), pp. 5217–5223.

Teixeira, M.R., Rosa, M.J. and Nyström, M., 2005. The role of membrane charge on nanofiltration performance. *Journal of Membrane Science*, 265(1–2), pp. 160–166.

Tsuru, T., Nakao, S.I. and Kimura, S., 1991. Calculation of ion rejection by extended Nernst–Planck equation with charged reverse osmosis membranes for single and mixed electrolyte solutions. *Journal of Chemical Engineering of Japan*, 24(4), pp. 511–517.

Van der Bruggen, B., Everaert, K., Wilms, D. and Vandecasteele, C., 2001. Application of nanofiltration for removal of pesticides, nitrate and hardness from ground water: Rejection properties and economic evaluation. *Journal of Membrane Science*, 193(2), pp. 239–248.

Van der Bruggen, B., Mänttäri, M. and Nyström, M., 2008. Drawbacks of applying nanofiltration and how to avoid them: A review. *Separation and Purification Technology*, 63(2), pp. 251–263.

Vatanpour, V., Esmaeili, M. and Farahani, M.H.D.A., 2014. Fouling reduction and retention increment of polyethersulfone nanofiltration membranes embedded by amine-functionalized multi-walled carbon nanotubes. *Journal of Membrane Science*, 466, pp. 70–81.

Vatanpour, V., Madaeni, S.S., Rajabi, L., Zinadini, S. and Derakhshan, A.A., 2012. Boehmite nanoparticles as a new nanofiller for preparation of antifouling mixed matrix membranes. *Journal of Membrane Science*, 401, pp. 132–143.

Wang, L., Fang, M., Liu, J., He, J., Li, J. and Lei, J., 2015. Layer-by-layer fabrication of high-performance polyamide/ZIF-8 nanocomposite membrane for nanofiltration applications. *ACS Applied Materials & Interfaces*, 7(43), pp. 24082–24093.

Wang, P., Wang, M., Liu, F., Ding, S., Wang, X., Du, G., Liu, J., Apel, P., Kluth, P., Trautmann, C. and Wang, Y., 2018. Ultrafast ion sieving using nanoporous polymeric membranes. *Nature Communications*, 9(1), pp. 1–9.

Wang, T., Yang, Y., Zheng, J., Zhang, Q. and Zhang, S., 2013. A novel highly permeable positively charged nanofiltration membrane based on a nanoporous hyper-crosslinked polyamide barrier layer. *Journal of Membrane Science*, 448, pp. 180–189.

Wei, X., Wang, Z., Fan, F., Wang, J. and Wang, S., 2010. Advanced treatment of a complex pharmaceutical wastewater by nanofiltration: Membrane foulant identification and cleaning. *Desalination*, 251(1–3), pp. 167–175.

Wei, X., Zhang, Q., Cao, S., Xu, X., Chen, Y., Liu, L., Yang, R., Chen, J. and Lv, B., 2021. Removal of pharmaceuticals and personal care products (PPCPs) and environmental estrogens (EEs) from water using positively charged hollow fiber nanofiltration membrane. *Environmental Science and Pollution Research*, 28(7), pp. 8486–8497.

Wen, Q., Yan, D., Liu, F., Wang, M., Ling, Y., Wang, P., Kluth, P., Schauries, D., Trautmann, C., Apel, P. and Guo, W., 2016. Highly selective ionic transport through subnanometer pores in polymer films. *Advanced Functional Materials*, 26(32), pp. 5796–5803.

White, L.S., 2006. Development of large-scale applications in organic solvent nanofiltration and pervaporation for chemical and refining processes. *Journal of Membrane Science*, 286(1–2), pp. 26–35.

Wilf, M., 2004, December. Fundamentals of RO-NF technology. In *International Conference on Desalination Costing*. Limassol.

Xie, H., Saito, T. and Hickner, M.A., 2011. Zeta potential of ion-conductive membranes by streaming current measurements. *Langmuir*, 27(8), pp. 4721–4727.

Yang, H.C., Xie, Y., Chan, H., Narayanan, B., Chen, L., Waldman, R.Z., Sankaranarayanan, S.K., Elam, J.W. and Darling, S.B., 2018. Crude-oil-repellent membranes by atomic layer deposition: Oxide interface engineering. *ACS Nano*, 12(8), pp. 8678–8685.

Yang, Y., Yang, X., Liang, L., Gao, Y., Cheng, H., Li, X., Zou, M., Ma, R., Yuan, Q. and Duan, X., 2019. Large-area graphene-nanomesh/carbon-nanotube hybrid membranes for ionic and molecular nanofiltration. *Science*, 364(6445), pp. 1057–1062.

Yaroshchuk, A.E., 1998. Rejection mechanisms of NF membranes. *Membrane Technology*, 1998(100), pp. 9–12.

Yoon, Y., Westerhoff, P., Snyder, S.A. and Wert, E.C., 2006. Nanofiltration and ultrafiltration of endocrine disrupting compounds, pharmaceuticals and personal care products. *Journal of Membrane Science*, 270(1–2), pp. 88–100.

Yoon, Y., Westerhoff, P., Snyder, S.A., Wert, E.C. and Yoon, J., 2007. Removal of endocrine disrupting compounds and pharmaceuticals by nanofiltration and ultrafiltration membranes. *Desalination*, 202(1–3), pp. 16–23.

Yu, Y., Zhao, C., Wang, Y., Fan, W. and Luan, Z., 2013. Effects of ion concentration and natural organic matter on arsenic (V) removal by nanofiltration under different transmembrane pressures. *Journal of Environmental Sciences*, 25(2), pp. 302–307.

Zaviska, F., Drogui, P., Grasmick, A., Azais, A. and Héran, M., 2013. Nanofiltration membrane bioreactor for removing pharmaceutical compounds. *Journal of Membrane Science*, 429, pp. 121–129.

Zhang, H., He, Q., Luo, J., Wan, Y. and Darling, S.B., 2020. Sharpening nanofiltration: Strategies for enhanced membrane selectivity. *ACS Applied Materials & Interfaces*, 12(36), pp. 39948–39966.

Zhang, H., Zhang, H., Li, X., Mai, Z. and Zhang, J., 2011. Nanofiltration (NF) membranes: The next generation separators for all vanadium redox flow batteries (VRBs)? *Energy & Environmental Science*, 4(5), pp. 1676–1679.

Zhang, Q., Fan, L., Yang, Z., Zhang, R., Liu, Y.N., He, M., Su, Y. and Jiang, Z., 2017. Loose nanofiltration membrane for dye/salt separation through interfacial polymerization with in-situ generated TiO2 nanoparticles. *Applied Surface Science*, 410, pp. 494–504.

Zhang, Y., Su, Y., Peng, J., Zhao, X., Liu, J., Zhao, J. and Jiang, Z., 2013. Composite nanofiltration membranes prepared by interfacial polymerization with natural material tannic acid and trimesoyl chloride. *Journal of Membrane Science*, 429, pp. 235–242.

Zhu, J., Guo, N., Zhang, Y., Yu, L. and Liu, J., 2014. Preparation and characterization of negatively charged PES nanofiltration membrane by blending with halloysite nanotubes grafted with poly (sodium 4-styrenesulfonate) via surface-initiated ATRP. *Journal of Membrane Science*, 465, pp. 91–99.

Zhu, J., Qin, L., Uliana, A., Hou, J., Wang, J., Zhang, Y., Li, X., Yuan, S., Li, J., Tian, M. and Lin, J., 2017. Elevated performance of thin film nanocomposite membranes enabled by modified hydrophilic MOFs for nanofiltration. *ACS Applied Materials & Interfaces*, 9(2), pp. 1975–1986.

4 Recent Advances in Biological Remediation of Volatile Organic Compounds (VOCs) and Heavy Metals

Khyati Arora, Vanshika Kumar, Shubham Jyoti Nayak, Aditya Surya, and Shobana Sugumar

CONTENTS

4.1 Introduction 50
4.2 Overview of VOC Treatment Methods 50
4.3 Biofiltration Method 51
    4.3.1 Use of Biofilter: *Health Risk Calculation*...52
4.4 Role of Biotrickling Filters in VOC Removal..53
    4.4.1 Basic Principle of Biotrickling Filtration 53
    4.4.2 BTF Mechanism, along with Limitations..53
    4.4.3 BTF System Combined with Other Techniques for Effective VOC Removal........54
    4.4.4 Use of Biotrickling Filter in VOC Elimination (Experiment)................54
    4.4.5 Inference 55
4.5 Bioscrubber Method 55
    4.5.1 Introduction to Bioscrubbers 55
    4.5.2 Bioscrubber Principle of Operations 56
    4.5.3 Bioscrubbing Process and Equipment Used...56
    4.5.4 Applications of Bioscrubbers..57
4.6 Use of Membrane Bioreactors 57
    4.6.1 Development of Anaerobic Membrane Bioreactor..................................58
    4.6.2 AnMBR Working and Configurations 59
    4.6.3 Side-Stream Configuration 59
    4.6.4 Submerged Membrane 60
    4.6.5 Membrane Operation 60
    4.6.6 Applications 60
    4.6.7 A Few Other Applications of AnMBR Are as Follows............................60
4.7 Bioremediation of Heavy Metals..60
4.8 Microbial Remediation of Heavy Metals 61
    4.8.1 General Mechanisms Followed for Microbial Bioremediation 61
    4.8.2 Bioremediation Mediated by Algae 61
    4.8.3 Bioremediation by Bacteria 61
    4.8.4 Bioremediation by Fungi 61
4.9 Phytoremediation...62
    4.9.1 Some Advantages of This Method 62

DOI: 10.1201/9781003165149-4

4.10 Combined Bioremediation 63
    4.10.1 Challenges 63
    4.10.2 Factors That Affect Combined Bioremediation Are 63
4.11 Conclusion 63
References 66

4.1 INTRODUCTION

The last couple of decades have witnessed a paradigm shift in global growth due to industrialisation and urbanisation. The world is accelerating towards innovation, but it is impacting the environment. With the hazardous disposal methods followed by large industries and manufacturers and even at an individual level, we have been successful in deteriorating the quality of the biosphere. Contamination in the environment can find its source as leaching of heavy metals, atmospheric pollutants, corrosion of metal, sediment resuspension to the soil, and groundwater.[1] This can further find its way into the food chain. Initially, several chemical methods were utilized to devoid and rid the soil of heavy metals and VOCs. Techniques such as filtration, reverse osmosis, biosorption, aerobic and anaerobic microbial-based degradation, flocculation, chemical precipitation, and coagulation were immensely explored.[2] In this book chapter/review, we discuss the various recent developments contributing to formulating newer methods of removing heavy metals and VOCs. An overview of them is the biofiltration method, bio trickling filters, bio scrubber methods, and bioremediation harnessing the microbes and plants.

4.2 OVERVIEW OF VOC TREATMENT METHODS

Some organic compounds (VOCs) and various other harmful air pollutants are produced in the polluted sewage. Due to regulatory policies and the need to manage these hazardous compounds, appropriate technology needs to be developed. Among the many traditional methods, biofiltration is gaining importance because of its environmentally safe process, which is also cost friendly. Performance records, laboratory tests, and written reviews indicate that biofiltration water efficiency of 90% VOC and odour can be easily obtained in a controlled environment.[3]

Nonetheless, as biodegradation and bioactivity are involved, a biofilter approach must fully understand this process. The temptation to remove VOC from the laboratory can work. Still, it requires the right design and the right working environment in a real work environment with many VOCs and natural resources. More biofilters have been installed worldwide, with a minimal VOC removal review by biofiltration produced from wastewater treatment plants.[4]

Practical applications of the volatile organic compound removal methods depend on the type of bio scrubber used for different pollutant particles. In the last 30 years, biological methods for VOC control became a properly organised field of process technology. Biotreatment began using biofilters for air pollution control, including different designs, configurations, and particulate materials for eliminating specific pollutants. Advanced types of filtering devices such as bio-trickling filters and bio-scrubbers have been developed for some time. For each device type, a range of design variations is currently accessible. There can be a combination of variations and processes (co-current or opposite current, fluid flow, etc.) and different microorganisms that reduce the VOC level, thus creating more engineering tools to cover the need for effective treatment. Compounds of organic matter and other pollutants.[5]

Using membrane-based bioreactor treatment, the conversion of natural waste into high-value products and non-hazardous building materials plays an essential role in controlling environmental pollution and better resource utilisation. Membrane bioreactors (MBR) are a superior wastewater treatment technology for waste recycling and solid free-standing because biomass is separated from treated water by filtering membranes. After discussing several detailed factors, applications, and limitations of the medical and diagnostic perspective, another study could bring to the surface that the membrane bioreactor can be transformed into the best possible medical technology.[5]

4.3 BIOFILTRATION METHOD

Biofiltration usually implies biology-based treatment to convert organic or inorganic pollutants into innocuous compounds in the gaseous phase. Although biological decomposition has been widely used in wastewater treatment and rehabilitation methods for soil and groundwater treatment, it was only in recent decades that biofiltration emerged as a treatment for VOC removal and its technology for industrial application. This process of using biofiltration is safe for the environment and capital. During the induction process, the flow of polluted air passes through the pulmonary media, where the contamination is absorbed by microorganisms and biodegraded. Bacteria harness energy from the VOCs and even produce endoergic mediators using the same, which results in the formation of carbon dioxide, water, and biomass. During the metabolism of H2S and NH3, products such as nitrogen oxide (NOX) and sulphur oxide (SOX) are produced.[3]

Biofilters are designed to control odours in composting plants or municipal wastewater treatment plants. Typically they are used in chemical production/processing, degradation, storing chemicals, cork dry heating, extraction of gas, film installation, fish film houses, investment bases from fertiliser works, harvesting, food and tobacco processing, steel and metal industries, and other sectors such as taste and smell, oil handling, tobacco processing, coffee roasting, printing stores, pet production, and European wastewater treatment plants. Biofilters are also widely used in the US and Japan. The biofiltration process can effectively remove contaminants up to 5,000 ppm if proper operating conditions are followed.[3]

Improving the biofiltration process commonly used to extract odorous chemicals controls agendas such as medium, temperature, and humidity, allowing biofiltration to increase the VOC elimination. Besides, in contrast with the physio-chemical process of purification (heat, extraction, and absorption), biofiltration depends on microbes' ability to degrade a wide range of chemicals, demonstrating economic and environmental benefits. Inside a biofilter, dissipated gas must be added to a layer of solid materials such as soil used for compost, peat for pressure, and tree bark. Therefore, air pollution was previously compounded into biomass by the action caused by microorganisms. The packaging volume of water remains constant.[4]

Most biofilters designed as a single open bed system to treat compost or odour gas have gradually evolved into eliminating VOC technology for commercial applications. The biofiltration process takes up the adulterated stream and dampens it and then it is sent to the biofilter. As the internal flow of the filter decreases, airborne contaminants are absorbed and synthesised into the body. This is more of a metabolism process for the bacteria.[6]

The primary hypothesis of using a bio trickling filtration system in conjunction with a biofilter (BTF/BF) is efficacious in the treatment of low-level VOCs where there is high H2S concentration. It is also thought that the BF phase works better than the BTF phase in VOC removal.[7]

Although biofilters may not be operated for years, bed water retention is one of the most critical and problematic performance limits. In many parts of the ecosystem, water is very abundant and easy to use. For biofilters, the bedding is usually non-abrasive and does not have a fluid flow. Therefore, it is difficult to identify or correct the online water content.[3]

There are two phenomena for the functioning of the biofiltration mechanism.[6]
They are:

1. Transfer pollutants from the air to the water section or support area—The transfer of contaminants from the gas phase to the liquid phase occurs according to physics laws. The filtration in the water is similar to that in the air and Henry's constant consistency. Typically, biofilters reach very high removal rates of soluble chemicals and water. Biofiltration patterns tend to support this because they can block cells and polysaccharides as they are treated with biofilm.
2. Biomass, end products, or carbon dioxide and water pollutants bioconversion – The contaminant decomposes in biofilm, which contains various substances that grow on the surface of a solid body.

Because of biofiltration's essential benefits, such as being eco-friendly, process efficiency, easy integration, increased efficiency, and cost, its use in eliminating air pollution has been widely considered recently. For example, the best way to reduce the concentration of certain chemicals such as styrene, toluene, and xylene at the lab level is by using biofiltration. It also contains combinations of various VOCs in the driver and industrial proportions. However, all technologies have limitations, and it is challenging to convert hydrophobic VOCs such as biofilter alkanes due to heavy transmission limits.[7]

To overcome this problem and reduce the contact time and construction costs required to increase the availability of chemicals, garbage gases should be treated before using this file system. The spray tower (ST) uses H2O as a solvent, a control strategy widely used to distribute a hydrophobic atmosphere. By this method, the risk factor for hydrophobic VOCs decreases with biofilter expansion. For example, there is evidence that EST can effectively remove air pollutants such as various particles and ethyl acetate from industrial pollutants. Therefore, combining biofiltration with ST is an excellent way to reduce VOC emissions and associated health risks.[6]

Use of a biofilter easily removes low levels of sulphur, amino compound and nitrogen compounds, AIHs, AH, and HH in public medical provisions, biosolids degrading, and waste treatment facilities. Although viruses are the engine of this process, there are very few studies examining the interaction between microorganisms in the biofilter and VOC in mixed removal. Besides, the microbial community varies according to the quantity and composition of VOCs. Therefore, it is essential to understand the social responsibility of viruses to remove various VOCs.[4]

Use of biofiltration for the removal of VOCs (experiment)[6] The testing site was chosen from the textile industry in the city of Guangzhou (Guangdong, China), which specialises in textile manufacturing. The production cycle comprises cleansing, scratching, adding corals, performing chlorination, colouring, and reviewing. These stages, mostly cleansing and drying, produce massive amounts of highly toxic wastewater. Advanced treatments use catalytic ozonation, biologically filtered air, and a membrane filter to remove dye for five days of oxygen demand (BOD5), the chemical oxygen demand (COD), and stable suspension stained water. However, contact with the natural contamination of workers in surfactants, detergents, and local airborne liquids has led to VOCs' release in the wastewater treatment process.

4.3.1 USE OF BIOFILTER: HEALTH RISK CALCULATION

Both the reduction in life (with ST-biofilter) and the treatment of plumbing VOCs effectively counter the odour management, chiefly when an odour is produced and excreted in the faeces. Also, prolonged exposure to released VOCs is associated with adverse impacts on public and technical life. Therefore, it is estimated that health risks, including the risk of non-cancer and cancer-risk VOCs, provide a basis for protecting air pollution in the public health arena, especially for workers. As a result, the biofilters could distinguish and calculate the inconsistencies released when cutting solid municipal waste. Therefore, the driver ST-biofilter scale's efficiency was tested online to compare the variability of VOCs in the treatment process using e-nose and PCA analytics. The results showed that the VOCs released from the water plant (inlet) and ST biofilter (outlet) were well separated from the odour control area, indicating an appreciable difference between TDWTP and VOC in the atmosphere. These results confirm that TDWTP is contaminated with unpleasant VOCs that require air purification. Besides, the more sensitive signals W1C, W5S, and W3C, used to describe benzene oxides, nitrogen, and aromatic amines, have been detected in samples compared with other sample sites, emitting oxidants and nitrogen and ammonia TD. Inlet samples are identified in the samples collected after the ST-biofilter treatment, indicating that ST-biofilter can successfully remove these odourless VOCs from TDWTP. The results show a detailed ST-biofilter treatment for the four VOC groups on the water treatment plant. In general, the technology used has demonstrated the successful elimination of four groups of limited VOCs associated with the results obtained using a biofilter.

Biological Remediation of Volatile Organic Compounds 53

Similarly, ethyl acetate (a type of NAOCCs) is much easier to remove than AHs during the filtration process. Further analysis proved that the RE of VOC fragrant VOCs in the treatment plant was massively different from the three given sample conditions, which may be due to the variability of the inputs and the various contaminant structures. Compared to the other three groups, NAOCCs can complete reaching 83.2% RE in the third sample system. Biofiltration techniques have been used worldwide in various ways to extract VOCs from polluted air streams, which is a technology that does not harm the environment and eliminates pollutants such as VOCs, abnormal chemicals, and other HAPs in addition to producing low-grade products used. Biofiltration operates at a low cost and is an accepted technology to treat pollutants containing VOCs, air toxins, and odours. More than 750 biofiltration equipment arrangements are installed in European countries to control the smell of polluted air and water. Current research on biofilters can be used as a powerful technology to remove VOCs compared to other available finishing methods. However, using biological processes to eliminate VOCs has its disadvantages.[8]

4.4 ROLE OF BIOTRICKLING FILTERS IN VOC REMOVAL

Biotrickling filters, also known as BTFs, have become an effective scent control in applied sciences. This method is based on non-perishable organisms that grow in biofilm attached to the packaging material. The odour flow occurs as a transfer of odour from a packed bed to a biofilm. Simultaneously, the water solution is trapped in the packaging sources to supply the water and nutrients needed by the insect community.[9] BTFs are well suited for biofilters. Compared to traditional methods, the included benefits are the low costs, empty sleep time (EBRT, hence being less than a foot), reduced heat loss, and low environmental impact, leading to high efficiency and visibility, high gas speed, and reasonable performance control.[10] Biotrickling filters display a significant decrease in the treatment of H2S, which has a soluble odour and active removal (REs) usually more than 99% in 2 to 10 seconds in treatment plants for wastewater and odour releases. However, WWTP emissions contain many high-quality and highly concentrated water-fearing organisms, essential for effective degradation. Although the effect of BTFs on the removal of H2S or industrial VOCs in high-energy areas has been investigated, many studies of extruded gases containing hydrophobic VOCs and sulphur compounds are commonly found in WWTPs – $1g\,m^3$ to $mg\,m^{-3}$.[10]

4.4.1 Basic Principle of Biotrickling Filtration

The air is full of pollutants built into biofilm by co- or counter-currently. The composition of the end-to-end systems suggests that the current combination is favourable because it reduces the imminent effect of air leakage, even though tests have failed to prove this to be true. For systems with mass transfer, the flow rate is not essential. Some barriers to device design are ignored until one or more operating modes are selected.[9] If there is a small amount of dust or tissue material in the air, perhaps grease will be released with a major purge. The medium feed contains nutrients. As is often the case in laboratory studies widely used in the field, it is converted from salts containing pesticides and pH buffer into a raw mixture of essential nitrogen and phosphorus fertilisers as well as potassium.[11]

4.4.2 BTF Mechanism, along with Limitations

The BTF system for VTs and deodorant is a complex combination of various physical, chemical, and biological conditions. When smoke passes through a BTF bed, waste and nutrients are transferred to microorganisms, where metabolic cell chemicals are usually absorbed. Microbes present in the packaging and the biofilm interact with the water phase. VOCs and air pollutants must first be transferred to the water phase and then to the biofilm's bacterial cells, where microorganisms eventually degrade them.[11] Previously, any nutrients (oxygen, nitrogen, phosphorus, etc.) needed for the decomposition of the vapor phase and bacteria's growth would dissolve in the water phase around the biofilm. Contaminants and O_2 are transferred directly to the water phase by a gradient of moisture or concentration, and this process is called mass transfer.[10] Biodegradation reactions occur when dissolved contaminants are formed and reduced by a biofilm by converting them into molecules that do not react positively with CO_2 or other chemicals. In the presence of nutrients, bacteria form biofilms via the energy released by the oxidation of these pollutants. In BTF, the biodegradation effect is controlled by transferring airborne contaminants from biofilm (distribution limit) and the biodegradation reaction (response limit). Once IL has been resolved, high EBRT means air pollution has a higher penetration rate and overall rate of flow. In this case, the water section around the biofilm is saturated with air pollution. Therefore, the reduction response is controlled by the response limit. At low EBRT, air pollution structures have a very high flow rate and a low penetration rate. In these cases, air pollution comes from the BTF regardless of the water category. Hence, the reduction is controlled by the distribution limit.[11] The separation of target contaminants takes place within the biofilm itself and allows the formulation of secondary metabolites. During production, the secondary metabolic target is the simultaneous distribution of significant pollutants and evolutionary processes as the key pollutants. Later, the final products are converted to the gas phase after the demolition of microorganisms or carbonate compounds in the water phase by VOC oxidation. Nevertheless, if secondary metabolites show more toxins than the first pollutants, their composition prevents decay processes during conversion.[9]

To get a successful BTF treatment, the impurities must be decomposing and harmless in the BTF. The most effective reduction was obtained using BTFs with highly soluble computing cells with low molecular weight and flexible binding structures. Chemicals with complex binding structures often require a lot of energy, sometimes not always available to microorganisms. As a result, the natural separation of these compounds is low or zero. It must be noted that organic compounds like -OH, -CHO, -CO, etc. have good biodegradability; phenols, chlorine hydrocarbons, PAHs, and high-grade HCs show a slow rate of decay. These, along with some anthropogenic compounds, do not corrode unless compounded with additional components like enzymes.[11]

4.4.3 BTF System Combined with Other Techniques for Effective VOC Removal

Combining the BTF system with other treatment options can achieve better exclusion efficiency than the BTF system. Microbes can produce many enzymes, like monooxygenase, accelerating antagonistic volatile organic compounds' degradation into normal molecules. It helps to promote the release of substrate oxidising enzymes into micro-micro-metabolic co-metabolites. The results showed an improvement in TCE removal and changes in the bacterial community with a magnetic field of strength of 60 mT. The stabilisation of this system has the same effect as removing impurities by incorporating phenol and sodium acetate as co-metabolites.[11]

Also, the different magnetic field forces profoundly affected the saprophyte clique in BTF systems and enhanced Ascomycota file mass, thereby increasing the TCE removal rate. As with the use of enzymes, early treatment of VOC antagonists improves the metabolism of viral impurities. For example, in treating certain fragrant chemicals such as styrene and chlorobenzene, UV treatment is often implemented before executing the BTF program. Ultra-violet energy directly removes hydrophobic VOCs, quickly stabilises, and decomposes mediators, a pre-treatment step before BTF treatment. As a result, weight transfer and reaction rate are enhanced by ultra-violet treatment, as it converts waste into soluble and destructive compounds. In addition to UV, various therapies such as plasma or photolysis therapies may be used.[11]

4.4.4 Use of Biotrickling Filter in VOC Elimination (Experiment)[10]

This study demonstrated the potency and high hydrophobic VOCs of BTFs and the effective removal of the organic sulphur odour. Most MeSH and VOC elimination was not observed in the

Biological Remediation of Volatile Organic Compounds 55

azoic trial, with significant input and output concentration differences. The results confirmed that the contamination was not removed through adsorption, adsorption, or photolysis, so only reducing germs caused the impurities to be released throughout the study.

4.4.5 INFERENCE

This study confirmed the BTF's ability to measure the treatment of VOCs and sulphur compounds at RE = 99% while simultaneously assisting the effective reduction of hydrophobic VOCs in doses such as mg m^3 and EBRT. These effects are due to increased KLa values noted by PUF. It has been hypothesised that either the collection of toxic metabolites in BTF or a low concentration of odours can cause irreversible damage to the bacterial community.

4.5 BIOSCRUBBER METHOD

Bioscrubber applications date back to the early 1980s when early equipment was installed and fitted for treating gasses from enamelling ovens (pollutants such as glycols, alcohols, resins, aromatic compounds, etc.), foundries (ammonia, phenols, formaldehyde, amines), and fat smelteries or incinerators.[14] In the 1990s, gases containing various compounds (methyl, H2S, dimethyl sulphide) from wastewater treatment plants were cleaned using bioscrubbers. Two different bio scrubbers can be 'fixed-film bio scrubbers' and 'suspended growth bio scrubbers'.[15]

4.5.1 INTRODUCTION TO BIOSCRUBBERS

A bioscrubber is a biological system/biofilter fitted with a wet scrubber that removes the waste gas by lowering the inlet temperature and removing dust within the system.

For waste gas treatment, the bio scrubbers used are characterised into two significant parts/steps. First is physically separating the volatile organic compounds into water inside the system. Then the subsequent step is biologically treating the water containing the VOCs in two-unit operations.

An absorber is used to clean the waste gas, and then inside a contactor of gas-liquid transfer of the pollutants into the aqueous phase from the gaseous phase occurs. For carrying out this waste gaseous treatment process, a gas-liquid contactor that's an absorption column fitted with a packed bed and gas and water counter-current flow is highly preferred. By biological treatment inside the bioreactor, the water coming out from the absorber, infected with dissolved pollutants, is regenerated. The bioreactor used in the bio scrubber system is mostly a tank that is aerated via air bubbles and contains activated sludge (suspended). The volume of this tank is way greater than the absorber. After cleaning the water, it is again recycled to the top of the absorber tank. A solution rich in nutrients is always added to the aqueous phase for ensuring microbial biodegradation activity. Control of the pH of the aqueous phase titrants (acid or alkali) can be added, and doing this helps in optimum absorption and optimum biological activity. In the aqueous phase, dissolved compounds and suspended biomass are accumulated, and then a stream of wastewater is produced. For replenishing the amount of water wasted and evaporated, freshwater is continuously added regularly.

In the aqueous phase, the dust and aerosols are also absorbed along with volatile organic compounds. Suppose a very high concentration of particles is present in the aqueous phase. In that case, a solids absorption process is carried out, and with the use of a filter or a secondary scrubber before the use of bioscrubber, the particles present in the gaseous phase are removed.[5]

There are various advantages of bioscrubber technology such as operation stability, controlling operating parameters such as dosage of pH and nutrients, small space requirement, and pressure drop. In bioscrubbers, packing material clogs that can be avoided, and high pollutant concentrations, large gas flow rates can be easily handled in these systems. The generation of toxic by-products can be maintained at low levels because the reaction products can be removed by washing.

There are two sides two a coin, and in the same way bioscrubbers too have their disadvantages, such as the generation of excess sludge and aqueous waste, slow-growing microbes getting washed out and sometimes getting inside the absorption column, in minimum residence time gaseous pollutants getting absorbed, and less water-soluble compound bio scrubbing not being suitable. The cost-effectiveness of bioscrubber applications for pollutants with Henry's coefficient below 0.01 was estimated, and this is one of the significant advantages.[15]

4.5.2 BIOSCRUBBER PRINCIPLE OF OPERATIONS

Physical and biochemical mechanisms are used for removal of the odorous compound in bioscrubbers that are:

1. Absorption – Gaseous to aqueous phase transfer of odorous compounds takes place. The mass contaminant transfer is an action of the surface area of contact, diffusivity coefficient, and contact time.
2. Adsorption – Compounds with high molecular weight and low water solubility can be physically absorbed into biological floccules.
3. Condensation – During the transfer of the warm odorous gaseous phase into the aqueous phase, condensation takes place, and then it is maintained at a low temperature
4. Bio-transformation or bio-degradation – For converting the odorous compounds inside the liquid phase, active microorganisms (heterotrophic or autotrophic ones) are responsible. For providing energy, cell growth, and synthesis, heterotrophic microbes require a source of organic carbon, carbon dioxide in the air stream provides carbon to autotrophic microbes, and cellular energy for respiration and growth comes via sulphide oxidation into either sulphur or sulfate.[16]

4.5.3 BIOSCRUBBING PROCESS AND EQUIPMENT USED

In an absorber unit, liquid-gas contact contaminants mass transfer is favoured from the gaseous phase to the surface area of the aqueous medium carried out. The packed tower absorbers are the best choice for bioscrubbers as others display negative elimination efficiencies for compounds with bad water solubility. A counter-current packed tower with its larger volume is much more suitable than cross-flow or co-current towers because they have higher efficiency of absorption and low-pressure drop, and costs of energy associated are very low. The packing materials used are saddles, Pall rings, and Raschigs.[5]

For absorption tower construction, the materials used should be corrosion resistant to liquids and gases. Materials used to build an outdoor absorber should be UV and thermal resistant; mechanical resistance to the temperatures and loading rates should also be present.[5]

Inside the bioreactor unit, the absorbed pollutant from the absorber must be degraded/destroyed. Due to continuous aeration, the conversion of the contaminant into CO_2 and H_2O biomass is carried out by active microbes. Another essential factor for bioreactor air supply is the bubble size. Fine bubble diffusers provide better reduction (above 99.5%) of both H2S and odours compared to bubble diffusers that are coarse and cause 95% reduction of odours and 92% H2S reduction. In the absorber, the bioreactor discharge is recycled and reused. The bioreactor's functional design is the same as wastewater treatment system sludge tanks. The hydraulic retention time is much longer in bioscrubbing, which equals sludge retention time, making the main difference. Therefore sludge retention is not required in bioscrubbers.[16] Most of the existing bioscrubbers remove single pollutants. Different improvements in the bioscrubber designs have taken place over time to improve operational flexibility and provide more effective odour control. Some of them are sorptive-slurry bioscrubbers, two-liquid phase bioscrubbers, anoxic bioscrubbers, airlift bioscrubbers, spray columns, or two-stage bioscrubbers.[3] *Thiobacillus* and *Hyphomicrobium* strains are efficient

Biological Remediation of Volatile Organic Compounds

degraders of compounds containing sulphur. Heterotrophic *Xanthomonas* remove H2S from the gas steam. Chlorobium limicola immobilised cells transform H_2S to sulphur through a reaction that is autotrophic. For maintaining the highest cleaning efficiency and removing inhibitory effects, constant pH should be maintained. An optimum pH between 8.5–9.0 and increased biological activity can be held; effective H_2S absorption can be achieved simultaneously. There is no work on temperature yet reported in the literature, and temperature plays a vital role in microbial activity.[17] Sublette et al. (1998) found that at temperatures between 25–35°C *thiobacillus denitrificans* efficiency of sulphide oxidation was at an optimum level, and below 16°C the efficiency decreased significantly. Without substrate supply, bioreactor biomass is highly sensitive to periods. Without the operation of a bioreactor for one or two days, it is useful to continue the supply of oxygen and substrate for maintaining the high activity of microbes. By sparing air into the bioreactor, oxygen can be added; for sufficient biological activity, at least a concentration of 1–2mg L-1 of oxygen is appropriate. Sludge accumulation in bioscrubbers is shallow compared to that of wastewater treatment plants. Also, it is less toxic; smaller flocs are formed in bioscrubbers than wastewater treatment systems.[16]

4.5.4 APPLICATIONS OF BIOSCRUBBERS

The significant advantage of a bioscrubber over a bio trickling filter is it can generate and contain substantially greater smaller units of microbial biomass and at the same time maintain increased specific substrate rates of utilisation.

A bioreactor can be started with activated sludge from wastewater plants, and the microbial growth depends on various environmental factors such as:

1. Temperature.
2. Ionic strength.
3. pH.
4. Toxic compounds present.
5. Substrate concentration.[16]

Bioscrubber technology development, advancement in removing odorous compounds, and controlling volatile organic compounds released from municipal and industrial wastewater can now be carried out with ease. The gaseous waste compounds that can be removed by the process of bioscrubbing using industrial-grade bioscrubbers are: Ammonia, odours from wastewater treatment plants, hydrogen sulphide, etc.[14]

4.6 USE OF MEMBRANE BIOREACTORS

Membrane bioreactor defines wastewater treatment procedures where a perm-selective membrane such as ultrafiltration or microfiltration is compiled with a biological process such as a suspended growth bioreactor.[18] The difference in a membrane bioreactor system arises from a 'polishing' process where the membrane used is a discrete tertiary treatment step where the active biomass does not reoccur in the biological approach. In today's time, all the commercial MBR systems use membrane as the filter that eliminates the solid particles formed during biological processes resulting in a disinfected, clarified product effluent. In simple words, an MBR system is a modern and developed version of a conventional activated sludge (CAS) system. The significant difference between the two systems is that a CAS system uses a secondary settlement tank for solid/liquid separation. Still, on the other hand, the MBR system uses a membrane for the same function. Due to this the MBR system has numerous advantages related to process control and product water quality over CAS.[18] As we observed, modern membrane bioreactors are composed of a suspended growth bioreactor unit and a perm-selective membrane. Now we will look at various membranes used for the treatment of pollutants in wastewater. This process is known as 'membrane-based separation' (MBS). MBS is a separation technique that is well known nowadays and has various water desalination applications as well as enabling recovery of valuables by toxic metal separation. The process of separation depends on the different types of membranes used in the system (MBR). These membranes are produced from different materials such as ceramics, polymers, zeolites, etc. and possess specific filtering features. They depend on various other factors such as pore size, surface charge, hydrophobicity/characteristics, and membrane morphology. The membranes are found in different modules such as microfiltration, ultrafiltration, nanofiltration, reverse osmosis, and forward osmosis. In these processes, the mode of separation differentiates from solution-diffusion to size-exclusive particle to molecular diffusion.[19] MF membranes consist of large pores (0.1–5 μm) compared to UF membranes, eliminating particles in the size of 0.1–10 μm. UF membranes with pore size in the range 0.01–0.1μm colloidal particles, biopolymers, macromolecules, viruses, and their sizes range from 0.01–0.2 μm, and it is based upon the principle of size exclusion. UF is mostly used for commercial treatment of wastewater, surfactants, industrial cleaning recovery, food processing, separation of proteins, etc. The fabrication of UF membranes is done from derivatives of cellulose, inorganic materials such as Al2O3, TiO2, ZrO, etc., with polymers such as polyacrylonitrile (PAN) polysulfone amide (PSA), polyethersulfone (PES), polyvinylidene fluoride (PVDF), etc.[20] NF membranes disallow molecules in a size between 0.001–0.01 μm, including most or organic materials, metallic salts, and biomacromolecules. The performance shown by NF falls between UF and RO. On the other hand, the membranes of RO are non-porous, formed from dense polymers, and contain pore sizes ranging from 0.0001–0.001 μm. They separate metal ions with low molecular weight. The most common application of RO is in treating paper and pulp mill effluents for producing potable water. In all of the membranes mentioned earlier, the water flux, chemical resistance, operating temperature, high PEP rejection, stability under stress, pressures applied, and engineering designs inform the choice of polymers to be used. In modern times ceramic and zeolite composite membranes are high-performance NF and RO membranes and are successfully used in PEP separation commercially.[19] Different configurations based on tubular, spiral-wound, hollow-fibre, and flat-sheet structures have been used in wastewater treatment applications for domestic and industrial purposes. The RO technology developed for water desalination studies had various membrane fouling issues, a requirement of high energy, concentration, and polarisation. This led to the development of FO by researchers, and in this the osmotic gradient plays an essential role in separation and mass transport. Therefore, FO is more preferred/suitable and is more energy efficient in treating feed with a high fouling tendency (e.g., landfill leachate), which would be more expensive by RO. Fertiliser dilution and fruit juice concentration are applications of FO. Initially, FO was treated as an efficient pre-treatment step in which the recovery of purified water could be made from a diluted draw solution.[21]

4.6.1 Development of Anaerobic Membrane Bioreactor

During the 1930s, both aerobic biological treatment and anaerobic biological treatment techniques were used to treat domestic and industrial wastewater. During this course, organic matters were converted into soluble forms such as $2H_2O$, CO_2, $4NH^+$, CH_4, $2NO^-$, etc. The end products used to differ depending upon the availability or unavailability of oxygen. In the 1960s, Dorr-Oliver commercialised activated sludge control with ultrafiltration. During the 1970s, the final effluent quality was mostly dependent on hydrodynamic conditions inside the sedimentation tank and sludge settling characteristics. Large volume sedimentation tanks offered several hs of residence for solid/liquid separation. In 1978 external cross-flow membranes were used to treat tank effluent, resulting in increased biomass concentration, BOD reduction by 85%–95%, 72% removal of nitrate, and orthophosphate reduction 24%–85%. In 1980, for solid/liquid separation, a secondary settling tank was used, limiting the effluent quality. The term anaerobic can be defined as

Biological Remediation of Volatile Organic Compounds

a biological treatment operated without oxygen, and with the help of a membrane, solid-liquid separation is carried out. These systems were first introduced in South Africa in the 1980s. The first commercially developed AnMBR was developed by Dorr-Oliver in the early stage of the 1980s and was initially known as the membrane anaerobic reactor system (MARS). By the 1990s, the AnMBR research activity increased with investigations for different materials to be used as a membrane. Membrane foulants started being characterised, and strategies for membrane cleaning and fouling management were laid out. By the 2000s AnMBR studies focused on filtration characteristics, system performance and membrane foulants characterisation, and success of membrane fouling made MBR a highly encouraged method for wastewater treatment. In 2009, statistical data was presented, which showed that municipal water treatment by AnMBR with COD 500 mg/L could recover 48% of methane with effluent COD being less than 40mg/L. By the late 2010s, submerged AnMBR treatment was deeply researched, and attempts were made to improve energy efficiency, increase and extend the application scope, and solve problems like membrane fouling. Since January 2018 to date, improvement in the hydrodynamic parameter of membranes, characteristics of membrane material, and effects of pollutants on the performance of AnMBR configurational development have been done, which broadens the area for further research.[22]

4.6.2 ANMBR WORKING AND CONFIGURATIONS

Different configurations such as hollow-fibre, flat-sheet, and tubular membranes are used in AnMBRs, with varying modules systems like submerged/immersed and external cross-flow. The system was designed based on two significant configurations, which were membrane design and operation:

1. External or side-stream configuration.
2. Submerged or immersed configuration.

In these, the membrane could be operated in a vacuum or under pressure.

In the external cross-flow configuration of the membrane, both the bioreactor and membrane are separate units, and for pushing the bioreactor effluent into the membrane unit for permeation, a pump is used. Plastic, ceramic, steel, etc. are the materials used for making these membranes.

Within plastic, the types considered for use are;

1. Polyethersulfone (PES).
2. Polyvinylidene Fluoride (PVDF).
3. Polyethylene terephthalate (PET).
4. Polytetrafluoroethylene (PTFE).

Plastic can be used to make any membrane, but ceramic is used to make flat sheets and sintered steel to make multi-tube membranes. The end-point is to know the feasibility of both side-stream and submerged configurations for arriving at optimum fouling strategies and to minimise the overall energy demand.[22]

4.6.3 SIDE-STREAM CONFIGURATION

In the membrane chamber, a trans-membrane is present. As a result, the cross-flow velocity completely and permanently disrupts filtration cake formation on the membrane's surface. This process utilises energy, and the remainder of the energy can be used for mixing suspension in the anaerobic reactor. The membrane's surface and the cross-flow velocity is kept at a range between 2–4 m.s-1.[23]

4.6.4 SUBMERGED MEMBRANE

This membrane operates under a vacuum and not under direct pressure, and in this the module is directly placed inside the liquid. Gravity or the vacuum pump is used to take out the permeate via the membrane. This can be used in various configurations, including directly immersed inside the bioreactor or in a different tank. It was observed from statistics that configurations of AnMBR treating wastewater was 3.7 kW.h/m3 for external cross-flow configurations, and 0.3 kW.h/m3 permeate was needed for submerged setup. The most critical challenge for this system comes from fouling mitigation, due to this elimination of gas sparging and replacing it with low energy-intensive processes.[23]

4.6.5 MEMBRANE OPERATION

Membrane surface scouring is vital for removing foulants, and it defers depending on the configuration of the reactor, conditions of operations, and influent wastewater.

The cross-flow unit uses a high-cross velocity for mitigating fouling, whereas submerged AnMBRs are dependent upon gas sparging. Membrane souring is an alternative to gas sparging in submerged systems and can be achieved by having a fluidised bed of GAC in direct contact with the membrane.[22]

1. Side-stream configuration/setting
2. Submerged or immersed membrane configuration/setting.

4.6.6 APPLICATIONS

Nitrogen removal by an anammox process can be carried out by membrane coupling with an anaerobic biological process, by which nitrite and ammonia in wastewater are converted into nitrogen gas. In an AnMBR non-woven fabric for the treatment of low-strength wastewater polytetrafluoroethylene (PTFE), composite membranes are being used. For treating sewage water, a cross-flow AnMBR where the COD loading rates of 1–2 kg COD m3d-1 were applied to AnMBR for 280 days. It was found out that the effluent COD was always lower than 40mg/l, but 30% of the inlet COD was not able to be removed, independent of the HRT, because of dissolved methane, sulphate reduction, and untreated COD in the permeate. The amount of methane that was recovered from the municipal wastewater decreased from 48% to 35% when HRT decreased from 12 to 6 hrs.

4.6.7 A FEW OTHER APPLICATIONS OF ANMBR ARE AS FOLLOWS

Renewable energy production, biofuel for household activity, conversion of biogas into electricity, lighting, heating, running a small-scale business from generated energy, bio-fertiliser for agricultural uses, pollution control, health, and sanitation improvement, recycle and reuse of wastewater, nitrogen removal by anammox process, etc.[24]

4.7 BIOREMEDIATION OF HEAVY METALS

Due to several human activities, the accumulation of hazardous heavy metals is prevalent, for instance: Lead (Pb), silver (Ag), cadmium (Cd), zinc (Zn), mercury (Hg), chromium (Cr), uranium (Ur), selenium (Se), arsenic (As), nickel (Ni), and gold (Au). These heavy metals can induce dwarfism due to reduced photosynthetic activity and essential enzymes, along with an imbalance of mineral nutrition.[25] When these enter the human system, they can be cytotoxic at low concentrations, becoming causative of cancers and other hazards.[26] This causes oxidative stress forming reactive oxygen species, leading to free radicals' generation while preventing the cells from repairing

Biological Remediation of Volatile Organic Compounds 61

damage.[27] Bioremediation is a prospective technology for tackling the problem of heavy-metal contamination. The core principle of bioremediation centres around the dynamic pH levels, redox reactions, and assessing the adsorption of pollutants and contaminants, thereby reducing natural contaminants' solubility. It restores the ecosystem by eliminating hazardous pollutants and harnessing the biological mechanisms innate to the microbes and plants[28]. The process can be ex-situ or in-situ. The effect of bioremediation relies upon a multitude of factors, such as prevalent microorganisms, environmental factors, and the degree of pollutants at the site of contamination. The process is successful by Phytoremediation and predominantly microbial remediation, based on the metabolic potential of microbes that change the pollutant particles into innocuous substances through the redox process.[29]

4.8 MICROBIAL REMEDIATION OF HEAVY METALS

4.8.1 General Mechanisms Followed for Microbial Bioremediation[30]

1. Binding of proteins and peptides (like metallothioneins, phytochelatins) by intracellular metal-binding or cell wall components to isolate toxic metals.
2. Blocking uptake of metals by altering their biochemical pathways.
3. Converting metals to innocuous forms using enzymes.
4. Reducing the intracellular concentration of metals with the help of efflux systems.

4.8.2 Bioremediation Mediated by Algae

The benefit of using algae for a bioremediation operation is that algae is photosynthetic with a faster growth rate than forest-originating biomass. It can thrive in many habitats, such as freshwater, marine areas, and soils.[31] Marine algae are more susceptible to bind to heavy metals due to the large biopolymer quantity. The heavy metals accumulated by several metals are independent and dependent on their metabolism. Examples of this are the brown and red algae seaweeds.[32]

4.8.3 Bioremediation by Bacteria

The microbial biomass is a possible solution to uptake heavy metals from various contaminated media. The intrinsic advantage is that microbial biomass is acquired as a residual product of fermentation. When the heavy metals are adsorbed, they become a part of this biomass matrix.[33] Several functional groups of bacteria like amide, carboxyl, sulfonate, hydroxyl, and phosphonate groups play a pivotal function in the consumption and uptake of metals from a base of aqueous solution.[34] Examples of such bacteria are – *Bacillus* and *Pseudomonas sp.* Using electrostatic forces arising from the anionic nature of bacteria, the surface allows binding to metal cations. Thus, comparing gram-negative and gram-positive bacteria, the latter is superior in metal ion trapping.[35]

4.8.4 Bioremediation by Fungi

The advantage of using fungi for bioremediation is that it can quickly adapt to ecological conditions such as varying nitrogen and carbon levels. It can tolerate and detoxify effluents with high metal toxicity and has higher receptivity towards physical and enzymatic contact.[36] Examples of this are Pleurotus *ostreatus, Pleurotus tuberregium, Pleurotus pulmonarius, Agaricus bisporus, Lentinula edodes,* and *Irpex lacteus.*[37]

In detoxification using fungi, the basic principle followed is that the contaminated region consisting of heavy metals has valence alteration and intracellular and extracellular precipitations. Fundamentally, mannuronic and guluronic acids persist in fungi cell walls, the carboxyl groups that contribute to proto sufficiency and lead to pronounced heavy metal biosorption.[38]

4.9 PHYTOREMEDIATION

In simple terms, phytoremediation is the application of genetically modified or wild type plants to eliminate pollutants from the soil. The approach of removing said pollutants may vary from transferring, stabilising, or even outright destroying the contaminants in the soil and even the groundwater.[39]

The large share of scientific and profit-oriented views of phytoremediation now centralises on concepts of phytoextraction along with phytodegradation. These scientific constructs utilise desired plant species grown on contaminated soils.

This brings up the question: Why would plants take up these apparent pollutants?

Consider here is that what seems to us to be a pollutant might be a vital nutrient for the plant. Here we refer very specifically to metals like Zn, Fe, Mn, Ni, and Mo.[40] These metals can be considered pollutants for humans in the context of concentrations that start to affect us by way of heavy metal poisoning. We must remember the concept of biomagnification while understanding why these metals classify as pollutants while they are nutrients to individual plants.[41] Plants have evolved to have unique and modified mechanisms to consume, shift, and store nutrients. Proteins moderate the motility of the metals across natural membranes with transport roles. Furthermore, fragile and intricate tools help maintain the physiological limits for the intracellular concentration of metal ions.[41] Plants take a certain amount of cadence for the uptake of heavy metals in contaminated soils. This relies on two considerations, (a) amount of biomass the plants produce and (b) bio-concentration factor, which is the ratio of metal concentration in the shoot tissue to the soil.[40]

Plants that survive in such contaminated soils can be classified into three categories:

1. Excluder Plants: These plants generally have a low potential for any extraction, as they have adapted to prevent pernicious metal uptake into root cells.
2. Accumulator Plants: These plants have adapted to have specific mechanisms for detoxifying high metal levels. These directly allow for bio-accumulation of relatively high levels of concentrations.
3. Indicator Plants: Tend to show poor control over metal uptake and transport process. In addition to accumulating toxic minerals in their tissues, plants are also seen to take up a range of harmful organic compounds, including some of the most copious environmental pollutants such as polychlorinated biphenyl (PCB), ammunition wastes (nitroaromatics such as trinitrotoluene (TNT) and glycerol trinitrate (GTN)), and halogenated hydrocarbons (trichloroethylene, TCE). Successive metabolism in plant tissues then mineralises or degrades such pollutants to non- or less-toxic compounds. These compounds have been tested for Phytoremediation by various agencies.[42]

TCE: USAF has used popular trees to contain TCE in groundwater. TNT: The US Army corps of engineers has experimented with plants like coontail and pondweed; these have been shown to reduce TNT concentrations to just 5% of the original concentrate. RDX: Submersed plants have been used to minimise RDX concentrations to 40%, and with the addition of certain microbes, a reduction of 80% was achieved.[43] Additionally, by way of hemofiltration, sunflowers have been used to remove radioactive contamination from water in a pond in an experiment at Chernobyl, Ukraine. These results are quite significant and have to be thought about with an earnest consideration for their implications for the future of sustainable development.[44]

4.9.1 SOME ADVANTAGES OF THIS METHOD

It's cheap. Only harvesting and weed control costs must be factored in. It's CO2 neutral. A profit can be made viable when the resulting biomass is used for heat and/or energy fabrication.

Biological Remediation of Volatile Organic Compounds 63

Even though these advantages make it sound great to invest in this technology, here are some factors proving why significant research needs to occur before this technology has a hope of becoming viable.[44] If the concentration is too high, the plants will inevitably die. It's very slow. In a sense, it takes several years and even decades to halve concentrations. Most plants have bioconcentration factors of less than one, which means it takes much longer than the average human lifespan to reduce contaminations to 50%. They are limited to sites with contamination in shallow soils, as the roots are unable to penetrate deep into the soil.[39]

Climatic factors play a huge role in plant growth, and locations must be studied appropriately. Introducing a new plant species can have ecological implications and also must be taken into account every time. Disposal of harvested plants can be cumbersome if they have high levels of heavy metals. Last, a large area of land is needed, and this land cannot be used for any other activity for years and years, which then is an economic trade-off that needs to be considered.[40]

4.10 COMBINED BIOREMEDIATION

Bioremediation is an up and coming, promising methodology to help combat highly prevalent issues in today's highly development-focused world. Although still a new technology, bioremediation has already been in media coverage when used to treat the Alaskan shoreline in the aftermath of the oil spill of Exxon Valdez in 1989.[46]

It is a natural method that is widely accepted and has a high potential to rehabilitate the soil. It particularly shines in the area of petroleum-contaminated soils, as demonstrated in various laboratory and field studies. It is an economical, non-invasive, and socially acceptable way to rehabilitate soil that is infested with contaminants.[47]

There have been various combinations used, such as the bacteria-fungi system, plant-microorganisms-earthworms.

4.10.1 CHALLENGES

1. Finding ways to make the strains competitive with native organisms.
2. Optimising methods to enlarge bioavailability to increase the bioremediation.
3. Researching and finding out optimal combinations for various types of contaminations.
4. Functioning of enzymes that microbes latently needs further research as enzymes are vital in the remediation of polycyclic aromatic hydrocarbons and heavy metals.

Some experiments implemented at a mesoscale level have effectively reduced hydrocarbons and metal-infested soil.[48]

4.10.2 FACTORS THAT AFFECT COMBINED BIOREMEDIATION ARE

1. Growth until critical mass is reached.
2. Mutations and horizontal gene transfer.
3. Solubility of contaminants.
4. Oxidation/Reduction potential.
5. Mass transfer limitations like miscibility in/with water.[49]

4.11 CONCLUSION

Technologies related to solving the pollution caused by volatile organic compounds and heavy metals were discussed. A lot of methods involving biological-based techniques are very promising. They are sustainable and much better for the environment than more conventional techniques. We saw the role of biotrickling filters and bioscrubbers and documented instances of how effective

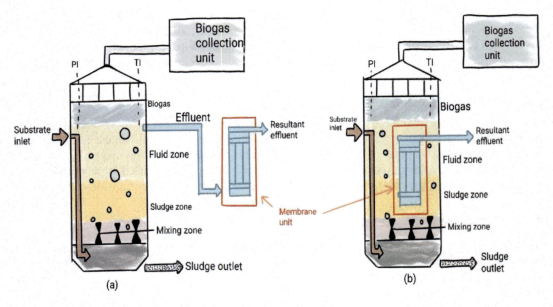

FIGURE 4.1 (a) Side-stream configuration/setting; (b) submerged or immersed membrane configuration/setting.[22]

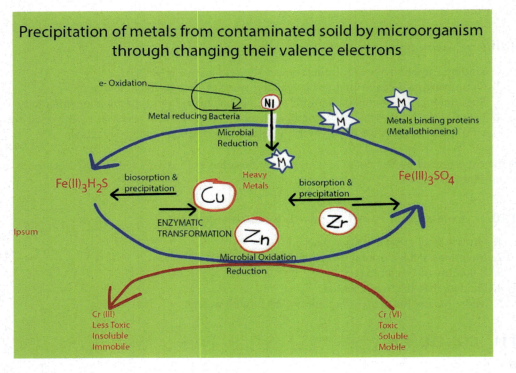

FIGURE 4.2 Mechanism of microbial remediation.[30]

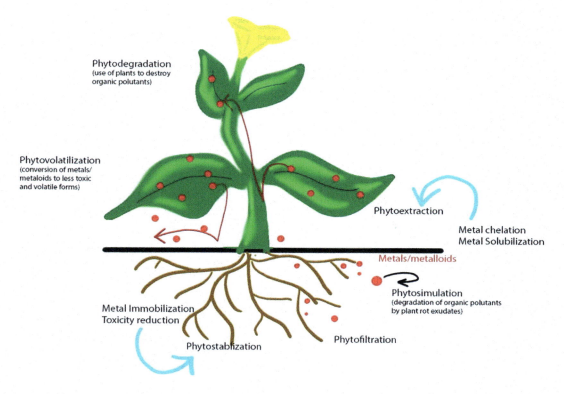

FIGURE 4.3 Process used in phytoremediation of heavy metals.[30]

they have been. Membrane bioreactors are a viable alternative for wastewater treatment and factors like different types of plastic and variations of usage arrangements in the membrane design and operation. With further research and development, it is safe to say it can be economical and environmentally sustainable to treat water contamination.[49] Another grave issue that needs to be addressed immediately is soil contamination with heavy metal and other dangerous compounds. It is evident why having healthy soil is very important. Our crops, food, and one of the first and essential links in the complex ecological food web, plants, and trees depend on soil.[50] If plants are unable to grow in soil due to contamination, that will gradually lead to grave consequences and we will not be able to reverse the damage. Phytoremediation is one of those technologies that has fantastic potential to help sustainably upkeep the soil's health. There are too many limitations for this technology to be deployed at a broad commercial scale today. Combined bioremediation is one of those methods that can be used to overcome some of the deficiencies; this type of remediation uses a combination of various organisms to increase efficiency.[47] At the base of all these technologies, there is one apparent common link, enzymes. Enzymes have had millions of years of evolution, while chemical catalysts or conventional methods are not more than 1,000 years old. Any human invention cannot overcome the sheer energy efficiency of enzymes. We must focus on using biotechnology-based solutions as they tend to be extremely sustainable compared to conventional mechanisms. More research and development will go a long way in making these technologies commercially viable within the next decade.[30]

REFERENCES

[1] Weerasundara, L., Amarasekara, R.W.K., Magana-Arachchi, D.N., Ziyath, A.M., Karunaratne, D.G.G.P., Goonetilleke, A. and Vithanage, M., 2017. Microorganisms and heavy metals associated with atmospheric deposition in a congested urban environment of a developing country: Sri Lanka. *Science of the Total Environment*, *584*, pp. 803–812.

[2] Kuppusamy, S., Thavamani, P., Venkateswarlu, K., Lee, Y.B., Naidu, R. and Megharaj, M., 2017. Remediation approaches for polycyclic aromatic hydrocarbons (PAHs) contaminated soils: Technological constraints, emerging trends and future directions. *Chemosphere*, *168*, pp. 944–968.

[3] Malakar, S., Saha, P.D., Baskaran, D. and Rajamanickam, R., 2017. Comparative study of biofiltration process for treatment of VOCs emission from petroleum refinery wastewater—A review. *Environmental Technology & Innovation*, *8*, pp. 441–461.

[4] Malhautier, L., Khammar, N., Bayle, S. and Fanlo, J.L., 2005. Biofiltration of volatile organic compounds. *Applied Microbiology and Biotechnology*, *68*(1), pp. 16–22.

[5] Van Groenestijn, J.W., 2001. Bioscrubbers. In *Bioreactors for Waste Gas Treatment* (pp. 133–162). Springer, Dordrecht.

[6] Liang, Z., Wang, J., Zhang, Y., Han, C., Ma, S., Chen, J., Li, G. and An, T., 2020. Removal of volatile organic compounds (VOCs) emitted from a textile dyeing wastewater treatment plant and the attenuation of respiratory health risks using a pilot-scale biofilter. *Journal of Cleaner Production*, *253*, p. 120019.

[7] Martinez, A., Rathibandla, S., Jones, K. and Cabezas, J., 2008. Biofiltration of wastewater lift station emissions: evaluation of VOC removal in the presence of H 2 S. *Clean Technologies and Environmental Policy*, *10*(1), pp. 81–87.

[8] Gostomski, P.A., Sisson, J.B. and Cherry, R.S., 1997. Water content dynamics in biofiltration: The role of humidity and microbial heat generation. *Journal of the Air & Waste Management Association*, *47*(9), pp. 936–944.

[9] Cox, H.H. and Deshusses, M.A., 2001. Biotrickling filters. In *Bioreactors for Waste Gas Treatment* (pp. 99–131). Springer, Dordrecht.

[10] Lebrero, R., Rodríguez, E., Estrada, J.M., García-Encina, P.A. and Muñoz, R., 2012. Odor abatement in biotrickling filters: Effect of the EBRT on methyl mercaptan and hydrophobic VOCs removal. *Bioresource Technology*, *109*, pp. 38–45.

[11] Wu, H., Yan, H., Quan, Y., Zhao, H., Jiang, N. and Yin, C., 2018. Recent progress and perspectives in biotrickling filters for VOCs and odorous gases treatment. *Journal of Environmental Management*, *222*, pp. 409–419.

[12] Iranpour, R., Cox, H.H., Deshusses, M.A. and Schroeder, E.D., 2005. Literature review of air pollution control biofilters and biotrickling filters for odor and volatile organic compound removal. *Environmental Progress*, *24*(3), pp. 254–267.

[13] López de León, L.R., Deaton, K.E. and Deshusses, M.A., 2018. Miniaturized biotrickling filters and capillary micro bioreactors for process intensification of VOC treatment with intended application to indoor air. *Environmental Science & Technology*, *53*(3), pp. 1518–1526.

[14] Ottengraf, S.P., 1987. Biological systems for waste gas elimination. *Trends in Biotechnology*, *5*(5), pp. 132–136.

[15] Mudliar, S., Giri, B., Padoley, K., Satpute, D., Dixit, R., Bhatt, P., Pandey, R., Juwarkar, A. and Vaidya, A., 2010. Bioreactors for treatment of VOCs and odours—A review. *Journal of Environmental Management*, *91*(5), pp. 1039–1054.

[16] Shareefdeen, Z., 2005. *Biotechnology for Odor and Air Pollution Control*. Springer Science & Business Media.

[17] Kohl, A.L. and Nielsen, R.B., 1997. *Gas Purification*. 5th ed. Gulf Publishing Company, Houston.

[18] Sublette, K.L., Kolhatkar, R. and Raterman, K., 1998. Technological aspects of the microbial treatment of sulfide-rich wastewaters: A case study. *Biodegradation*, *9*(3–4), pp. 259–271.

[19] Munk, P. and Aminabhavi, T.M., 2002. Macromolecules in solutions: Hydrodynamics of macromolecular solutions (Chapter 3.3). In *Introduction to Macromolecular Science*.

[20] Wang, J. and Bai, Z., 2017. Fe-based catalysts for heterogeneous catalytic ozonation of emerging contaminants in water and wastewater. *Chemical Engineering Journal*, *312*, pp. 79–98.

[21] Zhao, S., Zou, L., Tang, C.Y. and Mulcahy, D., 2012. Recent developments in forward osmosis: Opportunities and challenges. *Journal of Membrane Science*, *396*, pp. 1–21.

Biological Remediation of Volatile Organic Compounds

[22] Liao, B.Q., Kraemer, J.T. and Bagley, D.M., 2006. Anaerobic membrane bioreactors: Applications and research directions. *Critical Reviews in Environmental Science and Technology*, 36(6), pp. 489–530.

[23] Shoener, B.D., Zhong, C., Greiner, A.D., Khunjar, W.O., Hong, P.Y. and Guest, J.S., 2016. Design of anaerobic membrane bioreactors for the valorization of dilute organic carbon waste streams. *Energy & Environmental Science*, 9(3), pp. 1102–1112.

[24] Ho, J. and Sung, S., 2009. Anaerobic membrane bioreactor treatment of synthetic municipal wastewater at ambient temperature. *Water Environment Research*, 81(9), pp. 922–928.

[25] Nematian, M.A. and Kazemeini, F., 2013. Accumulation of Pb, Zn, Cu and Fe in plants and hyperaccumulator choice in Galali iron mine area, Iran. *International Journal of Agriculture and Crop Sciences*, 5(4), p. 426.

[26] Dixit, R., Malaviya, D., Pandiyan, K., Singh, U.B., Sahu, A., Shukla, R., Singh, B.P., Rai, J.P., Sharma, P.K., Lade, H. and Paul, D., 2015. Bioremediation of heavy metals from soil and aquatic environment: an overview of principles and criteria of fundamental processes. *Sustainability*, 7(2), pp. 2189–2212.

[27] Mani, S., 2015. Production of reactive oxygen species and its implication in human diseases. In *Free Radicals in Human Health and Disease* (pp. 3–15). Springer, New Delhi.

[28] Ayangbenro, A.S. and Babalola, O.O., 2017. A new strategy for heavy metal polluted environments: A review of microbial biosorbents. *International Journal of Environmental Research and Public Health*, 14(1), p. 94.

[29] Jan, A.T., Azam, M., Ali, A. and Haq, Q.M.R., 2014. Prospects for exploiting bacteria for bioremediation of metal pollution. *Critical Reviews in Environmental Science and Technology*, 44(5), pp. 519–560.

[30] Ojuederie, O.B. and Babalola, O.O., 2017. Microbial and plant-assisted bioremediation of heavy metal polluted environments: A review. *International Journal of Environmental Research and Public Health*, 14(12), p. 1504.

[31] Saratale, G.D., Saratale, R.G., Ghodake, G.S., Jiang, Y.Y., Chang, J.S., Shin, H.S. and Kumar, G., 2017. Solid state fermentative lignocellulolytic enzymes production, characterization and its application in the saccharification of rice waste biomass for ethanol production: An integrated biotechnological approach. *Journal of the Taiwan Institute of Chemical Engineers*, 76, pp. 51–58.

[32] Flores-Chaparro, C.E., Ruiz, L.F.C., de la Torre, M.C.A., Huerta-Diaz, M.A. and Rangel-Mendez, J.R., 2017. Biosorption removal of benzene and toluene by three dried macroalgae at different ionic strength and temperatures: Algae biochemical composition and kinetics. *Journal of Environmental Management*, 193, pp. 126–135.

[33] Dhanarani, S., Viswanathan, E., Piruthiviraj, P., Arivalagan, P. and Kaliannan, T., 2016. Comparative study on the biosorption of aluminum by free and immobilized cells of Bacillus safensis KTSMBNL 26 isolated from explosive contaminated soil. *Journal of the Taiwan Institute of Chemical Engineers*, 69, pp. 61–67.

[34] Arivalagan, P., Singaraj, D., Haridass, V. and Kaliannan, T., 2014. Removal of cadmium from aqueous solution by batch studies using Bacillus cereus. *Ecological Engineering*, 71, pp. 728–735.

[35] Karthik, C., Barathi, S., Pugazhendhi, A., Ramkumar, V.S., Thi, N.B.D. and Arulselvi, P.I., 2017. Evaluation of Cr (VI) reduction mechanism and removal by Cellulosimicrobium funkei strain AR8, a novel haloalkaliphilic bacterium. *Journal of Hazardous Materials*, 333, pp. 42–53.

[36] Congeevaram, S., Dhanarani, S., Park, J., Dexilin, M. and Thamaraiselvi, K., 2007. Biosorption of chromium and nickel by heavy metal resistant fungal and bacterial isolates. *Journal of Hazardous Materials*, 146(1–2), pp. 270–277.

[37] Rhodes, C.J., 2014. Mycoremediation (bioremediation with fungi)–growing mushrooms to clean the earth. *Chemical Speciation & Bioavailability*, 26(3), pp. 196–198.

[38] Raja, C.P., Jacob, J.M. and Balakrishnan, R.M., 2016. Selenium biosorption and recovery by marine Aspergillus terreus in an upflow bioreactor. *Journal of Environmental Engineering*, 142(9), p.C4015008.

[39] Nur-E-Alam, M., Mia, M.A.S., Ahmad, F. and Rahman, M.M., 2020. An overview of chromium removal techniques from tannery effluent. *Applied Water Science*, 10(9), pp. 1–22.

[40] Lasat, M.M., 2000. *The Use of Plants for the Removal of Toxic Metals from Contaminated Soils*. US Environmental Protection Agency.

[41] Clemens, S., 2001. Molecular mechanisms of plant metal tolerance and homeostasis. *Planta*, 212(4), pp. 475–486.

[42] Pueke, A.D. and Rennenberg, H., 2005. Phytoremediation: Molecular biology, requirements for application, environmental protection, public attention and feasibility. *EMBO J*, 6, pp. 497–501.

[43] Dietz, A.C. and Schnoor, J.L., 2001. Advances in Phytoremediation. *Environmental Health Perspectives*, *109*(suppl 1), pp. 163–168.

[44] Dietz, A.C. and Schnoor, J.L., 2001. Advances in phytoremediation. *Environmental Health Perspectives*, *109*(suppl 1), pp. 163–168.

[45] Hall, J.Á., 2002. Cellular mechanisms for heavy metal detoxification and tolerance. *Journal of Experimental Botany*, *53*(366), pp. 1–11.

[46] McGrath, S.P. and Zhao, F.J., 2003. Phytoextraction of metals and metalloids from contaminated soils. *Current Opinion in Biotechnology*, *14*(3), pp. 277–282.

[47] Boopathy, R., 2000. Factors limiting bioremediation technologies. *Bioresource Technology*, *74*(1), pp. 63–67.

[48] Liu, S.H., Zeng, G.M., Niu, Q.Y., Liu, Y., Zhou, L., Jiang, L.H., Tan, X.F., Xu, P., Zhang, C. and Cheng, M., 2017. Bioremediation mechanisms of combined pollution of PAHs and heavy metals by bacteria and fungi: A mini review. *Bioresource Technology*, *224*, pp. 25–33.

[49] Garbisu, C. and Alkorta, I., 1999. Utilization of genetically engineered microorganisms (GEMs) for bioremediation. *Journal of Chemical Technology & Biotechnology: International Research in Process, Environmental & Clean Technology*, *74*(7), pp. 599–606.

[50] Macci, C., Doni, S., Peruzzi, E., Ceccanti, B. and Masciandaro, G., 2012. Bioremediation of polluted soil through the combined application of plants, earthworms and organic matter. *Journal of Environmental Monitoring*, *14*(10), pp. 2710–2717.

5 Tailor-Made Microbial Wastewater Treatment

Shaon Ray Chaudhuri

CONTENTS

5.1 Wastewater Generation and Its Fate 69
5.2 Classical Approach for Wastewater Abatement 70
5.3 The Microbial Biofilm-Based Approach for Wastewater
Treatment 71
    5.3.1 Tailor-Made Microbial Consortium Development from Environmental
Origin 71
    5.3.2 Understanding the Biofilm Progression by the Developed Consortium for
Process Optimisation 75
    5.3.3 Microbial Biofilm Reactor for Remediation of Soluble Sulphate 75
    5.3.4 Microbial Biofilm Reactor for Remediation of Municipal Wastewater 75
    5.3.5 Microbial Biofilm Reactor for Remediation of Dairy Wastewater with
Byproduct Formation 78
    5.3.6 Microbial Biofilm Reactor for Petrochemical Wastewater Treatment 78
5.4 Conclusion 80
5.5 Acknowledgement 80
Notes 80
References 80

5.1 WASTEWATER GENERATION AND ITS FATE

Wastewater is water contaminated due to anthropogenic and natural activity that poses a threat to the environment. It is often referred to as sewage and can come from various sources, namely, domestic and community activities (municipal), industrial (food and beverage, tannery, refinery, petrochemical, mining, fertiliser factory, pesticide, textile) activities, agricultural runoff, storm water, aquaculture, and commercial sectors (Tchobanoglous et al., 2003). The wastewater characteristics are based on its properties, namely physical (colour, suspended as well as dissolved solids, turbidity, odour, oil and grease, and temperature), chemical (ionised metals like cadmium, chromium, iron, nickel, zinc, manganese; radicals like phosphate, nitrate, sulphate, chlorides; inorganic ions like sulphides; pH, alkalinity, acidity, hardness as well as salinity; chemical oxygen demand COD; biochemical oxygen demand BOD; total organic carbon TOC; total oxygen demand) and biological (bacteriological, namely coliform, faecal coliform, pathogens). These properties decide the treatment process for the wastewater.[1] A simple statistic reveals 36,400 million litres (MLD) of wastewater generation per day in India with a treatment capacity of 13,900 MLD.[2] This huge amount of untreated or improperly treated wastewater, if it seeps into the environment, causes undesirable events. In soil it causes bioaccumulation and hypersalinity, while in water it causes uncontrolled algal growth called eutrophication (Biswas et al., 2019; Gogoi et al., 2021). The later pollutes the freshwater (surface or ground water) sources, creating further atrocities due to scarcity of available drinking water sources. The existing treatment technologies are elaborate and labour

DOI: 10.1201/9781003165149-5

intense, requiring large area and energy with prolonged incubation time. Such a bottleneck ensures that the smaller industrial setups cannot adopt these existing technologies in most cases. This background explains the continuous need for enhancing the treatment capacity both in terms of the quantity (volume) and the quality (efficiency, performance, economics).

5.2 CLASSICAL APPROACH FOR WASTEWATER ABATEMENT

The conventional approach of wastewater treatment broadly involves physical, chemical, and biological treatment of wastewater.[3] These are used in combination based on the kind of pollutant to be removed. The physical treatment steps involve screening (Liu and Lipták, 1999), comminution, flow equalisation, sedimentation, flotation, and Granular-medium filtration. The chemical treatment mainly includes precipitation, adsorption, disinfection, and dichlorination. The biological methods are mostly activated sludge treatment, trickling filters, rotating biological contactors, aeration lagoons, pond stabilisation, and biological nutrient removal. Each process has its own advantages and limitations. The order in which these methods are used for wastewater treatment in different cases is based on the influent nature, quality, and volume as well as the available resources/facility. With the rapidly expanding anthropogenic activity, increasing volume of the wastewater generation, and growing complexity of its composition, there is a need to develop a sustainable, rapid, eco-friendly approach for upstream as well as downstream operations.

Most of the biological process of wastewater treatment involve activated sludge or suspended growth. In both cases there is a continuous requirement for biomass recharging as inoculum at regular intervals. Despite standard operating procedures, variation in system performance, when working under ambient condition, is demonstrated at operational scales. This variation in system performance is due to seasonal variation (temperature), bacterial biomass condition, and influent composition. Each organism has a range of temperatures that it can tolerate with optimum metabolic activity at a specific point. Any variation in the same results is slower performance due to delayed doubling time and suboptimum enzyme performance. As an example, the doubling times for *Bacillus thermophilus*, *Bacillus subtilis*, and *Escherichia coli* are 1.3 mins, 30 mins, and 20 mins respectively under optimum condition of growth. However, when there is deviation from the optimum condition, the time increases, and as a result their performance (decided by their metabolism) in their environment also declines. The slowing down of performance is a common phenomenon in systems running under ambient conditions during the winter seasons. Planktonic or free-floating cells are more susceptible to changes in environmental condition than the immobilised cells that remain protected with associated extracellular polymeric substances (EPS; Saha et al., 2018). Such a state of bacterial growth is called bacterial biofilm. Biofilms can be visualised with the arranged bacterial cells as well as the EPS using light microscopy and scanning electron microscopy (SEM) (Figure 5.1).

When the majority of the bacterial biomass is in the log phase of growth, the consortium would have its optimum performance. However, the presence of inhibitors (pesticides, heavy metals, antibiotics, xenobiotics, etc.) might slow down the performance by altering the cell physiology. This altered performance of the cell in response to the inhibitor like heavy metals is often associated with change in the cell morphology as depicted in the scanning electron micrographs of the bacterial cells in response to heavy metals in Figure 5.2.

These pollutants can often be part of the influent reaching the wastewater treatment plant. Based on the nature and concentration of these pollutants, the biomass may be forced to go into dormancy until the stress/pollutant is removed or reduced below the threshold level. The system might take time to revive back to its near normal performance. When the performance does not return to its normal level, it might be due to inactivation of certain members of the biomass that were more sensitive to the pollutant. The inactivated population might have been either directly involved in bioremediation or have produced some intermediate metabolites that were essential for the bioremediants in the biomass to complete the bioremediation (Mamta et al., 2020). In either case, the system does not return to its normal performance. In severe cases of accidental release of elevated

Tailor-Made Microbial Wastewater Treatment

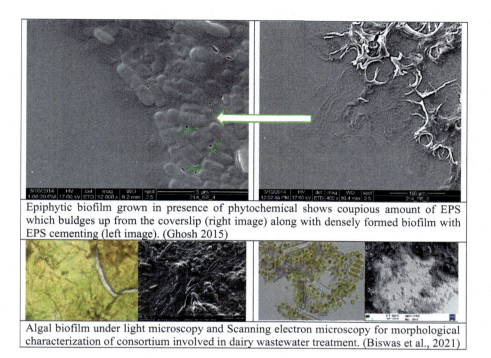

FIGURE 5.1 Microscopic visualisation of bacterial and algal biofilm.

concentration of pollutants from upstream operation, the total biomass is inactivated, requiring the complete replacement of it with fresh biomass. The process of replacing and recharging the biomass to attain optimum performance is time consuming as it is often acclimatised sludge or enriched suspension used for the purpose of conventional biological wastewater treatment. The acclimatisation itself requires substantial time, making biological wastewater treatment often a bottleneck for industrial scale operation in spite of being eco-friendly. Another limitation of biological treatment of wastewater is the bulk sludge formation that needs to be recovered, inactivated, and disposed of safely for sustaining the continuous operation. The sludge removal is a labour intense, energy intense (dewatering), cumbersome process.

To further improvise the process of biological wastewater treatment, a sludge-free system with one-time inoculation and a robust performance is required. This could be achieved through development of biofilm bioreactors using tailor-made microbial consortium specific for the wastewater (Saha, et al., 2018; Mamta et al., 2020; Chanda et al., 2020; Gogoi et al., 2021; Biswas et al., 2021).

5.3 THE MICROBIAL BIOFILM-BASED APPROACH FOR WASTEWATER TREATMENT

5.3.1 Tailor-Made Microbial Consortium Development from Environmental Origin

The first step towards development of a tailor-made microbial consortium is to characterise the wastewater to be treated in terms of its physicochemical properties (Debroy et al., 2012). Upon identification of the pollutants to be treated and the co-existing microbial inhibitors in the wastewater, it is possible to document from literature the types of microbes that are known to bioremediate these pollutants. Based on the documentation, the medium can be selected for culture-based isolation of the microbe from the environmental site (Ray Chaudhuri and Thakur, 2006).

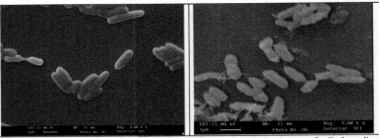

Scanning electron micrographs representing the morphological change of oil degrading *Pseudomonas aeruginosa* (GenBank Acc No. FJ788518) pre (left) and post (right) exposure to 4mM of Pb salt. The cells treated with Pb exhibited a rough, short and stout shape with distorted morphology.

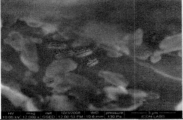 

Scanning electron micrographs of oil degrading *Pseudomonas mendocina* BWO (GenBank Acc No EU006700) displaying the morphological changes observed pre- and post-exposure to iron. The control cells were visualized by Environmental SEM where the cells were not prefixed and directly mounted on stubs for visualization. Upon treatment with iron, the cell elongates preventing further intake of metal inside through decreased surface area.

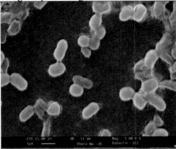 

SEM image of oil degrading *Acinetobacter grimontii* CL (GenBank Acc No EU006703) before and after exposure to silver salt. The micrographs were captured at 9000X magnification depicting elongation with arrested cell division of cells ensure inhibition of further metal uptake through decreased surface area.

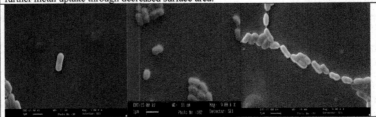

Scanning electron micrographs (magnification 9000X) of oil degrading *Acinetobacter hamolyticus* (GenBank Acc No EU006701) from East Kolkata Wetland before (left), and after exposure with cobalt (middle) and silver (right) salts. In case of cobalt, there is shrinkage (increased surface area) representing that the cells can accumulate further cobalt intracellularly. In presence of silver the cells shrink and develop a wooly coat (that prevents further entry of metal) indicating the stress response of the cells.

FIGURE 5.2 Scanning electron micrographs of bacterial cells with and without treatment with heavy metals.

Source: Mishra, 2009

Tailor-Made Microbial Wastewater Treatment

The site selection for isolation of the microbes is based on the following steps. The first step involves identifying environmental sites that receive the pollutant (to be removed) from point sources. The second step involves aseptic collection of the samples and physicochemical and biological characterisation of the collected samples. The third step involves identifying those sites that, in spite of receiving the point sources of the pollutant, show minimal concentration of the pollutant. That is an indication of presence of potential bioremediants at those sites from where the samples were collected. The fourth step would involve spreading the collected sample on the selected agar medium to obtain isolated colonies. The reason for selecting an enriched medium instead of the conventional minimal medium would be to ensure isolation of the major fraction of the microbes in the sample, which would be drastically reduced if the isolation was attempted on a minimal medium. The fifth step would involve testing the obtained isolates for survival on the agar medium with spiked concentration of the pollutants. For each round of streaking (resulting in isolate purification) the concentration of the spiked pollutant can be increased to test the tolerance of the isolates. The isolates that survive in the presence of a higher concentration of pollutants can then be acclimatised to minimal medium-based growth condition. As the sixth step, the isolates acclimatised to minimal condition can be grown in liquid minimal medium spiked with the pollutant till turbidity is observed. The concentration of the pollutant at the zero hour and after growth would be checked. The difference in concentration indicates the bioremedial ability of the isolates. The isolates such as those selected (Debroy et al., 2012) with higher tolerance can be compared for their bioremedial ability and, if desired, consortium can be developed, combining the isolates in desired proportion (Halder et al., 2020; Saha et al., 2018; Biswas et al., 2019; Gogoi et al., 2021; Biswas et al., 2021). Such consortium development ensures active involvement of each member with little presence of non-performing members in the consortium. This in turn ensures little sludge generation. The other steps carried out are the characterisation of the isolates/developed consortium for performance optimisation and long-term preservation.

Yet another approach would be to use liquid medium for enriching the consortium using samples collected from the selected sites. The consortium could be stabilised through repeated subculturing. The ability of bioremediation and performance optimisation can be carried out as per standard procedure (Ray Chaudhuri et al., 2016a, 2016b, 2017b, 2017c; Nasipuri et al., 2010, 2011; Chowdhury et al., 2011). The performance of the consortium can then be tested with actual effluent (Gogoi et al., 2021; Biswas et al., 2021; Halder et al., 2020; Biswas et al., 2019; Saha et al., 2018). In order to ensure minimum addition of COD through the medium components in wastewater, the minimal medium composition for sustaining the growth of the consortium can be determined as per reported procedure (Chanda et al., 2020; Saha et al., 2018; Ray Chaudhuri et al., 2013; Ray Chaudhuri, 2011; Shah, 2020b, 2021).

A third approach will be to use the existing well-characterised isolates from a laboratory or culture collection for consortium development. In such cases, the first step is to understand the mechanism preferred for the bioremediation or bioconversion of the pollutant. An example for the same is cited later. Here the purpose was to develop a consortium that could convert the nitrogenous components in the milk processing unit wastewater into ammonia (plant growth nutrient) instead of releasing it as nitrogen gas. The nitrogen gas release causes environmental pollution. The release of the untreated dairy wastewater (DWW) also causes environmental deterioration. The existing treatment technologies for DWW are elaborate and energy intense, which loses the nitrogenous component as nitrogen gas. Nitrogenous fertilisers are an integral part of plant nutrient requirement. Its production requires energy while only 30% of the applied fertiliser can be utilised by the plants. The rest leaches into the environment and pollutes the surrounding water and soil. So, the approach adopted here was to reuse the nitrogenous pollutants in the DWW by converting it and other pollutants to a form that the plants would preferentially uptake without damaging the environment. Here the microbial metabolic map available at Kyoto Encyclopedia of Genes and Genomes (KEGG) was used as a reference frame (Figure 5.3). The possible fate of the nitrogenous components (nitrate) could be either conversion of nitrate to nitrogen gas through

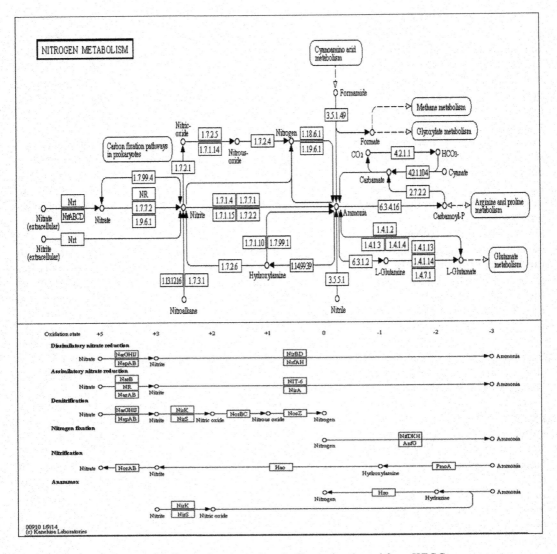

FIGURE 5.3 Nitrogen metabolism reference pathway of bacteria adopted from KEGG.
Source: www.genome.jp/kegg-bin/show_pathway?map00910.

denitrification/anammox reaction or to ammonia through nitrogen fixation/dissimilatory nitrogen reduction/assimilatory nitrate reduction (Ghoshal et al., 2014). The alternative fates are conversion of ammonia to nitrate through nitrification, to nitrogen gas via hydrazine, and to amino acids through anabolism. Based on this background information, the draft genome analysis data of well-characterised microbes (Ray Chaudhuri et al., 2013; Debroy et al., 2013a, 2013b, 2013c; Chatterjee et al., 2013; Ray Chaudhuri, 2016; Shah, 2020a) was used to select microbes that would show the presence of all genes required for conversion of nitrate to ammonia but would lack at least some of the genes required for the conversion of ammonia into nitrate or nitrogen gas. This would ensure that, provided there was no adverse environmental condition, these microbes would not lose the ammonia as nitrogen gas or it would not turn back to nitrate (hence slowing down the process of bioconversion of DWW into ammonia-rich liquid biofertiliser). The selected microbes can be

Tailor-Made Microbial Wastewater Treatment

then combined to develop consortium (Halder et al., 2020; Biswas et al., 2019; Halder, 2017; Ray Chaudhuri, 2020) that would be tested for real DWW conversion to liquid biofertiliser (Gogoi et al., 2021).

Based on the first approach, hexamine (HDV01) degrading *Micromonospora citrea* strain SRCHD01 (GenBank Accession Number MT995076.1) was isolated from hexamine containing industrial effluent with ability of 72.44% COD reduction within 48 hs at 37°C from simulated effluent at pH 7 with an initial COD of 3117mg/L. The isolate is gram positive; protease, amylase, lipase, oxidase, DNAse negative; catalase positive; strong biofilm former (Figure 5.4).

5.3.2 Understanding the Biofilm Progression by the Developed Consortium for Process Optimisation

After developing the appropriate consortium for specific wastewater treatment, the next approach is to ensure development of a stable system with sustained robust performance. In order to do so, the consortium is tested for its biofilm-forming ability (Martin et al., 2008) and biofilm progression (Gogoi et al., 2021; Figure 5.5a). Upon deciding the time for stable biofilm formation, the system performance optimisation is carried out under an immobilised condition (Saha et al., 2018) using methods like response surface methodology (Sarkar et al., 2021; Chandra et al., 2020; Figure 5.5b). Based on the optimisation parameters developed using a laboratory scale setup (Figure 5.5c), the full-scale reactors are designed and the process is implemented at an industrial (Figure 5.5d) scale (Gogoi et al., 2021). As examples, three case studies of pilot scale sludge free biofilm systems using tailormade bacterial consortium are detailed in the following sections.

5.3.3 Microbial Biofilm Reactor for Remediation of Soluble Sulphate

Sulphate is a major pollutant in wastewater from point sources like sewage, mining effluent, wastewater from tanneries, pulp mills, and textile mill effluents (to name a few). Physicochemical followed by biological treatment is the conventional process. Tailor-made consortium developed through enrichment (Nasipuri et al., 2010, 2011) in medium DSMZ 641 from different environmental sites under anaerobic condition (nitrogen and carbon dioxide) could effectively reduce soluble sulphate as single unit operation at laboratory scale within 12 to 96 hs (Nasipuri, 2010). The efficiency of the system was improved with biofilm formation achieving desired reduction within 12 hs. The most efficient consortium among these was found to be effective for tannery as well as mining effluent bioremediation with supplement media component addition. The process was scaled up to 1.32m³/day processing capacity in a biofilm reactor (Ray Chaudhuri et al., 2016a; Mukherjee et al., 2016a, 2016b) with desired reduction with 3.5 hs of incubation with one-time charging and consistent performance for over 18 months of operation (Ray Chaudhuri et al., 2011, 2013). It is essential to develop processes that could sustain bioremediation with the addition of minimum COD load to the wastewater. The minimal medium composition was worked out with maintained efficiency of the system (Chanda et al., 2020). The process is now ready for a pilot-scale trial of a sludge-free system.

5.3.4 Microbial Biofilm Reactor for Remediation of Municipal Wastewater

The current case study involves selectively combining three bacterial isolates from the environmental origin (rhizosphere of water lily; municipal sewage enriched biomass from radioactive nitrate reducing bioreactor; water of sewage canal) with the potential of simultaneous nitrate and phosphate removal from wastewater. All isolates were members of the genus Bacillus. The consortium includes one strong biofilm with two moderate biofilm formers. The latter two could continue to stick to the biofilm formed by the former to develop a system with sustained performance.

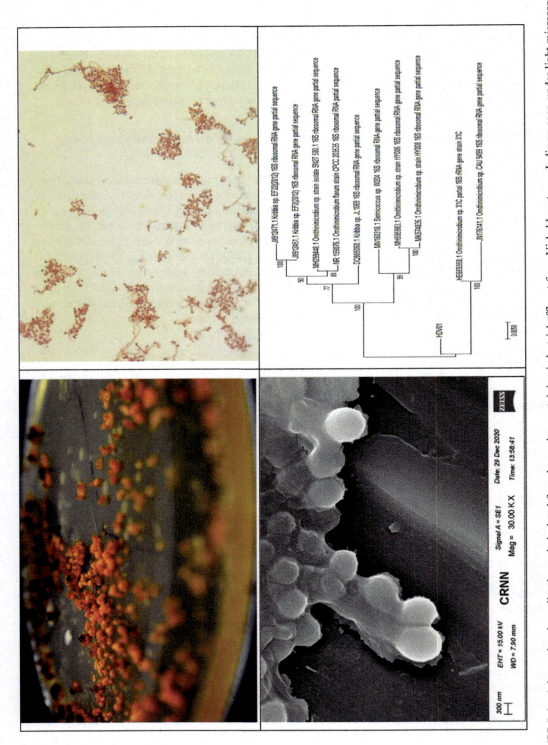

FIGURE 5.4 A hexamine degrading bacteria isolated from hexamine containing industrial effluent from Vishakhapatnam, India on agar, under light microscope, under scanning electron microscope and its phylogenetic position based on 16S rDNA sequence.

Tailor-Made Microbial Wastewater Treatment

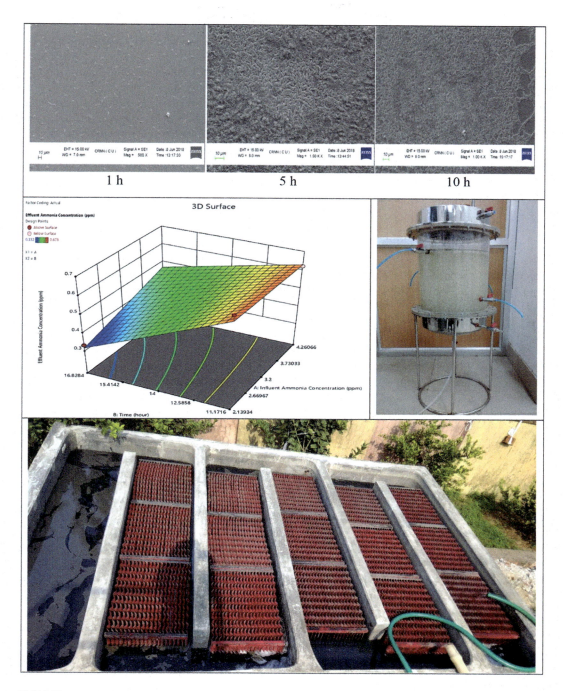

FIGURE 5.5 A Time course of biofilm formation by an ammonia removing isolate of *Bacillus albus* ARTU001 (GenBank Accession Number MK318642) through scanning electron microscopy (Ray Chaudhuri, 2021). 5b. Graphical representation of the process optimisation for tailormade consortium based aquaculture effluent treatment using RSM. 5c. Laboratory scale packed bed biofilm reactor used for process optimisation. 5d. Industrial scale biofilm reactor for DWW conversion to liquid biofertiliser with fiber reinforced plastic blocks as immobilisation matrix.

This consortium could bioremediate under both suspended and immobilised condition. The system can treat the municipal wastewater and agricultural runoff water within 2 hs of incubation with no added aeration, reducing the COD by 92% and BOD by 97%. This treated water was suitable for reuse in agriculture and aquaculture (Saha et al., 2018; Banerjee, 2017). Through this approach, the crisis of fresh water due to its misuse in non-portable application can be addressed to a major extent. The biomass in turn, with the accumulated nitrate and phosphate, could work as a biofertiliser with enhanced production of mung bean during field trials (Ray Chaudhuri et al., 2013; Ray Chaudhuri, 2011). The process was scaled up to 2.64m^3/day using single inoculum with sustained performance for more than 1.5 years. The scum-free system uses 55% less space and 80% less energy while releasing 90% less carbon dioxide equivalent (CO_2 e) gas due to its operation.

5.3.5 MICROBIAL BIOFILM REACTOR FOR REMEDIATION OF DAIRY WASTEWATER WITH BYPRODUCT FORMATION

Well-characterised bacterial isolates from the activated sludge of a dairy effluent treatment plant were selectively combined with environmental isolates with capability of reducing nitrogenous compounds, phosphates, proteins, and lipids to develop a consortium that could treat dairy effluent within 21 hs of hydraulic retention time. This treated effluent could be used for irrigation (Biswas et al., 2019). This process was 5.7 times faster than conventional treatment technology. It involves single unit operation instead of an eight-unit conventional setup.

There are algae bacterial mixed consortiums selectively enriched in algal medium from a wastewater-fed fish pond that could bioremediate the DWW producing lipid-rich algal biomass. The appropriate consortium development ensures DWW remediation with no added fresh water requirement along with economically important resource generation. The selectively enriched consortium could grow as biofilm, which can be easily recovered post bioremediation. This ensured reduction in the energy requirement for algal harvesting (Biswas et al., 2021).

A well-characterised bacterial consortium made using 6 isolates from environmental origins could convert DWW into liquid biofertiliser within 16 hs of incubation. The process is scum free. When compared to conventional system, it is 7.5 times faster and requires less space and less energy for operation, hence generating 90% less CO_2 e gas (Gogoi et al., 2021; Ray Chaudhuri et al., 2017a). The biofertiliser could enhance growth of economic crops like mung bean, black gram, sorghum Sudan grass, and lemongrass as shown in Figure 5.6.

5.3.6 MICROBIAL BIOFILM REACTOR FOR PETROCHEMICAL WASTEWATER TREATMENT

Bacterial isolates obtained from effluent slurry of the petrochemical industry were combined for development of consortium. The consortium could reduce the COD of petrochemical effluent to discharge level within 18 to 20 hs of incubation under ambient condition in a sludge-free, moving-bed biofilm reactor. The reactor design had multiple 5m^3 reactors in series. The process could be scaled up to 12m^3/day capacity with 96% BOD reduction. The released treated water was clean, scum/sludge free, and reusable for landscaping. The industrial system has been performing bioremediation continuously for more than four years with biofilm development done only once during the installation. The scum-free process is faster than conventional techniques and eco-friendly (as all microbes are from the environment), and treated water is suitable for landscaping purposes. The purchase of fresh water by the industry for maintenance activities (landscaping and storage for fire-fighting) has been replaced by reusing the treated water, resulting in savings due to the adoption of this system. The investment made on setting up the bioremedial unit could be recovered within three years of installation. The adoption of this tailor-made consortium-based biofilm system ensures compliance with the Environmental Protection Agency norms along with environmental protection (Ray Chaudhuri et al., 2020a).

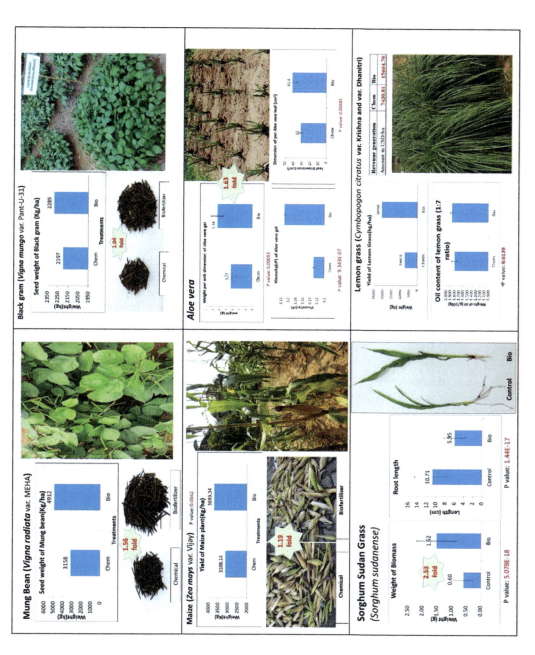

FIGURE 5.6 Photographs showing the growth enhancement in mung bean, black gram, maize, aloe vera, and lemongrass with the aforementioned liquid biofertiliser application. The growth enhancement in the case of sorghum Sudan grass with liquid biofertiliser application was compared with no fertiliser application as fodder crops are grown without fertiliser. Through this approach, fresh water misuse for agriculture can be stopped and the plant growth nutrients in DW can be completely reused instead of using chemical fertiliser (Gogoi et al., 2021).

5.4 CONCLUSION

This study documents the approach of developing tailor-made microbial consortium for wastewater treatment using different types of biofilm reactors. The systems are robust and sludge free with minimum stabilisation time to attain optimum performance. Such an approach is suitable for industrial scale application and can address the bottleneck of prolonged stabilisation time for activated sludge-based biotreatment process in most of the effluent treatment plants, be it from the food and beverage industry, sewage treatment plant, or petrochemical industry effluent. Through this approach and its further modifications, rapid, eco-friendly biological treatment systems can be developed not only for wastewater treatment but also for minimising pollution generated during different anthropogenic activities (Ray Chaudhuri et al., 2019; Ray Chaudhuri, 2010a, 2010b; Ray Chaudhuri et al., 2020b).

5.5 ACKNOWLEDGEMENT

The author acknowledges the granting agencies, namely the Ministry of Environment, Government of India (FAST Scheme); Department of Biotechnology, GoI (Twinning project); Biotechnology Industry Research Assistance Council (Biotechnology Ignition Grant), Department of Atomic Energy, GoI (BRNS, UGC-DAE CSR), Indian Council for Agricultural Research, GoI for funding the research whose outcome is depicted in this chapter. The author acknowledges the work done by the project trainees and the scholars associated with the Microbial Technology Group that yielded the results mentioned in this chapter. The author acknowledges the engineering input (bioreactor designing) received from Prof L. M. Gantayet for the reactor development for the industrial-scale setups mentioned in this chapter. The author acknowledges the administrative support of Maulana Abul Kalam Azad University of Technology, West Bengal and Tripura University, where she was associated during the work.

NOTES

1. https://books.google.co.in/books?id=EJ3eAAAAMAAJ
2. www.scribd.com/document/421191172/sewagepollution-pdf
3. https://books.google.co.in/books?id=EJ3eAAAAMAAJ

REFERENCES

Banerjee S. (2017). *Understanding the Effect of Plant Growth Promoting Bacteria (PGPB) Formulation on Nutritional Quality of Mung Bean Seeds*. PhD Thesis submitted to Maulana Abul Kalam Azad University of Technology, West Bengal, India.

Biswas T., Bhushan S., Prajapati S. K., Ray Chaudhuri S. (2021). An eco-friendly strategy for dairy wastewater remediation with high lipid microalgae-bacterial biomass production. *Journal of Environmental Management*. 286, 112196.

Biswas T., Chatterjee D., Barman S., Chakraborty A., Halder N., Banerjee S., Ray Chaudhuri S. (2019). Cultivable bacterial community analysis of dairy activated sludge for value addition to dairy wastewater. *Microbiology and Biotechnology Letters*. 47, 585–595pp.

Chanda C., Gogoi M., Mukherjee I., Ray Chaudhuri S. (2020). Minimal medium optimization for soluble sulfate removal by tailor-made sulfate reducing bacterial consortium. *Bioremediation Journal*. 24, 251–264pp.

Chatterjee D., Thakur A. R., Ray Chaudhuri S. (2013). Draft Genome of an ammonia producing *Acinetobacter* sp strain MCC2139 from dairy effluent. *Genome Announcement*. 1, e00410–13pp.

Chowdhury S., Thakur, A. R., Ray Chaudhuri S. (2011). Novel microbial consortium for laboratory scale lead removal from city effluent. *Journal of Environmental Science and Technology*. 4, 41–54pp.

Debroy S., Bhattacharjee A., Thakur A. R., Ray Chaudhuri S. (2013a). Draft Genome of a nitrate and phosphate accumulating *Bacillus sp* MCC0008. *Genome Announcement*. 1, e00189–12pp.

DebRoy S., Das S., Ghosh S., Banerjee S., Chatterjee D., Bhattacharjee A., Mukherjee I., Ray Chaudhuri S. (2012). Isolation of nitrate and phosphate removing bacteria from various environmental sites. *OnLine Journal of Biological Sciences*. 12, 62–71pp.

Debroy S., Mukherjee P., Roy S., Thakur A. R., Ray Chaudhuri S. (2013b). Draft Genome of a nitrate and phosphate removing *Bacillus sp* WBUNB009. *Genome Announcement*. 1, e00254–12pp.

Debroy S., Mukherjee P., Roy S., Thakur A. R., Ray Chaudhuri S. (2013c). Draft Genome of a phosphate accumulating *Bacillus* sp WBUNB004. *Genome Announcement*. 1(1), e00251–12pp.

Ghosh S. (2015). *Plant Microbe Interaction on Epiphytic Leaf Surface in Case of Traditional Indian Plants and Herbs*. PhD Thesis submitted to Maulana Abul Kalam Azad University of Technology, West Bengal, India. https://opac.wbut.ac.in/cgi-bin/koha/opacetail.pl?biblionumber=40304&query_desc=au%2Cwrdl%3A%20Sourav%20Ghosh.

Ghoshal T., Ghosh S., Saha A., Haldar N., Thakur A. R., Ray Chaudhuri S. (2014). Combination of conventional and in-silico approach for identifying industrially important isolates of *Aeromonas*. *OnLine Journal of Biological Sciences*. 14, 70–83.

Gogoi M., Biswas T., Biswal P., Saha T., Modak A., Gantayet L. M., Nath R., Mukherjee I., Thakur A. R., Sudarshan M., Ray Chaudhuri S. (2021). A novel strategy for microbial conversion of dairy wastewater into biofertilizer. *Journal of Cleaner Production*. 293, 126051pp.

Halder N. (2017). *Microbial Conversion of Dairy Waste Product (Effluent) into Ammonia*. PhD Thesis submitted to Maulana Abul Kalam Azad University of Technology, West Bengal, India. www.genome.jp/kegg-bin/show_pathway?map00910

Halder N., Gogoi M., Sharmin J., Gupta M., Banerjee S., Biswas T., Agarwala B. K., Gantayet L. M., Sudarshan M., Mukherjee I., Roy A., Ray Chaudhuri S. (2020). Microbial consortium-based conversion of dairy effluent into biofertilizer. *Journal of Hazardous, Toxic, and Radioactive Waste*. 24, 04019039-1-7pp.

Liu D. H. F., Lipták B. G. (Eds.). (1999). *Wastewater Treatment*. 1st ed. Boca Raton, FL: CRC Press. Taylor & Francis Group, 472pp. ISBN: 9780367399122.

Mamta B., Bhushan S., Ray Chaudhuri S., Simsek H., Prajapati S. K. (2020). Algae—and bacteria—driven technologies for pharmaceuticals remediation in wastewater. In: Shah M. P. (Eds.). *Removal of Toxic Pollutants through Microbiological and Tertiary Treatment*. 1st ed. Amsterdam: Elsevier, 373–408pp., ISBN-9780128210147.

Martin R., Soberon N., Vaneechoutte M., Florez A. B., Vazquez F., Suarez J. E. (2008). Characterization of indigenous vaginal lactobacilli from healthy women as probiotic candidates. *International Journal of Microbiology*. 11, 261–266pp.

Mishra M. (2009). *Bioremedial Studies Using Microbes from East Calcutta Wetland*. PhD Thesis submitted to West Bengal University of Technology, India. https://opac.wbut.ac.in/cgibin/koha/opacdetail.pl?biblionumber=40234&query_desc=au%2Cwrdl%3A%20Madhusmita%20Mishra.

Mukherjee I., Giri A., Barat P. (2016a). Quantitative characterization of Sulphate Reduction data Obtained from a Biofilm based bioreactor-Part II. In: Ray Chaudhuri S. (Ed.), *Life Science: Recent Innovations and Research*. New Delhi: Research Publishing House, 103–120pp. ISBN: 9879384443535.

Mukherjee I., Giri A., Sen C., Sebait R., Barat P. (2016b). Quantitative characterization of sulphate reduction data obtained from a biofilm based bioreactor—PART-I. In: Ray Chaudhuri S. (Ed.), *Life Science: Recent Innovations and Research*. New Delhi: Research Publishing House, 81–102pp. ISBN: 9879384443535.

Nasipuri P. (2010). *Isolation and Characterization of Efficient Sulfate Reducing Bacterial Consortia from Different Environmental Sites*. PhD Thesis submitted to West Bengal University of Technology, India.

Nasipuri P., Pandit G. G., Thakur A. R., Ray Chaudhuri S., (2010). Comparative study of soluble sulfate reduction by bacterial consortia from varied regions of India. *American Journal of Environmental Science*. 6, 152–158pp.

Nasipuri P., Pandit G. G., Thakur, A. R., Ray Chaudhuri S. (2011). Microbial consortia from taptapani hot water springs for mining effluent treatment. *American Journal of Microbiology*. 1, 23–29pp.

Ray Chaudhuri S. (2010a). *Hide Processing Methods and Compositions*. Indian Patent Application 863/KOL/2010 dt August 5, 2010; Granted Chinese Patent CN103080340B. https://patents.google.com/patent/CN103080340A/en, Granted US Patent US20120142073. https://patents.google.com/patent/US20120142073.

Ray Chaudhuri S. (2010b). *Microbial Enzymes as Detergent Additives*. Indian Patent Application 599/KOL/2010 dt June 1 2010; Granted US Patent US20120021489. https://patents.google.com/patent/US20120021489A1/en.

Ray Chaudhuri S. (2011). *Nitrate Reducing Microbial Consortium as Biofertilizer from Plant Growth Promotion.* Indian Patent Application 518/KOL/2011 dated April 11th 2011.

Ray Chaudhuri S. (2013). *Methods for Treating Sulphate Containing Water.* Indian Patent Application 1289/KOL/2013 dated November 13th 2013, Granted Indian Patent 341914 on 17th July 2020.

Ray Chaudhuri S. (2016). Draft genome sequence of an industrially important *Bacillus* sp. from Mandarmani coastal waters in Midnapur District, West Bengal, India. *Genome Announcement.* 4, e00867–16pp.

Ray Chaudhuri S. (2019). *Microbial Consortium and Process for Degumming of Ramie Fiber.* Indian Patent Application 201931048663 dated 27th November 2019.

Ray Chaudhuri S. (2020). *Microbial Combination for Environmental Protection and Agricultural Sustenance.* Indian Patent Application 203/KOL/2013 dated 21st Feb 2013. Granted Indian Patent 347939 on 28th September 2020.

Ray Chaudhuri S. (2021). *A Process and System for Ammonia Removal from Wastewater.* 202131002964 dated 21st January 2021, Indian Patent Application.

Ray Chaudhuri S., Gantayet L. M., Thakur A. R. (2017a). *Bio-fertilizer Production from Bacterial Consortium.* Indian Patent Application 201731003023 dated 27th January 2017.

Ray Chaudhuri S., Gantayet L. M., Thakur A. R. (2020a). *Formulation of Bacterial Consortium for Bioremediation of Petrochemical Wastewater.* Indian Patent Application 202031011766 dated 18th March 2020.

Ray Chaudhuri S., Gogoi M., Biswas T., Chatterjee S., Chanda C., Jamatia R., Modak A., Sett S. K., Mukherjee I. (2020b). Optimization of bio-chemical degumming of Ramie fiber for improved strength & luster. *Biotechnology Reports.* 28, e00532–1-7pp.

Ray Chaudhuri S., Mishra M., De S., Samal B., Saha A., Banerjee S., Chakraborty A., Chakraborty A., Pardhiya S., Gola D., Chakraborty J., Ghosh S., Jangid K., Mukerjee I., Sudarshan M., Nath R., Thakur A. R. (2017b). Microbe-based strategy for plant nutrient management. In: Farooq R., Ahmed Z. (Eds.), *Waste Water Treatment and Resource Recovery.* London: Intech, 38–55pp. ISBN: 9789535130451.

Ray Chaudhuri S., Mukherjee I., Datta D., Chanda C., Krishnan G. P., Bhatt S., Datta P., Bhushan S., Ghosh S., Bhattacharya P., Thakur A. R., Roy D., Barat P. (2016a). Developing tailor made microbial consortium for effluent remediation. In: Rahman O. A., Saleh H. E. M. (Eds.). *Nuclear Material Performance.* London: Intech, 17–35pp., ISBN: 9789535124481.

Ray Chaudhuri S., Saha A., Ghoshal T., Thakur A. R. (2013). Draft genome of an ammonia-producing aeromonas sp. MDS8 (Strain MCC2167) from sludge of dairy effluent treatment plant. *Genome Announcement.* 1, e00710–13pp.

Ray Chaudhuri S., Sharmin J., Banerjee S., Jayakrishnan U., Saha A., Mishra M., Ghosh M., Mukherjee I., Banerjee A., Jangid K., Sudarshan M., Chankraborty A., Ghosh S., Nath R., Banerjee M., Singh S., Saha A. K., Thakur A. R. (2016b). Novel microbial system developed from low level radioactive waste treatment plant for environmental sustenance. In: Saleh H. E. M., Rahman R. O. A. (Eds.), *Management of Hazardous Wastes.* London: Intech, 121–154pp., ISBN: 9789535126164.

Ray Chaudhuri S., Thakur A. R. (2006). Microbial genetic resource mapping of East Calcutta Wetland. *Current Science.* 91, 212–217pp.

Ray Chaudhuri S., Thakur A. R. (2011). *Self-Sustained Microbial De-Toxification of Soluble Sulfate from Environmental Effluent.* Indian Patent Application 789/KOL/2011 dated June 10th, 2011; Granted US patent US8398856B2. https://patents.google.com/patent/US8398856; US20120312743A1. https://patents.google.com/patent/US20120312743.

Ray Chaudhuri S., Thakur A. R., Mukherjee I. (2017c). *Microbial Consortium for Nitrate and Phosphate Sequestration for Environmental Sustenance.* Indian Patent Application 1179/KOL/2013 dated 16th Oct 2013, Granted Indian Patent 351564 on 13th Nov 2020; Granted Bangladesh Patent 1005753 dated 24th Oct 2017 filed on 14th Oct 2014.

Saha A., Bhushan S., Mukherjee P., Chanda C., Bhaumik M., Ghosh M., Sharmin J., Datta P., Banerjee S., Barat P., Thakur A. R., Gantayet L. M., Mukherjee I., Ray Chaudhuri S. (2018). Simultaneous sequestration of nitrate and phosphate from wastewater using a tailor-made bacterial consortium in biofilm bioreactor. *Journal of Chemical Technology and Biotechnology.* 93, 1279–1289pp.

Sarkar P., Biswas T., Chanda C., Saha A., Sudarshan M., Majumder C., Ray Chaudhuri S. (2021). Spent coffee waste conversion to value added products for pharmaceutical industry. In: Thatoi H., Das S. K., Mahaptr S. (Eds.), *Bioresource Utilization and Management Applications in Therapeutics, Biofuels, Agriculture, and Environmental Science.* Canada: Apple Academic Press (AAP), Inc., a Taylor & Francis Group. 471–487pp. ISBN: 9781771889339.

Shah M. P. (2020a). *Advanced Oxidation Processes for Effluent Treatment Plants*. Elsevier.

Shah M. P. (2020b). *Microbial Bioremediation and Biodegradation*. Singapore: Springer.

Shah M. P. (2021). *Removal of Emerging Contaminants through Microbial Processes*. Singapore: Springer.

Tchobanoglous G., Burton L. S., Stensel D. H. (2003). *Wastewater Engineering: Treatment and Reuse*. 4th ed. New York: Metcalf and Eddy, Inc. McGraw-Hill, 1848pp. ISBN: 007418780.

Waste-water Treatment Technologies: A General Review. (2003). *Economic and Social Commission for Western Asia*. United Nations, 121pp. https://books.google.co.in/books?id=EJ3eAAAAMAAJ.

www.scribd.com/document/421191172/sewagepollution-pdf.

6 Systematic Industrial Wastewater Treatment by Biomaterial Fabricated Nanofiltration Membrane

Puja Ghosh, Supriya Ghule, Nilesh S. Wagh, and Jaya Lakkakula

CONTENTS

6.1 Introduction 85
6.2 Biopolymers Used in Fabrication of Nanofiltration
Membrane 86
    6.2.1 Engaging Chitosan Biopolymer 86
    6.2.2 Engaging Cellulose Biopolymer... 88
    6.2.3 Engaging β-Cyclodextrin Biopolymer 89
    6.2.4 Engaging Alginate Biopolymer... 90
    6.2.5 Engaging Fibroin Biopolymer 92
6.3 Green NPs Used in Fabrication of Nanofiltration Membrane............................... 93
    6.3.1 Synthesis of TiO_2 NP from Living Sources in Nanocomposite Formation 93
6.4 Conclusion 96
References.. 100

6.1 INTRODUCTION

Industrial effluent is becoming a major concern for the environment as well as human health. Proper disposal and treatment of the effluents contaminating wastewater might result in a decline in environmental pollution and in health hazards. Commercially available filters and other physical and chemical treatments are already applied in many treatment plants for wastewater, but to remove the nanoscale pollutants present in it, new advancements have to be made. For this, researchers are shifting their focus towards developing fibrous membranes with nanosized pores, where both nanoscale pollutants and viruses can be removed [1]. However, the commercially available synthetically fabricated membranes are prone to clogging as they gradually age, and their disposal becomes another problem for the environment. Thereby, nano composite membranes, made up of biopolymers and other green technologies, are taken under consideration. Any polymeric compound extracted from any macro-organisms like honeybees or microorganisms like algae or trees are considered biopolymers, whereas other green technologies include nanomaterials like titanium dioxide (TiO_2) and silver synthesised from living beings to construct the nanofiltration membrane [2]. Their availability, efficiency, biocompatibility, and biodegradability as well as their renewability make them appropriate from both an economic and an environmental point of view [3]. This technology is drawing more attention due to the properties of the membrane, like reduced consumption of energy, truncated operational pressure, and exceptional rejection capacity [4]. Moreover, due to high area-to-volume ratio, nanofibers are preferred in wastewater treatment [5]. For the characterisation of the assembled membrane, different groups of researchers investigated

DOI: 10.1201/9781003165149-6

various parameters like water contact angle to measure the wettability and hydrophilicity of the surface of the membrane [6], rejection efficiency, water flux, durability, pore size, pore diameter, and thermal stability. To overcome the drawbacks of these individual membranes, different innovative approaches, including blending these biopolymers with other ingredients, have also been investigated. Another novel strategy of effluent removal can be carried out by introducing microorganisms (bacteria) onto the nanofibrous sheath. Different microorganisms have the potential to remove specific contaminants, thus when encapsulated with the nanocomposite material, they can induce the rejection efficiency [7].

In this chapter we listed a few biopolymers like alginate, chitosan, cellulose, cyclodextrin, fibrine, and zein, along with green nanoparticle (NP), to fabricate composite membranes with diverse properties to eliminate different varieties of effluents.

6.2 BIOPOLYMERS USED IN FABRICATION OF NANOFILTRATION MEMBRANE

6.2.1 ENGAGING CHITOSAN BIOPOLYMER

Chitin is a polysaccharide present in the exoskeletons of crustaceans. It is isolated from lobster, shrimp, crab, etc. Chitosan is a hydrophilic biopolymer that can be obtained industrially by hydrolysing the chitin (amino acetyl groups). It is a natural, biodegradable, non-toxic, poly-saccharide available in many forms like fine powder, beads, solution, flakes, and fibres [8]. Chitosan can be used as a good supporting material for developing filters.

An interfacial polymerisation of piperazine with trimesoyl chloride in the presence of H_2O was assisted by chitosan to produce an ultra-highly permeable and selective nanofilter (NF) for water treatment. Both chitosan and its NP establish an interlayer as they get slowly deposited, which provides the structure of the NF as well as enhances its performance. For the experiment, a 20-nm thickness of the polyamide layer was considered. Na_2SO_4 rejection was found to be 99.3% besides 45.2 Lm^{-2} h^{-1} bar^{-1} of pure water permeance. The synergistic effect of low water resistance nanochannels, large surfaces, and reduced thickness contribute to superior performance. This strategy was suggested to be very efficient and cost effective in wastewater treatment [9]. Wu et al. employed trimesoyl chloride and carboxylated chitosan to construct a polyesteramide nanofiltration membrane. The single-layer membrane was made up of 0.7% wt trimesoyl chloride and 3.5% wt carboxylated chitosan. The carboxylation of chitosan makes it water soluble, and it is otherwise soluble in water only in presence of acid. At the pressure of 0.6 MPa flux of 7.3 L/m^2h and rejection capacity of 66.3% for NaCl, 33.2% for $MgCl_2$, 65.7% $MgSO_4$, and 95% of Na_2SO_4 could be obtained. By repeating the cycles of the interfacial polymerisation reaction, they built a multiple layer structure, which showed a higher negatively charged surface and could reject the salt with more efficiency. However, because of the loose structure of a single-layer membrane, much more retention of reactive blue anionic dye could be observed. Moreover, the dye removal efficiency towards rhodamine B, reactive black 5 (> 90%), and indigo carmine were also reported [10].

He et al., developed a chitosan-based, hollow-fibre, positively charged composite nanofiltration membrane. For this, they employed trimesoyl chloride along with chitosan lactate and carried out interfacial polymerisation of these two with a hollow-fibre ultrafiltration membrane, made up of polyether sulphonate. A ramification derivative of chitosan is the chitosan lactate obtained from suspending chitosan in lactate acid followed by desiccation, freezing, and filtration. Various types of salts like Na_2SO_4 (22.3%), NaCl (27.6%), $MgSO_4$ (59.7%), $ZnCl_2$ (95.7%), and $MgCl_2$ were investigated to be rejected with the help of this NF membrane. The maximum rejection of 95.1% has been observed towards $MgCl_2$, with permeate flux of 10.3 L m-2 h-1 at 25°C and pressure of 0.4 MPa. They suggested that this membrane has higher potential to be used for industrial wastewater filtration [4].

The commercially available thin-film composite layer made up of polyamide is often susceptible to fouling through highly contaminated water. To overcome this, a similar membrane made up of chitosan biopolymer, forming a novel interfacial thin-film asymmetric nanofiltration membrane, has been developed. Additionally, silver NP has been incorporated due to its biocidal activity that mitigates the membrane fouling. The SEM characterisation studies indicated a more uniform, smoother surface coating of chitosan-impregnated silver nanofiber. This membrane showed an exceptional antifouling property with water flux of more than 100 $Lm^{-2}h^{-1}$, that is 2.5-fold higher than polyamide membrane. Investigation of the synergistic effect between chitosan and silver on filtration performance and porous support structure reports a 98% rejection of tannery wastewater and red-brown or organic dye and 40% rejection of NaCl [8].

To remove chromium from wastewater, a novel polyethersulfone-based nanofiltration membrane composed of graphene oxide and chitosan nanoplate has been employed. By modifying the graphene oxide with chitosan, this composite membrane has been built. The chitosan enhances the availability of highly active sites, which results in improvement of various properties of the composite membrane. The SEM image indicated a uniform distribution of graphene oxide and chitosan nanoplate over the membrane, which contributes to better antifouling performance. This membrane was then assessed for waster flux calculation and $CrSO_4$ and Na_2SO_4 rejection. Moreover, the rejection, water flux, and hydrophobicity of the composite membrane was found to be much higher than that of bare polyethersulfone membrane. Best performance was obtained only when 1% wt concentration of graphene oxide and chitosan nanoplate was added in the polyethersulfone membrane [11]. In a study, a nanofiltration bucky paper membrane utilising chitosan functionalised carbon nanotubes consisting of $COOH^-$ and NH_2 moieties was fabricated. The tensile strength of the membrane has been reported to be 17 ± 2 to 60 ± 2 MPa, with contact angle between $36°\pm3°$ and $105°\pm2°$, and surface area ranging from 12 ± 2 m²/g to 112 ± 4 m²/g. Moreover, the water permeability, magnetic properties, sonication time, salt rejection, surface charge, and morphology have also been noted. These variations in the parameters could be seen as the membrane was affected by the top layer of the multiwalled carbon nanotube. The highest tensile strength, along with the highest toughness, permeate flux (6.6 ± 0.21 L/m2), and elongation has been observed in case of the membrane that had COOH distributed in the top layer. But owing to the presence of crosslinking between the COOH and amide group present in the chitosan, the salt rejection capacity of this membrane gets disturbed. If the amino group in the top layer is balanced properly, then the highest salt rejection with the highest permeability can be observed. However, better rejection has been seen in the case of monovalent cations (Na^+) compared to higher variant cations (Mg^{2+}) [12]. Halakarni and his co-associates investigated selective separation and wastewater treatment through a chitosan-based and helical carbon functionalised loose nanofiltration membrane. The crosslinking of the membrane was done by employing sodium dodecyl sulphate. The increment in hydrophilicity ($44.4°\pm1$) was observed due to the presence of hydroxyl and amino groups in chitosan, through water contact angle measurement. Both anionic and cationic dye showed a more than 90% rejection and about an 8% rejection of $MgSO_4$ was reported at a pressure of 3 bar and flux of 40–60 $Lm^{-2}h^{-1}$. To observe the long-term stability profile, they used four types of real industrial wastewater (textile, molasses, sugar, pharma) and observed > 90% organic matter rejection throughout 100 h. After repeated operating cycles, no reduction in flux or rejection percentage was visualised, thus demonstrating how impactful and efficient this membrane is [13].

An inorganic hazardous pollutant, chromium is mainly found in wastewater coming out of the leather tanning industry. Zakmout et al. suggested that chromium can be removed from tannery effluent through nanofiltration. For this, they used a chitosan-modified membrane (NF270 and NF90), coating polyethersulfone with chitosan. Because of the adsorption capacity of chitosan towards metals, this nontoxic biopolymer has huge potential in heavy-metal removal. To achieve stability in the chitosan layer, it was crosslinked with glutaraldehyde through the amino groups present in chitosan. Although a high flux rate was observed at pH 6.1, efficiency of monovalent and divalent ions was much less. However, more than 99% removal of chromium was observed [14]. A study using a chitosan-based nanofiltration membrane demonstrated the rejection of NaCl and $CuSo_4$. The membrane was prepared upon polyethersulfone and a thin layer of chitosan modified sodium tripolyphosphate. Properties like high glass transition, good mechanics, and thermal properties make polyethersulfone a likely potential membrane for nanofiltration. The chitosan was coated or injected upon the polyethersulfone through pressure. A denser and more compact modified surface of the modified membranes was obtained. When 0.05% wt of sodium tripolyphosphate was taken for fabrication, a higher rejection capacity towards copper could be achieved. About 97% of NaCl rejection could be achieved through this membrane [15].

The layer-by-layer self-assembled nanofiltration membranes were fabricated to improve antibiological fouling and rejection of metal salt. For the membrane preparation, negatively charged carboxymethyl chitosan and positively charged glutaraldehyde and polyethyleneimine were assembled alternately on the polyethersulfone membrane, followed by coating with silver NP. Rejection of both metal salt ($Na_2SO_4 < NaCl < Na_2CO_3 < CaCl_2 < MgSO_4 < MgCl_2$) and heavy-metal salts like Ni^{2+} (83.47%), Cd^{2+} (82.69%), Cu^{2+} (84.04%), and Cr^{3+} (88.95%) were observed using homogeneous salt, whereas, in case of mixed salt, the rejection capacity increased to 87.22% for Ni^{2+}, 89.74% for Cd^{2+}, 90.70% for Cu^{2+}, and 88.95% for Cr^{3+}. The rejection capacity was further enhanced to 96.12% in Cu^{2+} and 94.38% in Cd^{2+} upon pH adjustment from 8 to 10 [16].

6.2.2 Engaging Cellulose Biopolymer

Cellulose is available abundantly and can be used as a good base for nanofilter techniques. Nano cellulose and its applications have gained high attraction in research and industrial areas because of its properties like high surface area, excellent mechanical properties, rich hydroxyl groups for modification, and environmental friendliness [17].

Weng et al. developed a cellulose nanofiltration membrane by extracting cellulose from bamboo for water purification, replacing petroleum-based polymer film. For fabrication of the membrane, 6% wt of cellulose and N-methylmorpholine-N-oxide were used, and it was prepared through a phase inversion strategy. This dense bamboo cellulose membrane was then modified by hydrolysis employing 1 mol/l NaOH and carboxymethylation employing 3% wt/v, to obtain a novel membrane. The pore size of the novel membrane was found to be 0.63 nm, and with 0.5MPa pressure the retention rate of methyl blue and methyl orange were noted to be 98.9% and 93% respectively. However, the stability of the novel membrane was found to be much less than that of pure bamboo cellulose membrane [18].

Anokhina et al. used cellulose solution to prepare composite nanofiltration membranes. DMSO was used as cosolvent in a mixture containing [Emim]OAc (1-Ethyl-3-methylimidazolium acetate), ionic liquid, and cellulose. DMSO lowers the dissolution time, and the least dissolution time could be obtained when a 1:1 ratio of [Emim]OAc and DMSO had been added. An increase in dissolution time was noted with further addition of DMSO. The fabrication of the membrane was done on poly (ethylene terephthalate) with 6, 8, 12, and 16% wt of cellulose. Depending on the DMSO concentration, 62% rejection of Remazol Brilliant Blue R and 42% rejection of orange-red could be achieved, whereas, at the highest concentration of cellulose (16% wt), the orange-red and Remazol Brilliant Blue R rejection factors were elevated up to 61% and 82% respectively [19].

Along with other harmful metals like arsenic, cadmium, and zinc, mercury industrialisation has led to the discharge of lead (Pb) into the wastewater. The heavy-metal contamination may lead to renal failure, convulsion, nausea, anaemia, and other chronic diseases. A recent investigation on Pb (II) removal from wastewater was performed, where the researchers used cellulose acetate-based complexation NF. Due to the metal rejection property along with the potential affinity for water permeation, biocompatibility, and eco-friendly nature, cellulose acetate has been chosen. But to overcome their poor thermal and mechanical properties, carbon-based graphene and graphene oxide were used. Therefore, for the fabrication of the filter by the dissolution casting method, graphene oxide modified with vinyl triethyoxysilane/cellulose acetate as well as gum arabic membranes were utilized. After characterisation of the NF, their thermal stability was analysed under nitrogen by thermogravimetry. The permeation flux obtained was $8.6 \, l \, m^{-2} \, h^{-1}$. At pH 9, rejection of 97.6% of Pb (II) was reported and the rate of adsorption also increased to $15.7 \, mg \, g^{-1}$ after a certain period [20].

Another green technology for fabrication of NF has been developed from bleached and unbleached rice straw. The cellulase nanofibers were isolated from the rice straws and thin-film ultrafiltration membranes were prepared. Higher water flux was obtained in the membrane produced from unbleached straw than that of bleached straw because of the high lignin content in the unbleached straw; it has higher porosity and flux. From SEM images it has been found that the pore size of bleached straw membrane was 128 ± 24 nm whereas the pore size of unbleached straw membrane was 106 ± 28 nm. Moreover, it has also been reported that the unbleached straw membrane depicted 112 % higher pore volume and 111% higher surface area than bleached straw membrane. From the water contact angle study, they saw that bleached straw membrane is more hydrophilic in nature due to the presence of the hydroxyl and carboxylic group. These types of NFs can be used in removing lime NPs and papermaking white water [21].

Mahdavi and Bagherifar fabricated a hybrid nanofiltration membrane with the help of SiO_2-poly (2-Acrylamido-2-methylpropane sulfonic acid) and cellulose acetate to eliminate ceftriaxone sodium from wastewater. The ceftriaxone sodium is an antibiotic, which, when it is dissolved in wastewater coming out from pharmaceutical industries, might cause adverse physiological consequences in living organisms, including humans. The used surface initiated the redox polymerisation and phase inversion method to develop this membrane. In the presence of silica NP the charge repulsion, pharmaceutical rejection, and size exclusion properties of neat cellulose acetate membrane are improved. About 96% rejection efficiency at pH 8 could be achieved through this membrane [22]. A group of researchers developed a polyamide nanofiltration membrane on cellulose nanofiber support. They used interfacial polymerisation to fabricate an arch bridge-structured membrane with elevated performance. This uncommon structure has been produced due to the Marangoni convection on extremely hydrated cellulose. The surface area of this arch bridge structured membrane was greater, therefore better water permeation of nearly 42.5 L/m^2h bar could be observed. About 99.1% desalination of Na_2SO_4 using this membrane has been reported [23].

6.2.3 ENGAGING β-CYCLODEXTRIN BIOPOLYMER

β-cyclodextrin is made up of seven glucopyranose units and is a cyclic derivative of starch. β-cyclodextrin has been explored as the encapsulating medium for water pollution treatment. The external part of the cyclodextrin is hydrophilic, and the internal part is hydrophobic obtained by degradation of the starch and production by intramolecular transglycosylation [24].

For the removal of methylene blue, a nanocomposite material composed of sodium alginate, β-cyclodextrin, activated charcoal, and magnetic iron oxide polymer was studied. As β-cyclodextrin based polymers are very effective in dye elimination from wastewater, addition of β-cyclodextrin to any composite enhances its adsorption capacity. The ability of sodium alginate to crosslink with various polyvalent ions and absorb antibiotics, dyes, and heavy metals makes it an interesting candidate for NF. Again, the activated charcoal and magnetic iron oxide further elevate the stability and strength of the NF in addition to lowering the filtration time. The mean pore diameter of the membrane was analysed to be 1.47 nm. With the increase in temperature, the adsorptive removal rate also gets increased. The water soluble hazardous cationic dye can be removed by 99.53% within 90 mins at a pH of 6. Not only adsorption of methylene blue but also good separation properties and extraordinary regeneration ability make this membrane very appropriate for wastewater treatment [25].

Machut and his co-workers synthesised gold doped TiO_2 with the help of modified cyclodextrin via microwave irradiation technique to carry out dye degradation. TiO_2 has been taken under consideration due to good stability, non-toxicity, and cost-effectiveness. In addition to this, the gold NP helps in the uplifting of photocatalytic properties of the membrane. Here, the cyclodextrin acts as both stabilising agent and reducing agent in the synthesis of gold NP. Photodegradation of methyl orange dye has been demonstrated in this experiment in the presence of UV irradiation, and it has been observed that, after one h of irradiation, nearly zero concentration of methyl orange was left in the wastewater [26].

Another study to remove 4-aminoazobenzene dye and methylene blue dye was carried out by Guo et al., in which they prepared a novel composite nanomembrane using β-cyclodextrin-based polymers and ε-polycaprolactone *via* electrospinning method. The cavity molecular structure of ε-polycaprolactone produced a host-guest interaction, which contributes to enhancement in adsorption uptake, stability, and mechanical strength of the membrane. The innumerable cavities present in the ε-polycaprolactone also guarantee the selective adsorption capacity of azo dyes. The 4-aminoazobenzene dye-removal efficiency was found to be 24.1 mg/l, whereas at eight cycles only 78% of methylene blue could be removed [27].

In a study, graphene quantum dots and β-cyclodextrin were polymerised to produce a high-flux nanofiltration membrane. Due to the host-guest interaction and unique hydrophobic cavities (presence of hydrophobic methylene group and hydrophilic hydroxyl group at internal and external sites respectively), β-cyclodextrin proves to be very effective in water treatment. Moreover, the quantum dots contribute to improving the permeability of the membrane. The surface properties of the membrane can be modulated by non-specific H-bonding between graphene quantum dots and β-cyclodextrin. When the concentration of the quantum dots was increased from 0 to 0.5% wt the water flux also was elevated from 122.2 to 474.7 L/m^2h at 0.1 MPa pressure. Both congo red and Eriochrome black T dye rejection were reported to be more than 93%. The extraordinary chlorine resistance, antifouling property, and superiority in the filtration performance of this membrane make it promising for dye rejection from wastewater [28].

Several micropollutants like polycyclic aromatic hydrocarbons, polychlorinated biphenyls, and other pharmaceutical residues contaminate water bodies and adversely affect aquatic plants and animals. Jurecska et al. developed a NF of 1.5–3.5 mm thickness using β-cyclodextrin. Due to the presence of hydrophilic outer surface and cylindrical hydrophobic inner chamber, β-cyclodextrin acts as a molecular capsule and holds different complexes without forming any chemical bond. Along with this they added some inorganic chemical compounds (NH_4HCO_3, $NaHCO_3$, $NaCl$) to enhance the adsorption capacity and ethanol to enhance regeneration capacity. Maximum efficiency was obtained by using 30% m/m polymer beads of β-cyclodextrin, 12mmol NH_4HCO_3 and 70% m/m polyethylene of ultra-high molecular weight. Absorption capacity of 4.5 µmol/g has been seen, thus suggesting a promising method of water treatment [29].

For metal ions removal from wastewater, a group of researchers developed a magnetic membrane by blending polyvinylidene fluoride with β-cyclodextrin via phase conversion strategy and induction of magnetic field. Polyvinylidene fluoride was used here due to its chemical inertness, better thermal stability, and excellent mechanical strength. To improve the antifouling ability, the hydrophobicity of the membrane has been elevated by blending it with Fe3O4 magnetic particles and graphene oxide modified by β-cyclodextrin. Properties like anti-pollution performance, pore size (28 nm), porosity, hydrophobicity, and water flux are improved in this fabricated membrane. The zeta potential of the membrane also is increased from −31.25 mV to −38.11 mV. The resultant membrane showed high efficiency (75%) in elimination of Cu^{2+} from wastewater. It has also been observed that the absorption capacity (0.94 mg/g) of Cu^{2+} did not get altered even after five cycles, which suggests that, for a longer time membrane recycling could be done [30].

6.2.4 ENGAGING ALGINATE BIOPOLYMER

Alginate is anionic polymer obtained from brown seaweed, and it is extensively investigated due to its biocompatibility, relatively low cost, and low toxicity. It has also been used for many biomedical applications. Mild gelation can be done by addition of divalent cations such as Ca^{2+}. Alginate hydrogels can be made by various cross-linking methods [31]. The previously mentioned properties are perfect for nanofiber base synthesis.

An alginate green and cheap biodegradable bio-membrane has been prepared through polymer casting and crosslinking sodium alginate in presence of calcium chloride. To improve the membrane stability as well as mechanical strength, crosslinking of sodium alginate is essential. As a support, alumina discs, non-woven polyesters, cellulose, and polyacrylonitrile have been added. The pore size of 50 nm and thickness of 1.7 μm were detected through SEM images and AFM consecutively. The presence of strong hydrophilic alginate makes the contact angle 25°. An exceptional chemical stability and organic solvent nanofiltration performance of this membrane has been reported. The membrane permeance varied from 0.08 L/m2 to 1.8 L/m2 depending on the different parameters. A promising level of Brilliant Blue and Methyl Orange adsorption has been investigated [32].

Ravikumar and his co-associates experimented on Congo Red removal from the solutions or effluents by using novel nano bio composite beads. In this, *Klebsiella* sp. (adapted to 50mg/L of Congo Red) with nano zerovalent iron (NZVI) green synthesised were immobilised together in beads of sodium alginate. In this work, all process and components are kept environment friendly. Analytical techniques like Brunauer Emmett and Teller (BET) method for studying surface area and pore volume of nano bio composite (NB) beads and the values obtained were 1.25m2/g and 0.000556cm3/g respectively. A batch reactor was used to remove Congo Red using NB beads. The experimental factors like interaction time, pH, dye concentration, and weight of bead were optimised. From the experiments, optimised conditions for Congo Red are a concentration of 50mg/L, pH of 5.0, NB bead weight of 20% w/v, and interaction time for maximum removal of is recorded at 180 min. The maximum removal (%) we got was 97.53 ± 0.4% and removal capacity was obtained at 126 ± 1.2mg/g. The combination of three-factor Box–Behnken design (BBD) with quadratic programming and RSM were used in optimisation of the conditions for removal of Congo Red. Applications of NB beads in real life were tested by taking lake water, groundwater, and tap water samples. They got 95.53 ± 0.4%, 95.88 ± 0.6%, and 95.82 ± 0.3% removal of dye respectively. The effluent solution showed decrease in toxicity compared to untreated solution when observed during interaction with algae and bacteria in the environment. This result showed the novel nano-bio composite efficiency in dye removal with consequent detoxification of effluent solution [33].

In another study, a nanocomposite consisting of sodium alginate, polyethersulfone, polyvinyl alcohol, and graphene oxide was fabricated by immersion precipitation technique upon blending with hydrogel. In the presence of a $CaCl_2$ and boric acid solution, the crosslinking takes place between the nanocomposite and epoxy groups, carboxylate ions, and hydroxyl and carboxyl group. The nanocomposite has the property of improving both porosity and permeability of the membrane. Moreover, the antifouling property and permeability of the membrane seems to improve by incorporation of 1% wt of hydrogel. The hydrogel inclusion can also provide good mechanical stability, regeneration ability, reusability, adsorption capacity, and selectivity. Therefore, here both nanocomposite (1% wt) and hydrogel (1% wt) were exploited for NF fabrication. The resultant hybrid membrane induced negative charge on the surface of the membrane to bring about 83% of Lanasol Blue 3R dye rejection and better solution flux, and because of the presence of hydrophilic functional group in the hydrogel, it also improved antifouling properties [34].

Bhangi and Ray have developed a nanocomposite copolymer made up of alginate and nano silver chloride for removal of synthetic dyes from wastewater. For this, they copolymerised hydroxyethyl methacrylate and acrylic acid in the presence of sodium alginate and water, followed by coprecipitation of silver chloride NPs in the matrix of polymer. The silver NP has high surface area density and surface plasmon resonance properties in the presence of UV; these make it a great candidate for removal of toxic compound from water. Further, prevention of agglomeration of silver NPs can be carried out by the polymer gel. Both the distribution of the particles and their size can possibly be controlled by altering the structural network of the polymer. Adsorption of brilliant Cresyl Blue coming out from the textile industry was investigated and the result showed > 95% removal of the dye [35].

A double-layered nanomembrane has been formulated utilising alginate and maize-derived cellulose nanowhiskers of 3–12 nm diameter. To improve the mechanical properties and filtration efficiency of the membrane this formulation has been taken under consideration. Further, the performance of the membrane can be enhanced by functionalising cellulose. Here the cellulose was employed as barrier layer, whereas alginate was electro-spun as a substrate. Polyethylene oxide enhanced the electro-spinnability, and for crosslinking aqueous calcium washing was implemented. Both membrane productivity and antifouling property were intensified through the cross linking of the middle layer with the top layer. Water contaminates including 10–100 nm NPs have also been reported to be removed by this nanomembrane by water retention and the size exclusion principle. As the pH of the solution is increased, the percentage of chromium rejection also increases. At pH of 11 the chromium rejection was found to be nearly 80%, which revels that this membrane can be used for efficient water treatment [3].

A recent study demonstrated that activated carbon impregnated alginate coated with nanohydroxypeptide can remove uranium from wastewater. The naturally occurring, mineral calcium apatite, hydroxyapatite NPs were used because of their proven biocompatibility. Furthermore, the activated carbon was employed due to its high purity, cost-effectiveness, high-radiation stability, high sorption efficiency, and selective adsorption. For 7 hs using the batch method, applying pH (5–7), the uranium removal efficiency was investigated. The maximum uptake of uranium was observed at pH 6. Very high efficiency of nearly 92% has been achieved for uranium removal with maximum sorption capacity of 18.66 mg g−1 [36].

Chen et al. crosslinked epichlorohydrin with sodium alginate to synthesise a nanocomposite membrane. In aqueous solution, due to the presence of hydrophilic carboxyl and hydroxyl groups in sodium alginate, it swells, which leads to the decrease in mechanical strength and selectivity of the membrane. To overcome this, epichlorohydrin was employed. When sodium alginate was 1.0% and epichlorohydrin were used at a temperature of 60°C, optimum performance of the NF could be observed after 18 h of crosslinking. The following were visualised using this membrane at pressure of −232.8 mV MPa-1: 11.1% rejection $MgCl_2$ at flux of 48.9 L/m2h, 29.9 % rejection of KCL at flux of 68.3 L/m²h, 35.2% rejection of NaCl at flux of 58.1 L/m2h, 65.8% rejection of MgSO4 at flux of 45.9 L/m²h, 87.2% rejection K2SO4 at flux of 28.5 L/m2h, and the highest rejection of 88.7% towards Na2SO4 at flux of 21.4 [37].

6.2.5 ENGAGING FIBROIN BIOPOLYMER

Fibrin is a tough protein arranged in long fibrous chains. It is formed from the fibrinogen molecule (soluble protein produced by the **liver** and present in blood plasma). Fibrin (Fbn)-based nanostructures can provide a suitable matrix for processes like water filtration and effluent treatment [38].

Bombyx Mori cocoon derived silk fibre is composed of fibroin biopolymer. Due to the good biocompatibility, high rate of water flux, and high mechanical strength, silk nanofibers are selected for water purification. Sivakumar et al. developed a composite membrane using silk nanofiber along with graphene oxide to remove salt (Na_2SO_4) and dyes present in industrial wastewater using the green chemical method as observed in Figure 6.1. Uniform pore size (~10–20 nm) and more active sites were obtained in the composite membrane compared to individual membranes. The hydrogen bond between 1D silk nanofiber and 2D graphene oxide helps in ensuring specific rejection of Na_2SO_4 at 95%. Moreover, satisfactory dye removal of 97% was also visualised. Besides constant rejection value, this composite structure has been reported to enhance the flux rate (43 LMH) as well as provide good stability. Thus this composite membrane has huge potential in industrial water treatment [1].

A group of researchers used silk fibroin to fabricate an eco-friendly NF via green electrospinning method, which can perform filtration. Particulate matter 2.5 can be removed by this strategy. A natural protein, silk fibroin has aqueous processibility, zero toxicity, good biocompatibility, and robust strength. For preparation of the NF, degumming of the silkworm cocoons were implemented, followed by fabrication using polyethylene oxide. A high filtration efficiency of 99.99% along with only 75 Pa air resistance make this membrane highly likely to be used at an industry level. Moreover, after being used, these membranes can be disposed of [39]. In an experiment, to demonstrate copper removal from wastewater, silk fibroin was blended with nylon-6 to fabricate

Biomaterial Fabricated Nanofiltration Membrane

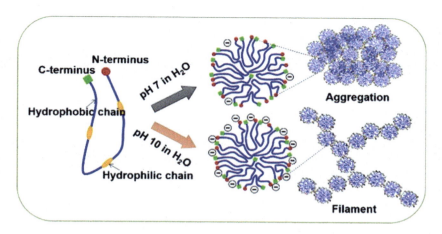

FIGURE 6.1 Synthesis of silk fibre nano filtration membrane at different pH.

Source: Reprinted with permission from Sivakumar, Mani & Liu, Ding-Kun & Chiao, Yu-Hsuan & Hung, Wei-Song. (2020). Synergistic effect of one-dimensional silk nanofiber and two-dimensional graphene oxide composite membrane for enhanced water purification. Journal of Membrane Science. 606. 118142. doi: 10.1016/j.memsci.2020.118142

a NF. For the same, the electrospinning method was applied along with incorporation of calcium phosphate crystals to modify the surface, which lead to an increment in the affinity towards divalent heavy metals. To get better mechanical and biomedical properties, nylone-6 was used, which forms a hydrogen bond with silk fibroin. An extraordinary fibrous structure with 250 ± 50 nm diameter was formed, and adsorption studies were carried out in both a continuous system and a batch system. At a pH of 5, the highest Cu (II) adsorption capacity was observed due to increase in metal ion binding sites in the membrane. The result of the absorption studies showed that through a continuous system only 33% removal of Cu (II) could be achieved, whereas through a batch system 77% removal of Cu (II) could be achieved [40].

To adsorb heavy metal ions like Cu^{2+} from industrial wastewater, a group of researchers prepared a composite biofilter using silk fibroin and cellulose acetate via the electrospinning method. The functional group present in the nanofibers such as carboxyl, amido, amidoxime, carbonyl, hydroxyl, and others provides an electrostatic force or, by chelation, absorbs the toxic heavy metals. Here silk fibroin has been used because of the presence of a large amount of functional groups that are present in their amino acids (tyrosine, serine, alanine, acetic acid). Further, the cellulose acetate has been used because of its excellent compatibility and good hydrolytic stability. The anti-felting property of the 100–600 nm diameter membrane has been caried out by treatment with 100% ethanol. The blended membrane showed higher affinity towards Cu^{2+} than the membrane made up of individual pure element. The blended membrane having 20% of cellulose acetate showed the maximum affinity of 22.8 mg/g towards Cu^{2+}, thus indicating the synergy between silk fibroin and cellulose acetate. It has also been noted that adsorption efficiency might be affected by the initial concentration and running time of Cu^{2+}, and the efficiency is directly proportional to the initial concentration of Cu^{2+} [41].

6.3 GREEN NPS USED IN FABRICATION OF NANOFILTRATION MEMBRANE

6.3.1 Synthesis of TiO_2 NP from Living Sources in Nanocomposite Formation

Ranjitha and co-researchers worked on the synthesis of a nanocomposite made of reduced graphene oxide (rGO) and TiO_2/CO_3O_4. The rGO-TiO_2/CO_3O_4 hybrid nanocomposite was synthesised successfully via the co-precipitation method. Graphene oxide (GO) was prepared by a modified

Hummer's method, then reduced using *Shuteria involucrata* leaf extract solution and dispersed in deionised water (50ml). After that it was sonicated for 4 h. The average size of nanocomposite crystals decreased from 45 nm to 18 nm after the rGO addition. The size decrease helped in the photocatalytic activity of nanocomposite rGO/TiO$_2$/CO$_3$O$_4$. The band gap energy estimated for TiO$_2$/CO$_3$O$_4$ and rGO/TiO$_2$/CO$_3$O$_4$, rGO/TiO$_2$/CO$_3$O$_4$ had low band gap energy 2.74eV. These nanocomposites were able to degrade Methylene Blue and Crystal Violet by the photocatalytic method. The decolourisation percentage of Methylene Blue (MB) dye solution was high for small duration of time (210 min) compared to Crystal Violet (CV) dye, which takes double time compared to MB. This result concludes that the commercialisation of the rGO/TiO$_2$/CO$_3$O$_4$ catalyst may be useful for treating various dyes in industries [42].

A group of researchers synthesised TiO$_2$ nanosheet using *Aloe Vera* leaves for the removal of Rhodamin B (RhB). Decolourisation of RhB was performed by photocatalytic method as observed in Figure 6.2. The photocatalytic efficiency in the presence of UV light detected 58% in 50 min. TiO$_2$ nanosheet was non-uniform sheet-like crystals in the range of 50–200 nm. The effect of pH was studied, and at pH range (pH > 6), RhB was strongly adsorbed on surface of the TiO$_2$ nanosheet. Above pH 6, the photocatalyst was negatively charged and the RhB dye is cationic dye, which helped in adsorption. The decolourisation of dye was done by raising the pH from 9 to 10. Raising the RhB concentration from 10 to 50 ppm led to decrease in catalytic efficiency from 54% to 30.3%. The maximum removal efficiency of RhB was observed 54% at RhB concentration 20 ppm [43–45].

For elimination of chromium (VI) from the wastewater coming out from the tannery industry, a study has been carried out to develop an antibacterial, self-cleaning, and eco-friendly membrane. To fabricate the membrane, *Cajanus cajan* seed extract has been taken to synthesise titanium dioxide NP (15.89 nm) and blended with a hydrophobic membrane consisting of polyvinylidene

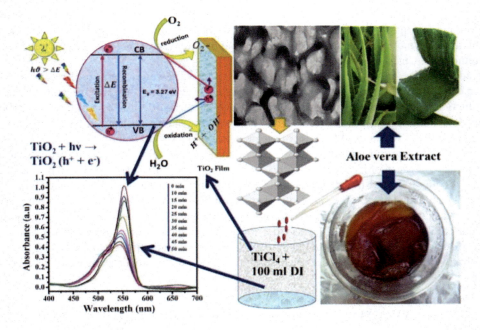

FIGURE 6.2 Synthesis of TiO$_2$ nanosheet using *Aloe Vera* leaves for the removal of Rhodamin B.

Source: Reprinted with permission from Rakesh K. S., Gaurav H., S.R. Sabhajeet, S. Sikarward, Rahule, Sandhya S. Materials Science & Engineering B: Green synthesis of TiO2 nanosheet by chemical method for the removal of Rhodamin B from industrial waste. Materials Science and Engineering: B 258, 114577 (2020), https://doi.org/10.1016/j.mseb.2020.114577

fluoride. The presence of TiO_2 makes the membrane hydrophilic in nature besides enhancing thermodynamic properties. The resultant membrane not only rejects the chromium (VI) but also reduces it into chromium (III) in the presence of sunlight. The result of the study showed that, with the increase in flux, 85.59% rejection and 92% reduction of chromium could be achieved if 0.02% wt of TiO_2 NP was incorporated into polyvinylidene fluoride membrane [2].

A group of researchers has worked on synthesis and application of algae-TiO_2/Ag hybrid bio-nano material for removing Cr (VI) from wastewater by reducing it to its less toxic form Cr (III). The membrane was prepared by mounting algae cells (*Chlorella vulgaris*) on the ultrafine TiO_2/Ag chitosan hybrid nanofiber mat, as microalgae and TiO_2 face difficulty in recovering from solution. Chitosan nanofibers were created using the one-step electro spinning method. The TiO_2/Ag NF showed absorbance of Cr (VI): 6% and algae- TiO_2/Ag NF: 11%. The nanofiber obtained was able to maintain its shape in the water. The ultrafine nanofiber's average diameter was about 36.7 nm. Metal ions (Ag) improve the photocatalytic activity of TiO2; both NPs together act as electron trap. Algae can photo generate reactive radicals that help induce the degradation of pollutant in visible light. Reactive oxygen species (ROS) that are photo generated are quantified in different reaction systems first with algae, then with pure TiO_2, TiO_2/Ag nanofiber, and last algae-TiO_2/Ag hybrid materials. The changes in antioxidant enzymes activity in the algae cells were observed and noted after irradiation. After the reaction, more than 50% of Cr (VI) was removed from the system with algae- TiO_2/Ag NF. This nanofiber mat was easily recovered and maintained good photocatalytic activity throughout the five successive cycles [46].

Sethy along with his co-researchers have done work on synthesis of TiO_2 NPs along with extract of *Syzygium cumini* for removal of lead (Pb) present in explosive industrial wastewater. These TiO_2 NPs (green synthesised) were tested for removal of lead from industrial wastewater by photo catalytic activity. Synthesised TiO_2 NPs have spherical morphology, an anatase phase, and a large surface area of 105 m2/g. TiO_2 NPs have 10.50 nm pore size, pore volume 0.278 cm3/g, and average diameter of 18 nm in size. To determine the lead concentration, (ICP) inductive coupled plasma spectroscopy was used. The obtained results for chemical oxygen demand (COD) were 75.5% and 82.53% removal of lead (Pb^{2+}). Kinetic study results showed that photo-catalytic and COD removal of lead from industrial wastewater both follow first-order kinetics [47].

A group of researchers has worked on TiO_2 NPs, polyvinylidine fluoride hollow fibre membrane (PDVF) fabrication, and high-density algae (*Chlorella vulgaris*) cultivation for wastewater treatment. PVDF hollow-fibre membrane was fabricated by phase inversion method. Pore size of membrane was 20 nm, inner density 0.6 mm and outer density 1 mm. Algal membrane bioreactors removed 78% phosphorous from the solution, which was having initial phosphorous concentration of 3.5 to 8.6 mg/l. The membrane resistance is lower in TiO_2/PVDF (49% approximately) than the PVDF membrane. Also, water flux is increased from 106 to 134 L/m². That helped in enhancing hydrophilicity of membrane [48].

Chao and his associates worked on laccase immobilisation on Titania NPs by extracting crude enzyme (laccase) from a culture of *P. ostreatus* and using this combination for the degradation of micro-pollutants. Carbamazepine and bisphenol-A micro-pollutants were selected because they frequently are detected in water bodies. Effective degradation of micro-pollutants in wastewater can be done by enzymatic treatment, but pure enzymes are costly, so a new technique is introduced in this work in that crude enzyme extracts from *P. ostreatus* were immobilised on a functionalised TiO_2 NP surface. The extracted crude enzyme is confirmed to have laccase by comparing the bands on SDS PAGE results of sample 1: crude enzyme and sample2: purified commercial laccase. Both samples were then immobilised onto TiO_2 NPs. Crude enzyme extract from *P. ostreatus* gave these results: Laccase loading (μg/mgNP): 0.8 ± 0.03, apparent activity (U/mgNP): 0.019 ± 0.001, and activity recovery (%): 125.8 ± 15.2 and its recovery percentage from the solution after the process is much more than the purified enzyme. The drawback of crude enzymes having other components that interfere with laccase immobilisation was solved by using ten-time high dilution and more than 80% laccase from crude extract immobilised onto the NPs. Reusability of the biocatalytic NPs was investigated by observing five continuous degradation cycles of bisphenol-A and carbamazepine mixture in solution. Removal of bisphenol-A observed for the solution: 90% within only 6 hs. Recalcitrant carbamazepine degradation was nearly 10% even after 48 hs. For carbamazepine, degradation was greatly enhanced in the presence of bisphenol-A; nearly 40% removal of carbamazepine was observed within 24 hs of treatment achieved due to redox mediator p coumaric acid (oxidisation products of bisphenol-A) that helped in carbamazepine degradation. In this work, use of enzyme extracts for cost-effective practical applications was achieved [49–52].

Ricardo and co-associates synthesised Fe-TiO$_2$ NPs that have photocatalytic application in wastewater treatment. This work is focused on studying the synthesis and structural characterisation of the NPs (Fe-doped TiO$_2$) synthesised by green synthesis. The main aim of this was to determine the effect of the doping element's introduction into the physicochemical, optical, structural, and morphological properties. Iron ions have doped on anatase TiO$_2$ NPs and are synthesised by using green chemistry. In that method aqueous extract of lemongrass (*Cymbopogon citratus*) was used, which was obtained by Soxhlet extraction. The TiO$_2$ anatase phase doped with Fe^{3+} at 350°C and 550°C respectively. The scanning electron microscopy and energy-dispersive x-ray (SEM-EDS) shows clusters of NP and impregnation efficiencies of between 66.5 to 58.4%. (TEM) transmission electron microscopy shows final sizes of particle range between 7 and 26 nm depending on the presence or absence of dopant. The (PL) photoluminescence and (CL) cathodoluminescence studies of the undoped and doped NPs show signal of luminescence attributed to surface oxygen vacancies (visible and PL emission 350–800 nm and CL emission 380–700 nm). The six bands (A1g + 2B1g + 3Eg) found by Raman spectroscopy and the x-ray diffraction (XRD) pattern confirm that TiO$_2$ synthesised is only anatase phase, commonly used as a catalyst in wastewater treatment, specifically in processes of heterogeneous photo catalysis [53].

A group of researchers performed green synthesis of nanocomposite of TiO$_2$ NPs/pristine pomegranate peel extracts and used it for disinfecting water with its antimicrobial activity. In this study pristine pomegranate peel (PPP) extract was used for green synthesis of (TiO$_2$ NPs) titanium dioxide NPs and nanocomposite formation (PPP-TiO$_2$). The results of dynamic light scattering (DLS) showed that TiO$_2$ NPs has 100% peak intensity with 620 nm Z-average value, PDI = 0.178, which affirm the homogeneity and monodisperisty of TiO$_2$ NPs synthesised. DLS result of PPP-TiO$_2$ show presence of both PPP and TiO$_2$ in the composite. The mean value was found to be 1,230 nm for hydrodynamic diameter (Z-average) of PPP-TiO$_2$. The values of ζ-potential were PPP-TiO$_2$ NPs (−11.4mV) have higher stability compared to TiO$_2$ NPs (−6.96mV). The well diffusion method was used to test biological activity. The antimicrobial activity of PPP-TiO$_2$ was 1.5 times higher than TiO$_2$ NPs and PPP. In addition, the results showed maximum inhibition against *S.aureus*. Biological oxygen demand (BOD) indicated the organic matter and microbial communities in water samples. The results showed lower values of BOD for samples that contain PPP-TiO$_2$ compared to TiO$_2$ NPs [54].

6.4 CONCLUSION

In this chapter, use of bio-nano filtration techniques and methods for industrial effluent treatment are discussed (Table 6.1). There are many industries present like textile, pharma, paper, food, cosmetics, pesticides, and some others that contribute to a major part of water pollution when they release their effluent untreated in water bodies. Major water pollutants found in water are heavy metals that can harm the environment. Active pharmaceutical ingredients were also detected in the wastewater. Due to this, resistance development became an issue. One azo dye named Procion Red was mainly found in textile wastewater. This dye is used commercially frequently because of its unique properties including high molecular weight and less degradability. But upon contamination it can interrupt the ecosystem along with contributing to carcinogenic and toxic effects. Also dyes like Methylene Blue and Congo Red, heavy metals like chromium, arsenic, cadmium, zinc, mercury, and lead being leached into the wastewater have been detected by different industries. The heavy-metal

Biomaterial Fabricated Nanofiltration Membrane

contamination caused health problems like renal failure, convulsion, nausea, anaemia, and other chronic diseases like cancer and some mutations also. The effect of these effluents on the environment occur due to their resistance to biodegradation, and they eventually ends up in food chains of aquatic life forms that cause long-term effects.

The bio nano filtration techniques are effective and can be used for a long time for removal of these effluents from water bodies. Molecules like chitosan, cellulose, and alginate were used in some nano filter techniques because they are eco-friendly and can be used as a good base material. Techniques like electrospinning, encapsulation of bacteria, doping of molecules like gold or iron, immobilisation of enzymes, and many more were found very effective for removal of effluents as well as some unwanted microbes. The sources for these nano materials were mostly plant and microbes, and the use of biological methods to obtain nano material is cost effective as well as environmentally friendly. Some frequently studied nanofibers were TiO_2 nanofiber composites like algae-decorated TiO_2/Ag hybrid nanofiber membrane, gold-doped TiO_2 assisted by modified cyclodextrins, and Fe-doped TiO_2. Molecules like histidine, laccase enzyme, or eggshell components were also experimented with, and they were surprisingly effective for water treatment. For checking efficiency of these materials, various parameters were studied like reaction time, effects of pH and temperature, concentration of effluent components, and recovery of nano fibre membranes, and most of them were found effective for effluent treatment processes.

TABLE 6.1
Fabricated nanofiltration membranes and their role in rejection of various industrial effluents.

No.	NFs	Organisms	Effluents	Characterisation	Ref.
1.	TiO_2 Nanofiber composites from *Cajanus cajan* seed extract	*Cajanus cajan*	Chromium (VI)	X-ray diffraction, contact angle, Zeta/nanosize analyser, thermo-gravimetric analysis, differential scanning calorimetry, FTIR, DRS	Arif et al., 2020
2.	TiO_2/PVDF (polyvinylidine fluoride hollow-fibre membrane)	*Chlorella vulgaris*	Phosphorous	SEM, EDS	Hu et al., 2015
3.	Algae decorated TiO_2/Ag hybrid nanofiber membrane	*Chlorella vulgaris*	Cr (VI)	SEM, TEM	Wang et al., 2016
4.	Laccase on Titania NPs	*P. ostreatus*	Bisphenol-A and carbamazepine	ELECTROPHORESIS AND SDS PAGE	Chao et al., 2017
5.	rGO-TiO_2/CO_3O_4 hybrid nanocomposite		Methylene Blue and Crystal Violet	FESEM, XRD, EDS, UV-Vis spectroscopy	Ranjitha et al., 2019
6.	Fe-doped TiO_2	Extract of lemongrass (*Cymbopogoncitratus*)	Used in processes of heterogeneous photo catalysis	(SEM-EDS) AND (TEM)	Ricardo et al., 2018–2019
7.	Nanocomposite of TiO_2 NPs/pristine	Pomegranate peel extracts	S. aureus and some other species	SEM, DLS, XRD, zeta potential (ζ-potential)	Muna et al., 2015
8.	TiO_2 nanosheet	Aloe-vera leaves	Rhodamin B. (RhB)	SEM, TEM	Rakesh 2020
9.	TiO_2 NPs	*Syzygium cumini*	Lead (pb)	XRD, SEM, DLS, FTIR	Sethy et al., 2019
10.	Gold-doped TiO_2 assisted by modified cyclodextrins	-	Methyl Orange degradation	UV-Vis spectroscopy, XRD, TEM, and ICP-OES	Machut et al., 2020

(Continued)

TABLE 6.1 *(Continued)*
Fabricated nanofiltration membranes and their role in rejection of various industrial effluents.

No.	NFs	Organisms	Effluents	Characterisation	Ref.
11.	Cyclodextrin containing NF	-	Pharmaceutical residues	Total organic carbon analyser and ibuprofen containing model solution	Jurecska et al., 2014
12.	ε-polycaprolactone (PCL) and beta-cyclodextrin-based polymer fibres	-	4-aminoazobenzene dye and Methylene Blue	FE-SEM, TG-DSC, EDXS, FTIR, XRD	Guo et al., 2019
13.	Magnetic/activated charcoal/β-cyclodextrin/alginate polymer nanocomposite	-	Methylene Blue	HRTEM, SEM, EDX, FTIR, VSM, XRD	Yadav et al., 2020
14.	β-cyclodextrin-modified/PVDF blend magnetic membranes	-	Cu^{2+}	FT-IR, Raman, XRD, TEM, XPS	Zhang et al., 2019
15.	Nanofiltration membranes from β-cyclodextrin and graphene quantum dots	-	Eriochrome Black T and Congo Red	TEM, DLS, ATR-FTIR, AFM	Xue et al., 2020
16.	Cellulose acetate based complexation NF membranes	-	Pb (II) from wastewater	FTIR, TGA, SEM, TEM, AFM	Idress et al., 2021
17.	Cellulose solution along with [Emim] OAc-DMSO mixture	-	Orange Red II dye and Remazol Brilliant Blue R dye	SEM	Anokhina et al., 2017
18.	Cellulosic nanofiber from unbleached rice straw	Rice straws	Lime NPs and papermaking white water	TEM, SEM	Hassan et al., 2020
19.	Cellulose acetate/ SiO_2-poly (2-Acrylamido-2-methylpropane sulfonic acid) hybrid nanofiltration membrane	-	Ceftriaxone sodium	FE-SEM, water contact angle, thermogravimetric analysis, FTIR, TEM	Mahdavi & Bagherifar, 2018
20.	Bamboo cellulose NF	Bamboo	Methyl Orange and Methyl Blue dye	XRD, FT-ID, SEM	Weng et al., 2017
21.	Salt reinforced hydrophilic bacterial cellulose nanofiber composites	Bacteria	Desalination (Na_2SO_4)	FTIR, SEM, TEM, Zeta potential	Teng et al., 2020
22.	Chitosan based NF	-	Na_2SO_4	-	Hao et al., 2020
23.	Chitosan-based polyesteramide nanofiltration	-	$NaCl$, $MgSO_4$, Na_2SO_4, $MgCl_2$, Rhodamine B dye, Reactive Black dye, indigo carmine dye.	ATR-FTIR, Zeta potential, DSA10-MK2 contact angle meter	Wu et al., 2020

Biomaterial Fabricated Nanofiltration Membrane

TABLE 6.1 *(Continued)*
Fabricated nanofiltration membranes and their role in rejection of various industrial effluents.

No.	NFs	Organisms	Effluents	Characterisation	Ref.
24.	Tripolyphosphate crosslinked chitosan/ polyethersulfone composite nanofiltration	-	(Heavy metal) Copper ion rejection	FTIR-ATR, zeta potential, contact angle, FE-SEM, AFM	Afsarian & Mansourpanah, 2018
25.	Chitosan modified membrane	-	Chromium, Na^+, K^+, Mg^{2+}, Ca^{2+}	FTIR-ATR, SEM	Zakmout et al., 2020
26.	Carboxymethyl chitosan-based NF	-	Heavy metals and salts	FTIR, XRD, SEM, AFM, zeta potential	Xiong et al., 2020
27.	Chitosan lactate-based hollow-fibre nanofiltration membrane	-	$MgCl_2$, $ZnCl_2$, $MgSO_4$, NaCl, Na_2So_4	SEM, ATR-IR	He et al., 2019
28.	Chitosan containing functionalised multiwall carbon nanotubes functionalised with COOH and NH_2	-	Desalination (NaCl and $MgSO_4$)	SEM, Zeta potential, UV-vis-NIR spectroscopy, contact angle measurement, micrometric ASAP-2020	Alshahrani et al., 2020
29.	Helical carbon functionalised chitosan-based loose nanofiltration	*Parthenium hysterophorus*	Organic pollutants	ATR-IR, FE-SEM, Contact angle, Zeta potential	Halakarni et al., 2020
30.	Chitosan and silver-based NF		NaCl, red brown/ organic dye and tannery wastewater	ATR-IR, FESEM, AFM, UV-Vis spectroscopy and contact angle measurement	Maalige et al., 2021
31.	Graphene oxide and chitosan-based membrane	-	Na_2So_4 and $CrSO_4$	SEM, FTIR, AFM, Contact angle, porosity measurement	Bagheripour et al., 2017
32.	Novel Nano-bio (nano zerovalent iron and *Klebsiella* sp.) composite beads	*Klebsiella* sp. (adapted to 50mg/L of Congo Red)	Congo Red	SEM, BET	Ravikumar et al., 2019
33.	Alginate membrane	Brown seaweed	Methyl Orange, Brilliant Blue	SEM, FTIR, AFM, contact angle	Aburabie et al., 2019
34.	Alginate/polysulfone composite using epichlorohydrin crosslinking	-	Na_2SO_4, K_2SO_4, $MgSO_4$, NaCl, KCL, $MgCl_2$	SEM, AFM	Chen et al., 2010
35.	Nanofibrous alginate membrane coated with cellulose nanowhiskers	Maize cellulose	Chromium	SEM, TEM,	Mokhena et al., 2017
36.	Polyvinyl alcohol-graphene oxide-sodium alginate nanocomposite hydrogel blended PES nanofiltration	-	Lanasol Blue 3R dye	SEM, AFM, ATR-FTIR, water contact angle, overall porosity, and mean pore radius measurement	Amiri et al., 2020

(Continued)

TABLE 6.1 *(Continued)*

Fabricated nanofiltration membranes and their role in rejection of various industrial effluents.

No.	NFs	Organisms	Effluents	Characterisation	Ref.
37.	Nanohydroxyapatite coated activated carbon impregnated alginate	-	Uranium	SEM-EDS and XRD	Saha et al., 2020
38.	Nano silver chloride and alginate incorporated composite copolymer	-	Brilliant Cresyl Blue	FTIR, UV-Vis, TEM-EDAX, XRD, XPS	Bhangi & Ray, 2019
39.	Silk fibroin-based green NF	Bombyx Mori	PM2.5	FE-SEM, FTIR	Gao et al., 2018
40.	Silk fibroin/nylon-6 blend NF matrix	Bombyx Mori	Copper removal	ATR-FTIR, SEM, EDXS, XPS	Yalcin et al., 2015
41.	Silk nanofiber with graphene oxide composite membrane	Bombyx Mori	Na2SO4 salt rejection and dye removal (Rose Bengal, Brilliant Blue R, Amido Black 10B, and Orange G)	FTIR, TGA, UV, FE-SEM, water contact angle, zeta potential, and positron annihilation for free volume analyses	Sivakumar et al., 2020
42.	Silk fibroin/cellulose acetate blend nanofibers	Bombyx mori	Heavy metal ions (Cu^{2+})	SEM, FTIR, atomic absorption spectrometry (AAS), atomic emission spectrometry (AES), inductively coupled plasma emission spectrometry (ICP-ES), and electrochemical and spectrometric method	Zhou et al., 2011

REFERENCES

1. Sivakumar, M., Liu, D.-K., Chiao, Y.-H. & Hung, W.-S. Synergistic effect of one-dimensional silk nanofiber and two-dimensional graphene oxide composite membrane for enhanced water purification. *Journal of Membrane Science*, 606, 118142 (2020). https://doi.org/10.1016/j.memsci.2020.118142

2. Arif, Z., Sethy, N.K., Mishra, P.K., et al. Development of eco-friendly, self-cleaning, antibacterial membrane for the elimination of chromium (VI) from tannery wastewater. *International Journal of Environmental Science Technology*, 17, 4265–4280 (2020). https://doi.org/10.1007/s13762-020-02753-6

3. Mokhena, T.C., Jacobs, N.V. & Luyt, A.S. Nanofibrous alginate membrane coated with Cellulose nanowhiskers for water purification. *Cellulose*, 25, 417–427 (2018). https://doi.org/10.1007/s10570-017-1541-1

4. He, Y., Miao, J., Chen, S., Zhang, R., Zhang, L., Tang, H. & Yang, H. Preparation and characterization of a novel positively charged composite hollow fiber nanofiltration membrane based on chitosan lactate. *RSC Advances*, 9, 4361–4369 (2019). https://doi.org/10.1039/C8RA09855G

5. Yeo, J.H., Kim, M., Lee, H., Cho, J. & Park, J. Facile and novel eco-friendly poly (Vinyl Alcohol) nanofilters using the photocatalytic property of titanium dioxide. *ACS Omega*, 5(10), 5026–5033 (2020). https://doi.org/10.1021/acsomega.9b03944

6. Nidhi Maalige, R., Aruchamy, K., Polishetti, V., Halakarni, M., Mahto, A., Mondal, D. & Kotrappanavar, N.S. Restructuring thin film composite membrane interfaces using biopolymer as a sustainable alternative to prevent organic fouling. *Carbohydrate Polymers*, 254, 117297 (2021). https://doi.org/10.1016/j.carbpol.2020.117297

7. Keskin, N.O.S., Celebioglu, A., Sarioglu, O.F., Uyar, T. & Tekinay, T. Encapsulation of living bacteria in electrospun cyclodextrin ultrathin fibers for bioremediation of heavy metals and reactive dye from wastewater. *Colloids and Surfaces B: Biointerfaces*, 161, 169–176 (2017). https://doi.org/10.1016/j.colsurfb.2017.10.047

8. Kas, H.S. Chitosan: Properties, preparations and application to microparticulate systems. *Journal of Microencapsulation*, 14(6), 689–711 (2009). https://doi.org/10.3109/02652049709006820

9. Hao, Y., Li, Q., He, B., Liao, B., Li, X.H., Hu, M., Ji, Y.-H., Cui, Z., Younas, M. & Li, J. An ultra highly permeable and selective nanofiltration membrane mediated by in-situ formed interlayer. *Journal of Materials Chemistry A*, 8, 5275–5283 (2020). https://doi.org/10.1039/C9TA12258C

10. Wu, D., Zhang, X., Chen, Y., et al. Thin film composite polyesteramide nanofiltration membranes fabricated from carboxylated chitosan and trimesoyl chloride. *Korean Journal of Chemical Engineering*, 37, 307–321 (2020). https://doi.org/10.1007/s11814-019-0426-4

11. Bagheripour, E., Moghadassi, A.R., Hosseini, S.M., Van der Bruggen, B. & Parvizian, F. Novel composite graphene oxide/chitosan nanoplates incorporated into PES based nanofiltration membrane: Chromium removal and antifouling enhancement. *Journal of Industrial and Engineering Chemistry*, 62, 311–320 (2018). https://doi.org/10.1016/j.jiec.2018.01.009

12. Alshahrani, A., Alsohaimi, I., Alshehri, S., Alawady, A., El-Aassar, M., Nghiem, L. & In Het Panhuis, M. Nanofiltration membranes prepared from pristine and functionalised multiwall carbon nanotubes/biopolymer composites for water treatment applications. *Journal of Materials Research and Technology*, 9, 9080–9092 (2020). https://doi.org/10.1016/j.jmrt.2020.06.055

13. Mahaveer, H., Ashesh, M., Kanakaraj, A., Dibyendu, M. & SaNna, K.N. Developing helical carbon functionalized chitosan-based loose nanofiltration membranes for selective separation and wastewater treatment. *Chemical Engineering Journal*, 127911 (2020). https://doi.org/10.1016/j.cej.2020.127911

14. Zakmout, A., Sadi, F., Portugal, C., Crespo, J.G. & Velizarov, S. Tannery effluent treatment by nanofiltration, reverse osmosis and chitosan modified membranes. *Membranes*, 10(12), 378 (2020). https://doi.org/10.3390/membranes10120378

15. Afsarian, Z. & Mansourpanah, Y. Surface and pore modification of tripolyphosphate-crosslinked chitosan/polyethersulfone composite nanofiltration membrane; characterization and performance evaluation. *Korean Journal of Chemical Engineering*, 35, 1867–1877 (2018). https://doi.org/10.1007/s11814-018-0085-x

16. Xiong, C., Huang, Z., Ouyang, Z., et al. Improvement of the separation and antibiological fouling performance using layer-by-layer self-assembled nanofiltration membranes. *Journal of Coatings Technology and Research*, 17, 731–746 (2020). https://doi.org/10.1007/s11998-019-00298-z

17. Dufresne, A. Nanocellulose: A new ageless bionanomaterial. *Materials Today*, 16(6), 220–227 (2013). https://doi.org/10.1016/j.mattod.2013.06.004

18. Weng, R., Chen, L., Xiao, H., et al. Preparation and characterization of cellulose nanofiltration membrane through hydrolysis followed by carboxymethylation. *Fibers and Polymers*, 18, 1235–1242 (2017). https://doi.org/10.1007/s12221-017-7200-1

19. Anokhina, T.S., Pleshivtseva, T.S., Ignatenko, V.Y., et al. Fabrication of composite nanofiltration membranes from cellulose solutions in an [Emim]OAc–DMSO mixture. *Petroleum Chemistry*, 57, 477–482 (2017). https://doi.org/10.1134/S0965544117060020

20. Idress, H., Zaidi, S., Sabir, A., Shafiq, M., Khan, R.U., Harito, C., Hassan, S. & Walsh, F.C. Cellulose acetate based Complexation-NF membranes for the removal of Pb(II) from waste water. *Scientific Reports*, 11(1), 1806 (2021). https://doi.org/10.1038/s41598-020-80384-0

21. Hassan, M.L., Fadel, S.M., Abouzeid, R.E., et al. Water purification ultrafiltration membranes using nanofibers from unbleached and bleached rice straw. *Scientific Reports*, 10, 11278 (2020). https://doi.org/10.1038/s41598-020-67909-3

22. Mahdavi, H. & Bagherifar, R. Cellulose acetate/SiO2-poly(2-Acrylamido-2-methylpropane sulfonic acid) hybrid nanofiltration membrane: Application in removal of ceftriaxone sodium. *Journal of the Iranian Chemical Society*, 15, 2839–2849 (2018). https://doi.org/10.1007/s13738-018-1470-4

23. Teng, X., Fang, W., Liang, Y., et al. High-performance polyamide nanofiltration membrane with arch-bridge structure on a highly hydrated cellulose nanofiber support. *Science China Materials*, 63, 2570–2581 (2020). https://doi.org/10.1007/s40843-020-1335-x

24. Rajput, K.N., Patel, K.C. & Trivedi, U.B. *β-Cyclodextrin Production by Cyclodextrin Glucanotransferase from Alkaliphile Microbacterium Terrae KNR 9 Using Different Starch Substrates*. Hindawi Publishing Corporation Biotechnology Research International, 2016, https://doi.org/10.1155/2016/2034359

25. Yadav, S., Asthana, A., Chakraborty, R., Jain, B., Singh, A.K., Carabineiro, S.A.C. & Susan, A.B.H. Cationic dye removal using novel magnetic/activated charcoal/β-cyclodextrin/alginate polymer nanocomposite. *Nanomaterials*, 10, 170 (2020). https://doi.org/10.3390/nano10010170
26. Machut, C., Kania, N., Léger, B., Wyrwalski, F., Noël, S., Addad, A., MonfliEr, E. & Ponchel, A. Fast microwave synthesis of gold-doped TiO_2 assisted by modified cyclodextrins for photocatalytic degradation of dye and hydrogen production. *Catalysts*, 10, 801 (2020). https://doi.org/10.3390/catal10070801
27. Guo, R., Wang, R., Yin, J., Jiao, T., Huang, H., Zhao, X., Zhang, L., Li, Q., Zhou, J. & Peng, Q. Fabrication and highly efficient dye removal characterization of beta-cyclodextrin-based composite polymer fibers by electrospinning. *Nanomaterials*, 9, 127 (2019). https://doi.org/10.3390/nano9010127
28. Xue, J., Shen, J., Zhang, R., Wang, F., Liang, S., You, X., Yu, Q., Hao, Y., Su, Y. & Jiang, Z. High-flux nanofiltration membranes prepared with β-cyclodextrin and graphene quantum dots. *Journal of Membrane Science*, 612, 118465 (2020). https://doi.org/10.1016/j.memsci.2020.118465
29. Jurecska, L., Dobosy, P., Barkács, K., Fenyvesi, É. & Záray, G. Characterization of cyclodextrin containing nanofilters for removal of pharmaceutical residues. *Journal of Pharmaceutical and Biomedical Analysis*, 98, 90–93 (2014). https://doi.org/10.1016/j.jpba.2014.05.007
30. Zhang, R., Li, Y., Zhu, X., et al. Application of β-cyclodextrin-modified/PVDF blend magnetic membranes for direct metal ions removal from wastewater. *Journal of Inorganic and Organometallic Polymers and Materials*, 30, 2692–2707 (2020). https://doi.org/10.1007/s10904-019-01416-5
31. Lee, K.Y. & David, J.M. Alginate: Properties and biomedical applications. *Progress in Polymer Science*, 37, 106–126 (2012). https://doi.org/10.1016/j.progpolymsci.2011.06.003
32. Aburabie, J., Puspasari, T. & Peinemann, K.-V. Alginate-based membranes: Paving the way for green organic solvent nanofiltration. *Journal of Membrane Science*, 596, 117615 (2019). https://doi.org/10.1016/j.memsci.2019.117615
33. Ravikumar, K., Soupam, D., Osborneb, J.W., ChaNdrasekaran, N. & Amitava, M. Novel nano-bio (Nano Zerovalent Iron and Klebsiella sp.) composite beads for Congo red removal using response surface methodology. *Journal of Environmental Chemical Engineering*, 7, 103413 (2019). https://doi.org/10.1016/j.jece.2019.103413
34. Amiri, S., Asghari, A., Vatanpour, V. & Rajabi, M. Fabrication and characterization of a novel polyvinyl alcohol-graphene oxide-sodium alginate nanocomposite hydrogel blended PES nanofiltration membrane for improved water purification. *Separation and Purification Technology*, 250, 117216 (2020). https://doi.org/10.1016/j.seppur.2020.117216
35. Bhangi, B.K. & Ray, S.K. Nano silver chloride and alginate incorporated composite copolymer adsorbent for adsorption of a synthetic dye from water in a fixed bed column and its photocatalytic reduction. *International Journal of Biological Macromolecules*, 144, 801–812 (2020). https://doi.org/10.1016/j.ijbiomac.2019.09.070
36. Saha, S., Basu, H., Rout, S., Pimple, M.V. & Singhal, R.K. Nano-hydroxyapatite coated activated carbon impregnated alginate: A new hybrid sorbent for uranium removal from potable water. *Journal of Environmental Chemical Engineering*, 8(4), 103999 (2020). https://doi.org/10.1016/j.jece.2020.103999
37. Chen, X., Wang, D., Wang, W., Su, Y. & Gao, C. A novel composite nanofiltration (NF) membrane prepared from sodium alginate/polysulfone by epichlorohydrin cross-linking. *Desalination and Water Treatment*, 30(1–3), 146–153 (2011). https://doi.org/10.5004/dwt.2011.1941
38. Rajangam, T. & An, S.S.A. Fibrinogen and fibrin based micro and nano scaffolds incorporated with drugs, proteins, cells and genes for therapeutic biomedical applications. *International Journal of Nanomedicine*, 8, 3641–3662 (2013). https://doi.org/10.2147/IJN.S43945
39. Gao, X., Gou, J., Zhang, L., Duan, S. & Li, C. A silk fibroin based green nano-filter for air filtration. *RSC Advances*, 8, 8181–8189 (2018). https://doi.org/10.1039/C7RA12879G
40. Yalçın, E., Gedikli, S., Çabuk, A., et al. Silk fibroin/nylon-6 blend nano filter matrix for copper removal from aqueous solution. *Clean Technologies and Environmental Policy*, 17, 921–934 (2015). https://doi.org/10.1007/s10098-014-0845-1
41. Zhou, W., He, J., Cui, S., et al. Preparation of electrospun silk fibroin/Cellulose Acetate blend nanofibers and their applications to heavy metal ions adsorption. *Fibers Polymers*, 12, 431–437 (2011). https://doi.org/10.1007/s12221-011-0431-7
42. Ranjitha, R., Renganathanb, V., Shen-Ming, C., Selvana, N.S. & Rajama, P.S. Green synthesis of reduced graphene oxide supported TiO2/Co3O4 nanocomposite for photocatalytic degradation of methylene blue and crystal violet. *Ceramics International*, 45, 12926–12933 (2019). https://doi.org/10.1016/j.ceramint.2019.03.219

43. Sonker, R.K., Hitkari, G., Sabhajeet, S.R., Sikarwar, S., Rahule & Singh, S. Green synthesis of TiO_2 nanosheet by chemical method for the removal of Rhodamin B from industrial waste. *Materials Science and Engineering: B*, 258, 114577 (2020). https://doi.org/10.1016/j.mseb.2020.114577

44. Shah, M.P. *Removal of Emerging Contaminants through Microbial Processes*. Springer, 2021.

45. Shah, M.P. *Advanced Oxidation Processes for Effluent Treatment Plants*. Elsevier, 2020.

46. Wang, L., Zhang, C., Gao, F., Mailhot, G. & Pan, G. Algae decorated TiO2/Ag hybrid nanofiber membrane with enhanced photocatalytic activity for Cr(VI) removal under visible light. *Chemical Engineering Journal*, 314, 622–630 (2016). http://doi.org/10.1016/j.cej.2016.12.020

47. Sethy, N.K., Arif, Z., Mishra, P.M. & Kumar, P. Green synthesis of TiO_2 nanoparticles from Syzygium cumini extract for photo-catalytic removal of lead (Pb) in explosive industrial wastewater. *Green Process Synthesis*, 9, 171–181 (2020). https://doi.org/10.1515/gps-2020-0018

48. Hu, W., Yin, J., Deng, B. & Hu, Z. Application of nano TiO2 modified hollow fibre membrane in algal membrane bioreactors for high density algal cultivation and wastewater polishing. *Bioresource Technology*, 193, 135–141 (2015). http://doi.org/10.1016/j.biortech.2015.06.070

49. Chao, J., Luong, N.N., Hou, J., Faisal, I.H. & Vicki, C. Direct immobilization of laccase on Titania nanoparticles from crude enzyme extracts of P. ostreatus culture for micro-pollutant degradation. *Separation and Purification Technology*, 178, 215–223 (2017). http://doi.org/10.1016/j.seppur.2017.01.043

50. Shah, M.P. *Removal of Emerging Contaminants through Microbial Processes*. Springer, 2021.

51. Shah, M.P. *Advanced Oxidation Processes for Effluent Treatment Plants*. Elsevier, 2020.

52. Shah, M.P. *Microbial Bioremediation and Biodegradation*. Springer, 2020.

53. Ricardo, A.S., Herrera, A.P., Maestre, D. & Cremades, A. Fe-TiO_2 nanoparticles synthesized by green chemistry for potential application in waste water photocatalytic treatment. *Hindawi Journal of Nanotechnology*, 2019, 1–11 (2019). https://doi.org/10.1155/2019/4571848

54. Abu-Daloa, M., Jaradata, A., Albissb, B.A. & Al-Rawashdeh, N.A.F. Green synthesis of TiO2 NPs/ pristine pomegranate peel extracts nanocomposite and its antimicrobial activity for water disinfection. *Journal of Environmental Chemical Engineering*, 7, 103370 (2019). https://doi.org/10.1016/j.jece.2019.103370

7 Biological-Based Methods for the Removal of VOCs and Heavy Metals

Amrin Pathan and Anupama Shrivastav*

CONTENTS

7.1 Introduction 105
    7.1.1 Heavy Metals in Industrial Wastewater 105
    7.1.2 VOCs in Wastewater and in Air 106
7.2 Biological Treatment for Heavy-Metals Removal 106
    7.2.1 Remediation of Heavy Metals 107
    7.2.2 Biofiltration Methods 107
        7.2.2.1 Component of Biofiltration 108
        7.2.2.2 Advancement within the Biofilters Effectiveness by Hereditary
                 Alteration of the Microorganisms 109
    7.2.3 Bioelectrochemical System 110
        7.2.3.1 Concept and Principle 110
    7.2.4 Biosorption 110
7.3 Biological Treatment for VOCs Removal 112
    7.3.1 Organically Actuated Froth for VOCs Removal 112
    7.3.2 Hollow-fibre Film Bioreactors 113
    7.3.3 Biotrickling Filters 113
        7.3.3.1 BTF Instruments and Limitations 114
        7.3.3.2 Microorganisms in BTF Systems 115
        7.3.3.3 Other Procedures Combined with BTF Systems 116
    7.3.4 Bioscrubbers 116
    7.3.5 Biofiltration 117
    7.3.6 NTP and UV-BTF Coordinates Innovation for Cl-VOCs Removal 119
7.4 Conclusion 120
References 120

7.1 INTRODUCTION

7.1.1 HEAVY METALS IN INDUSTRIAL WASTEWATER

Heavy metals are components having nuclear weights between 63.5 and 200.6 and a particular gravity more prominent than 5.0. Living beings require a small amount of heavy metals, like vanadium, iron, cobalt, copper, manganese, molybdenum, zinc, and strontium. Intemperate levels of fundamental metals, in any case, can be inconvenient to the life form. Non-essential heavy metals of specific concern to surface water frameworks are cadmium, chromium, mercury, lead, arsenic, and antimony. Heavy metals that are moderately copious within the Earth's hull and habitually utilised in mechanical forms or agribusiness are harmful to humans. These can make critical modifications to the biochemical cycles of living things.

DOI: 10.1201/9781003165149-7

Most of the point sources of heavy metal toxins are mechanical wastewater from mining, metal handling, tanneries, pharmaceuticals, pesticides, natural chemicals, elastic and plastics, amble and wood items, etc. The heavy metals are transported by runoff water and sully water sources downstream from the mechanical location. All living things including microorganisms, plants, and creatures depend on water for life. Heavy metals can attach to the surface of microorganisms and may indeed enter the interior of the cell. Inside the microorganism, the heavy metals can be chemically changed as the microorganism employs chemical responses to process food.

The objective of this chapter is to analyse different organic strategies utilised for the removal of heavy metals from mechanical wastewater and the change within the productivity of the forms by rising strategies such as the application of hereditary design for the treatment of heavy metal.

7.1.2 VOCs in Wastewater and in Air

VOCs are omnipresent within the ambient air and they are vital antecedents to ozone arrangement and secondary organic pressurised canned products. VOCs are criminal and they are radiated from different sources such as the chemical industry, activity emanation, oil and gas abuse, and so on. Some VOC species are too discuss toxics with different wellbeing impacts including acute and inveterate dangers depending on their chemical compositions. In a normal petroleum refinery emanating treatment plant (ETP) act as conclusion of the pipe treatment office for different squander water streams radiated from the refinery prepare units & offsite/utility offices. In this way ETP gets diverse unstable natural compounds (VOC) and inorganics like NH3, H2S, and so on related to the effluents, and in this it way becomes an outflow source for these substances. As the VOCs have a greater bubbling point or higher vapor weight at room temperature, they vanish effectively and a few are dangerous to the environment and people. VOCs moreover create ozone after responding with oxides of nitrogen, which in turn shapes exhaust clouds that are destructive to people and vegetation. In petroleum refining, VOCs appear in rough oil and are disseminated to the environment by means of leaks/venting from capacity, channels, fittings, types of gear, loading/unloading offices, etc., and a portion goes to fluid wastewater delivered from prepared units and offsite facilities. Thus, the VOC in a refinery to a great extent depends upon the sort of rough handling, refining, capacity commitments, etc. Hence, the full VOC outflow regularly ranges from 50–1000 tons per million tons of rough oil handled [1]. In addition, hundreds of unstable natural compounds are found in a refinery and are classified as alkanes, alkenes, aromatics, and cyclic hydrocarbons. In spite of the fact that there are various VOCs in an ordinary petroleum refinery wastewater, common VOCs are methyl tertiary-butyl ether, benzene, biphenyl, cresols, xylene, cumene, ethylbenzene, hexane, naphthalene, phenol, styrene, toluene 1, 3-butadiene 2, and 4-trimethylpentane [2,3].

7.2 BIOLOGICAL TREATMENT FOR HEAVY-METALS REMOVAL

Several physico-chemical strategies have been broadly utilised for evacuation of heavy metals from mechanical wastewater, such as particle trade, actuated charcoal, chemical precipitation, chemical lessening and adsorption, etc. The customary strategies utilised for the treatment of heavy metals from mechanical wastewater show a few restrictions. There are still a few common issues related with these strategies, such as that they are cost-expensive and can themselves create other issues, which has constrained their mechanical applications.

Among the accessible treatment forms, these days the application of natural forms is steadily getting force due to the taking after reasons:

- Chemicals' necessity for the entire treatment prepare is reduced.
- Fewer working costs.
- Eco-friendly and cost-effective elective of customary techniques.
- Productive at lower levels of contamination.

Biological-Based Methods for Removal of VOCs and Heavy Metals

7.2.1 Remediation of Heavy Metals

The vital evacuation of heavy metals ought to be the avoidance of heavy metals entering the human body. Lessening debilitation and industry pollution is one of the issues that can be focused on. Other ways heavy metals are entering our bodies are through nourishment. Phytoremediation and intercropping are ways in which heavy metals can be retained and removed from the soils, silt, and waters. Hyperaccumulator plants are planted within the soils to evacuate heavy metals. These sorts of plants have root frameworks that have a specific take-up, where the contaminant is translocated and bioaccumulated, and after that the plants are entirely corrupted to evacuate the heavy metals. Phytostabilisation and phytoextraction happen for inorganic compounds. Phytoextraction is the form where the heavy metals are translocated from the roots and into the shoots. Phytostabilisation is comprised of as it were parts of the plant to require portion in heavy metal removal. Organic compounds are expelled through the forms phytodegradation, rhizofiltration, and rhizodegradation. Rhizofiltration includes the adsorption or precipitation of the poisons that are in arrangement and surround the root and permeate the root, once they are damp. Rhizodegradation involves microbial movement where the contaminants are debased in the rhizosphere and made strides by the roots. Plant species may be noteworthy when it comes to choosing the plant as distinctive species take up distinctive heavy metals. Other factors include the soil's pH, natural matter display, sum of phosphorous in the soil, root zone, the chelating specialist included in the soil that influences the bioavailability of the metals such as ethylenediaminetetraacetic acid (EDTA) and temperature, which impacts the vegetative take-up through the root length. Hyperaccumulator plants don't show signs of toxicity when they assimilate heavy metals. On the other hand, the strategy of intercropping comprises growing two distinctive species of plants at the same time. Plant biomass is improved, and the collection of heavy metals is aided.

However, the biofilters are the most recent and the foremost promising advancement in natural forms for the treatment of heavy-metals-sullied mechanical wastewater.

7.2.2 Biofiltration Methods

Microorganisms settled into a permeable medium utilised within the biofiltration prepare to break down poisons shown within the wastewater stream. The microorganisms grow in a biofilm on the surface of the medium or are suspended within the water stage encompassing the medium particles. The channel bed medium comprises generally inactive substances that guarantee huge surface connection zones and extra supplement supply. The general viability of a biofilter is administered by the properties and characteristics of the bolster medium, which incorporates porosity, degree of compaction, water maintenance capabilities, and the capacity to have microbial populaces. Basic biofilter operational and execution parameters incorporate the microbial vaccination, medium pH, temperature, the medium dampness, and supplement substance.

In a biofiltration framework, the biodegradable poisons are expelled due to organic corruption instead of physical straining as with the case in an ordinary channel. With the advancement of the filtration handle, microorganisms (oxygen consuming, anaerobic, facultative, microbes, parasites, green growth, and protozoa) are steadily created on the surface of the channel media and frame an organic film or thin layer known as biofilm. The significant point for the fruitful operation of the biofilter is to control and keep up a solid biomass on the surface of the filter.

Since the execution of the biofilter generally depends on the microbial exercises, a steady source of substrates (natural substance and supplements) is required for its reliable and viable operation, in spite of the fact that a few chemoautotrophic microbes may utilise inorganic chemicals as vitality source. The expulsion productivity of biofilters is controlled by a few parameters like pH, temperature, O_2 substance, introductory concentration of poisons, etc. The evacuation proficiency may be progressed by chemical alteration of the channel media or hereditary alteration of microorganisms.

7.2.2.1 Component of Biofiltration

The basic instruments, which permit biofilters to work and which must be controlled to guarantee victory, are complex. The biofilter contains a permeable medium whose surface is covered with water and microorganisms. The contaminant may shape complexes with natural compounds within the water and may be adsorbed by the bolster medium. Eventually, biotransformation changes over the contaminant to biomass, metabolic by-products, or carbon dioxide and water. The biodegradation is carried out by a complex environment of degraders, competitors, and predators that are mostly organised into a biofilm. There are three fundamental organic forms that can happen in a biofilter:

- Connection of microorganisms.
- Development of microorganisms.
- Rot and separation of microorganisms.

The components by which microorganisms can join and colonise on the surface of the channel media of a biofilter are transportation of microorganisms, starting grip, firm connection, and colonisation. The transportation of microorganisms to the surface of the channel media is controlled by four fundamental forms:

- Dissemination (Brownian motion).
- Convection.
- Sedimentation due to gravity.
- Dynamic portability of the microorganisms.

Long before the microorganisms reach the surface, beginning attachment happens that may be reversible or irreversible depending upon the entire interaction vitality, which is the whole of van der Waal's drive and electrostatic constraint. The forms of film connection and colonisation of microorganisms depend on influent characteristics (such as natural sort and concentration) and surface properties of the channel media. Afterward, parameters are taken into consideration to assess the connection of microorganisms on the surface of the channel media:

- The steric effect.
- Hydrophobicity of the microorganisms.
- Contact point.
- Electrophoretic portability values.

The components that impact the rate of substrate utilisation inside a biofilm are substrate mass transport to the biofilm, dissemination of the substrate into the biofilm, and utilisation energy inside the biofilm. Adsorption and biodegradation perform at the same time in biofilters to evacuate biodegradable and water-dissolvable perilous natural atoms, which come about in synchronous adsorption and biodegradation. In spite of the fact that the component of heavy metals expulsion by biofilter contrasts from that of the natural chemicals expulsion, overall the conditions with respect to development of biomass within the biofilm are the same. As it was a special case, in this case there's no biodegradation in heavy metals expulsion. The instrument of heavy metals expulsion from sullied water in a biofilter is as takes after. The non-biodegradable water-dissolvable heavy metals are either oxidised or decreased by the microorganisms and deliver a less solvent species. The less dissolvable frame of these metals that are shaped due to microbial responses are adsorbed or precipitated/co-precipitated on the surface of the adsorbent and the additional cellular protein of the microorganisms within the biolayer. The methylation of metals is additionally another vital course for bioremediation of heavy metals in water. In spite of the fact that the microbial activity

Biological-Based Methods for Removal of VOCs and Heavy Metals 109

on metal particle change is still a matter of inquiry, it is accepted that there are two ways. In one way oxidation or the decrease of heavy-metal particles takes additional cellular proteins where the metal particles don't enter into the bacterial cell. With the other way, the metal particles are transported into the microbial cells by transmembrane proteins and are changed over to other less soluble forms by metabolic activities of proteins within the cells by ensuing excretion from the cells, however, both ways are plasmid intervened. Whether the microbial activity on a metal particle is performed one way or by both the ways could be a matter of research.

7.2.2.2 Advancement within the Biofilters Effectiveness by Hereditary Alteration of the Microorganisms

Bioremediation is the change or corruption of contaminants into non-hazardous or less dangerous chemicals. Microscopic organisms are by and large utilised for bioremediation, but parasites, green growth, and plants have too been utilised. There are three classifications of bioremediation:

- Biotransformation: The change of contaminant particles into less- or non-hazardous molecules.
- Biodegradation: The breakdown of natural substances into littler natural or inorganic molecules.
- Mineralisation: The total biodegradation of natural materials into inorganic constituents.

These three sorts of bioremediation can happen either ex situ or in situ. There are both points of interest and impediments related to ex situ and in situ forms. In the previous case, the contaminants are expelled and set in a contained environment, which makes the remediation handle quicker by permitting simpler observation and keeping up of conditions and advancement. In any case, the expulsion of the contaminant from the sullied location is time consuming, expensive, and possibly unsafe, which are the major impediments of the method. In differentiate, the in situ preparation does not require the expulsion of the contaminant from the sullied location. Instead, either biostimulation or bioaugmentation is connected. The first is the expansion of nutrients, oxygen, or other electron givers or acceptors to the facilitated location in order to extend the populace or movement of normally occurring microorganisms accessible for remediation, whereas the second mentioned is the expansion of microorganisms that can biotransform or biodegrade contaminants.

Bioremediation innovation includes the utilisation of microorganisms to diminish, kill, contain, or change into prevalent contaminants shown in soils, silt, and water. *Staphylococcus, Bacillus, Alcaligens, Escherichia, Pseudomonas, Citrobacteria, Klebsilla, and Rhodococcus* are the living beings that are commonly utilised in bioremediation. This includes biochemical responses or pathways in a living being that result in movement, development, and generation of that life form. Chemical forms included in the microbial digestion system are composed of reactants, contaminants, oxygen, or other electron acceptors, which change over metabolites to well-characterised items. A key factor to the remediation of metals is that metals are non-biodegradable but can be changed through sorption, methylation, and complexation and changes in valence state. Although utilising bioremediation could be an awesome thought, very regularly the contaminants are also toxic to the dynamic organisms included within the bioremediation. This issue can make it exceptionally troublesome to keep the rate of bioremediation high. A solution to this issue is hereditarily built organisms that are safe for the particular conditions of the sullied location and conjointly have bioremediation properties. Bioremediation is picking up significance recently as an interchange innovation for evacuation of essential toxins in soil and water, which require compelling strategies of purification.

The designed microorganisms are more specific. The change in expulsion productivity for all the cases is recognisable. It is trusted that, in combination, these methods would progress channel proficiency.

7.2.3 BIOELECTROCHEMICAL SYSTEM

A one-of-a-kind approach based on the integration of electrochemistry and microbiology has as of late developed. It is commonly alluded to as electrode-assisted bioremediation or electro-bioremediation or bioelectrochemical remediation of heavy-metal-sullied situations. The reactor frameworks utilised to attain this are alluded to as bioelectrochemical frameworks (BESs). BESs combine microbial and electrochemical forms to change over the chemical vitality stored in biodegradable natural matter into power, hydrogen, or other valuable chemicals or to drive forms such as saltwater desalination and the expulsion of different contaminants.

Electrochemical treatment forms work on the rule of electrolysis in which, by giving vitality from an connected source, metal oxidation at the anode and metal reduction at the cathode are encouraged at particular electric possibilities. In spite of the fact that this technique is successful, it requires a lot of support, care, and vitality input and in this way includes significant operational costs. To overcome a few of the disadvantages of this approach, analysts began investigating BESs, which utilise microorganisms to intervene, encourage, or catalyse the redox responses at both or either of the cathodes [4]. It is possible that the microorganisms are compelling biocatalysts to impact the portability of metal particles in natural and built or planned frameworks under controlled conditions [5, 6]. BESs open up an opportunity for the improvement of a novel and valuable biotechnology based on the utilisation of microorganisms in electrochemical frameworks for the evacuation and recovery of heavy metals from the sullied soil, dregs, and water environments.

7.2.3.1 Concept and Principle

BESs comprise the anode and cathodes that are ordinarily isolated by an ion-selective layer. Microorganisms that have extracellular electron exchange capabilities are for the most part utilised to catalyse either the substrate oxidation or diminishment responses at the anodes in BESs. Such microorganisms are alluded to as electroactive microorganisms. They have the capacity to connect and intervene in electron exchange to the solid-state electron acceptors and from benefactors such as cathodes to support their respiratory or metabolic exercises. BESs can be worked in two diverse modes, microbial fuel cell (MFC) and microbial electrolysis cell (MEC) depending upon the target handle. Microbial substrate oxidation response at the anode is ordinarily utilised for the treatment of natural matter containing wastewater, while substrate diminishment responses at the cathode are particularly utilised for the generation of decreased items such as hydrogen and biochemicals [7] or the remediation of obstinate contaminants and decrease of oxidised shapes of heavy metals displayed in several sorts of wastes, prepare streams, wastewaters, and leachate arrangements. It can moreover be utilised for the lessening of other electron acceptors such as perchlorate, nitrobenzene, azo colours, and nitrate [8, 9]. In BESs, the electrons delivered by the microbial oxidation of natural matter at the anode can be utilised to drive or encourage the diminishment of heavy metals at the cathode (Figure 7.1). There are two conceivable outcomes for the 56 Coordinates Microbial Fuel Cells for Wastewater Treatment lessening of heavy metals at the cathode in BESs. To begin with, in the event that the lessening potential of the heavy metal at the cathode is higher (or more positive) than the oxidation potential of the electron giver at the anode, the response will continue suddenly and create a net positive cell voltage in MFC mode operation. The metals that can be decreased at the cathode of MFCs incorporate Cu^{2+}/Cu^0, Se^{4+}/Se^0, Cr^{6+}/Cr^{3+}, and V^{5+}/V^{4+}. The moment plausibility is through utilising MECs when the diminishment potential of the metal is lower (or more negative) than the oxidation potential of electron benefactor; an outside vitality should be connected to encourage the metal diminishment response at the cathode.

7.2.4 BIOSORPTION

The component of biosorption of heavy metals perhaps happens in two steps: The primary step happens on the cell surface, counting physical adsorption, particle trade, and surface complexation.

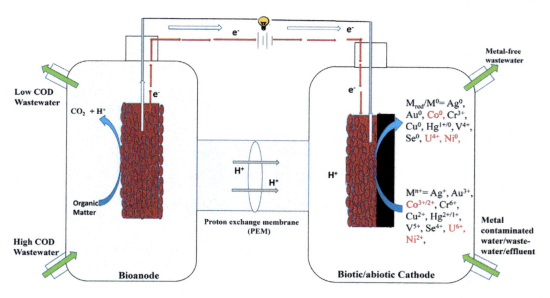

FIGURE 7.1 Schematic of a two-chambered bioelectrochemical system (BES) showing the principle of heavy-metals removal [10].

This step is driven by the chemical osmotic slope of the natural cytoplasmic layer [11], and it too can be done by dead natural cells. The important steps are the inside aggregation of cells and surface microprecipitation, both of which expend cellular vitality and require the interest of metabolic processes happening within the uncommon state of organic cells [12]. The intuitive adsorption mechanism is shown in Figure 7.2.

In spite of the fact that the introductory comes about of inquire about on biosorption of heavy metals by numerous natural materials have been accomplished, a single organic adsorbent is difficult to meet the needs of down to earth application. In this way, modern heavy-metals treatment strategies based on biomass have been developing. Among them, hereditarily built microorganisms are exceedingly anticipated. Be that as it may, the hereditarily altered microorganism is difficult to realise, due to the huge sum of building and the tall working conditions it requires. Most importantly, the existing hereditary information isn't sufficient to completely foresee how the hereditarily engineered microorganisms will influence the human living environment in the future. In this manner, agreeing to the current natural, financial, and innovative status, the utilisation of generally straightforward chemical or physical strategies to move forward the adsorption effectiveness of biosorbents is of awesome centrality for the treatment of heavy-metal wastewater. Chemical adjustment may be by and large separated into two categories: Surface modification and inner adjustment of the natural cell. The biggest reason for surface alteration is to expel the debasements on the cell surface and increment the heavy-metal official location on the surface of the natural cell or alter the charge on the cell surface. Inner adjustment is more complex, which includes changes within the inside structure or composition of natural cells, counting protein expression, enzymatic movement, or self-generated nanomaterials inside the cell. Invigorating or restraining the movement of chemicals included in heavy-metal transport or collection inside the cell will have a great impact on the decontamination of heavy metals.

The combination of biomass and other materials cannot as it was fathoming the issue with the lacking of mechanical quality of free suspended biomass in building application and the trouble of solid-liquid division, but moreover at the same time apply two sorts of materials perform on the adsorption [14, 15]. There may moreover be synergistic impacts within the assimilation of heavy

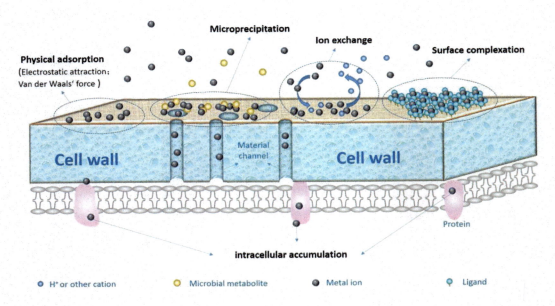

FIGURE 7.2 Adsorption mechanism of biological cells for heavy metals removal [13].

metals beneath fitting fabric choice and working conditions. The commonly utilised biological-bound materials are alginate, enacted carbon, and different polymer materials. In later a long time, a couple of studies have combined biomass with nanoparticles to form a unused fabric that has amazing adsorption properties for heavy metals [16, 17].

In expansion, multi-biological combinations have moreover been demonstrated to have great heavy-metal expulsion and steadiness in complex heavy-metal-sullied situations. The use of a multi-biological combination frequently has an advantageous and synergistic impact. A strain in a multi-biological combination can alter the natural oxygen substance and pH esteem through the organic digestion system to supply superior adsorption conditions for the other strain. The point of this work is to display the state of the craftsmanship of three strategies on heavy-metals biosorption. Fabric planning, test investigate strategies, treatment impact, and expulsion component were displayed and talked about, and their potential application and improvement were anticipated. Methodically compare these strategies, we given a point of view approximately biosorption for heavy metals expulsion. It is hoped that this audit may be accommodating to advancing investigations in this field.

7.3 BIOLOGICAL TREATMENT FOR VOCS REMOVAL

Despite the viability of physical–chemical strategies in wrecking VOCs and rotten gasses, the drawbacks of tall taken a toll and auxiliary contamination are characteristic of these techniques.

7.3.1 Organically Actuated Froth for VOCs Removal

Prepare designers at the Orange County Water District (OCWD) in Wellspring Valley, CA, USA have created a ceaselessly renewable biologically actuated foam (BAF) reactor that quickly 'eats' VOCs, changing over the contaminants into carbon dioxide, water, and characteristic biological build-up. As the late protected handle employments a vapour-phase reactor and BAF to annihilate the contaminants. The manufactured 'biofilm' delivered amid this handle boosts transport of VOCs and oxygen and stops microbial fouling issues customarily related to settled film forms, say

Biological-Based Methods for Removal of VOCs and Heavy Metals

the engineers. A major advantage of the BAF reactor is that, not at all like a fixed-film reactor, the microbe's concentration is controllable. In case microscopic organisms within the framework do multiply out of control it is simple to turn the framework off, close, it down, and start the reactor up once more. The downtime is brief, as there ought not to extricate the strong pressing, clean it, put it back and reinoculate the packing.

7.3.2 HOLLOW-FIBRE FILM BIOREACTORS

A hollow-fibre film-employed novel bioreactor was created for the coupled transcription/translation framework utilising T7RNA polymcrase and Escberichiaroli S30 extricate. The expansive surface zone per the response volume of the reactor guaranteed quick mass exchanges of substrates into rhc response blend and of squanders out from it over the layer by theit- atomic dissemination. The flux was expansive, sufficient to preserve nucleotide concentrations for more than 3 h, which expanded the protein union greatly. In expansion, the T eliminator grouping, downstream from the columnist qualities, was found to extend the integrated protein significantly, especially when the item of polymerase chain response. PCR was utilised as one template. Implementation of this finding and utilisation of the bioreactor created increased the efficiency of protein by the in $z\&v$ coordinate expression from the PCR template.

7.3.3 BIOTRICKLING FILTERS

In BTF frameworks, waste gas being treated is carried through a pressed bed, which is persistently or irregularly watered with a watery arrangement containing fundamental supplements required by the natural framework. Microorganisms from an outside source are immunised on the surface of the pressing fabric and shape a biofilm. Toxins are at first retained by a fluid film that encompasses the biofilm and are at that point corrupted by the biofilm. The BTF framework has the points of interest of low working and capital costs, lower weight drop amid long-term operation, and capability to expectation treat acidic corruption items of VOCs and acidic rotten gasses [18]. In any case,

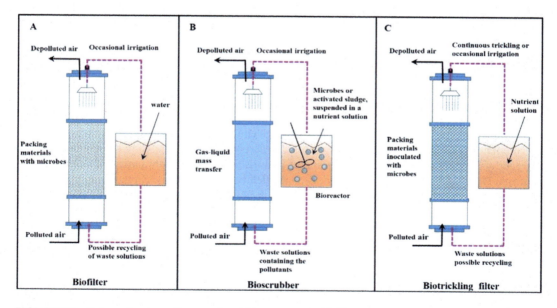

FIGURE 7.3 Schematic view of conventional bioreactors. A: biofilter (BF); B: bioscrubber; C: biotrickling filter (BTF) [19].

BTFs are moderately complex in development and operation and amass overabundant biomass. Comparison of BFs, BTFs, and bioscrubbers uncovered that BTFs have way better capacity in treating obstinate VOCs and acidic or soluble compounds. Additionally, water administration is the major advantage of BTFs, when compared with the BF frameworks, permitting characterised control of pH, supplement supply, and expulsion of poisonous metabolites, hence accomplishing higher toxins end rates. When compared with bioscrubbers, BTFs show quick biodegradation of toxins owing to enhancement of poisons with extracellular polymeric substances acting as surfactants, coming about in expansive numbers of immobilised microorganisms coming into contact with the pollutants.

In spite of the fact that the development and operation of a BTF may be complex, these frameworks as a normal frame of natural treatment innovation have succeeded in treating numerous sorts of toxins. Hence, when combining this advantage with it's taken a toll adequacy, BTF innovation is an appealing choice for controlling VOCs an emanation from different mechanical forms. In any case, a few confinements in BTF forms (such as abundance biomass amassing, low mass exchange rate, etc.) may influence the toxin expulsion execution of BTFs. In a BTF framework, the plan, working condition, mass exchange, pressing materials, and microorganisms are basic impacts of BTF forms and can altogether influence the expulsion of vaporous toxins.

Numerous strategies have been examined for upgrading the expulsion execution of BTFs, such as poison pre-treatment, parasitic BTF, surfactant expansion, and other strategies. Basically, these strategies are outlined to optimise the basic parameters of BTFs. In this way, examination of these basic parameters is fundamental, since they are the base for utilising BTFs and progressing the toxin evacuation.

7.3.3.1 BTF Instruments and Limitations

A BTF framework for VOCs and musty gas evacuation is ordinarily a complex combination of distinctive physicochemical and organic marvels. As the vapor passes through the BTF bed, contaminants and supplements are exchanged to the microorganisms, where the cellular digestion system breaks down the chemicals into less difficult components. Microorganisms will be immobilised on the surface of the pressing fabric and the surface of biofilms is secured by a water stage. The VOCs and musty toxins within the contaminated discuss to begin with got to be transferred to the water stage layer, and after that moved to microbial cells within the biofilm where they are eventually debased by the microorganisms.

At first, the vapor-phase contaminants and any supplements required for microbial development (oxygen, nitrogen, phosphorus, etc.) must be broken up within the water stage encompassing the biofilm. The poison vapours and oxygen are transported in muggy discuss or specifically into the water stage by concentration angle contrasts, and this handle is shown as mass exchange. Along these lines, biodegradation responses happen, amid which the broken-down contaminants are acclimatised and corrupted by the biofilm through change to CO_2 or other ineffectively unstable intermediates with simple chemical structures. In the proximity of supplements, the vitality discharged by the oxidation of these contaminants is utilised by the microorganisms to create a biofilm. In a BTF, the biodegradation impact is controlled both by mass exchange of vaporous toxins from discuss to biofilm (dissemination restriction) and by the biodegradation responses (response impediment). In a past study [20], the relationship between end capacities (EC) and channel stack (IL) was talked about at three purge bed maintenance times (EBRT). When IL was settled, longer EBRT implied the vaporous poison characteristics were a high channel concentration and low add up to stream rate. In this case, the water stage encompassing the biofilm was immersed by vaporous toxins; in this way, the debasement response was controlled by the response restriction. At shorter EBRT, the vaporous poison characteristics added up to high stream rate and low gulf concentration. Beneath these conditions, the vaporous poison left the BTF without satisfactory contact with the water stage; in this way, the debasement was controlled by dissemination impediment [20].

Biological-Based Methods for Removal of VOCs and Heavy Metals 115

The real biodegradation of target contaminants happens inside the biofilm, and the arrangement of auxiliary metabolites is conceivable. In this event, auxiliary metabolites will experience the same concurrent dissemination, biodegradation, and sorption forms as the essential toxins. At long last, the conclusion items (such as CO_2) after microbial corruption are exchanged to the gas stage or shape carbonate in the fluid stage via VOC oxidation. In any case, in the event that the auxiliary metabolites appear more harmful than the first poisons, their arrangement will halt the biodegradation forms; for illustration, generation of p-benzoquinone amid the biodegradation of chlorophenol or 4-chloronitrobenzene and collection of tert-butyl liquor amid the change of methyl tert-butyl ether (MTBE) [21, 22].

For viable BTF treatment, the contaminants must be biodegradable and non-toxic or have no poisonous quality to BTF. The foremost fruitful toxin expulsion utilising BTFs has been accomplished for low atomic weight and exceedingly solvent natural compounds with straightforward bond structures. Compounds with complex bond structures more often than not require more energy, which, now and then, isn't continuously accessible to the microorganisms, and as a result, small or no biodegradation of these compounds happens. It must be noted that natural compounds such as alcohols, aldehydes, ketones, and a few basic aromatics have amazing biodegradability; phenols, chlorinated hydrocarbons, PAHs, and exceedingly halogenated hydrocarbons show direct to moderate debasement, and certain anthropogenic compounds may not biodegrade at all unless a few extra components, such as vital proteins, are included.

7.3.3.2 Microorganisms in BTF Systems

The adequacy of a BTF depends on the capacity of the microorganisms to biodegrade toxins. Microorganisms vary in their corruption execution in treating a particular toxin. The biofilms in BTFs more often than not comprise high numbers of microbes and low numbers of organisms, hence, most articles focus on the investigation of bacterial communities [23]. Microbes of the genera Pseudomonas, Bacillus, Staphylococcus and Rhodococcus are most regularly found in BTFs. Pseudomonas has been recognised as the dominant species of the bacterial populace in several bioreactors used to expel nitrogen, H_2S and numerous VOCs [24–26]. Bacillus can happen simultaneously under high-impact nitrification-denitrification conditions, Staphylococcus can decrease nitrate to nitrite [27], and Rhodococcus has the capacity to metabolise destructive natural toxins, including toluene, naphthalene, herbicides, and other compounds.

In spite of the fact that toxin corruption in BTFs is more often than not ascribed to bacteria, now and then, organisms may play a vital part. A few studies have demonstrated that organisms display greater VOC evacuation performance than microscopic organisms. Moreover, fungi permit simple assimilation of numerous VOCs from the bulk gas phase because of their filamentous structures with ethereal mycelia and a large surface range. In the meantime, the resistance of fungi to low stickiness favours mass exchange of hydrophobic VOCs from the gas phase to parasitic surfaces. Compared with microbes, parasites can better resist the situations of low stickiness and great sharpness. Therefore, utilising parasites as the most corruption microorganism in BTFs has great potential.

Filamentous organisms in the genera Scedosporium, Paecilomyces, Cladosporium, and Cladophialophora, white-rot parasites, and yeasts of the type Exophiala are regularly gotten in BTFs [28]. Jin et al. [29] demonstrated that a fungi-dominated BTF showed great potential to resist stun loads within the treatment of alpha-pinene and might quickly accomplish full execution after a three-to-seven-day starvation period. Besides, organisms can evacuate profoundly concentrated VOCs in exceedingly acidic conditions. However, parasites have a few downsides. Compared to microscopic organisms, fungi have by and large lower metabolic rates that make the start-up periods of BTFs much longer, and their filamentous structure frequently leads to clogging in BTFs.

To make strides, the evacuation execution of BTF, bacteria-dominated BTF, and fungi-dominated BTF can be combined. Cheng et al. [30] set up three distinctive BTFs (bacterial BTF, parasitic BTF, and bacterial-and-fungal BTF) to ponder the distinction with the treatment of toluene. The bacterial-and-fungi BTF come to an RE of 90%, whereas the RE of the contagious BTF and the bacterial BTF were 60% and 20%, respectively.

7.3.3.3 Other Procedures Combined with BTF Systems

Combination of a BTF framework with other sorts of treatment can achieve much superior evacuation execution than the use of a BTF system alone. A few microorganisms can create distinctive sorts of enzymes such as monooxygenase, which can upgrade the debasement of refractory VOCs to straightforward atoms. Co-metabolism can offer assistance to the microorganisms to invigorate the discharge of substrate oxidising proteins. Quan et al. [31] utilised an inactive attractive field combined with BTF to treat TCE waste gas and compared this combined technique to a single BTF. The resulting demonstrated progressed TCE removal performance and change within the bacterial community beneath 60 mT of magnetic field escalated. Attractive field fortifying can create similar effects of poison evacuation with the expansion of phenol and sodium acetate as co-metabolic substrate. Besides, distinctive magnetic field power altogether influenced the contagious community in the BTF frameworks and progressed the wealth of the phylum Ascomycota, thus expanding the TCE expulsion rate [32]. Comparative to using proteins, pre-treatment of headstrong VOCs may advance the metabolism of toxins by microorganisms. For example, in the treatment of a few fragrant compounds such as styrene and chlorobenzene, UV pre-treatment is ordinarily utilised earlier in a BTF framework. The UV vitality can straightforwardly change the hydrophobic and stable VOCs into water-soluble and effectively biodegradable intermediates, making it a productive pre-treatment step before the BTF treatment. As a result, both mass exchange and response rate are improved by UV pre-treatment since it changes toxins into more solvent and biodegradable compounds. Other than UV, plasma or other photolysis pre-treatment methodologies can also be connected.

7.3.4 BIOSCRUBBERS

In bioscrubbers, the sullied gas is treated by means of two ways. In the first step, the sullied stream is mixed with water in a reactor pressed with inactive media, coming about in assimilation of contaminant to the fluid stage. The fluid is at that point put into an enacted slime reactor or any organic unit where the contaminants are organically corrupted. The treated liquid effluent from the bio-reactor (after clarification) is re-circulated to the primary reactor [33]. Subsequently, the reactor permits the bioscrubber to treat higher concentrations of VOCs than biofilters. Since retention and biodegradation happen in two different reactors, these reactors can be optimised. Though biotrickling channels and bioscrubbers are often more prominent administrator control over pH, nutrient and wash out of corruption by-products since of nearness of fluid stage.

The main disadvantage of bioscrubbers over biofilters is transfer of abundance slime/profluent. As the system depends on retention, increasingly water dissolvable poisons can be effectively treated. Since a huge portion of the VOCs that show in ETP or STP emanations are modestly hydrophobic in nature, bioscrubbing at that point ended up less prevalent [33]. In a bioscrubber, the inlet VOC or odour stacking for the most part is < 5 g m^{-3} [34].

Rene at al. [35] examined the BTEX expulsion in a fungi-dominated biofilter and in general evacuation productivity found it to be 35–97% beneath different operating conditions. Zamir et al. [36] conducted a compost biofilter try with toluene after immunisation with an extraordinary sort of white-rot organism Phanerochaetechrysosporium and found 92% diminished effectiveness. Li et al. [37] created a styrene biofilter test containing PU froth as media and the evacuation effectiveness was $> 96\%$. Chen et al. [38] planned a biofilter for toluene evacuation obtaining suspended biofilm. The expulsion proficiency obtained was $> 90.2\%$ after 14 days of start-up time and 128 days of working time. Rene et al. [39] conducted a compost biofilter test with benzene and toluene

Biological-Based Methods for Removal of VOCs and Heavy Metals 117

stacking by shifting channel concentrations and the expulsion was 72.7% for benzene and 81.1% for toluene individually. Natarajan et al. [40] were taking care of an ethylbenzene and xylene blend in a biofilter having blended microbial culture with tree bark media. The expulsion effectiveness of 58–78% and 68–89% were recorded for xylene and ethylbenzene for a nonstop 96 days of biofilter operation. Rahul et al. [41] conducted a biofilter operation with corncob as channel media with BTEX and saw more than 99% of removal. Gallastegui et al. [42] worked a toluene and p-xylene biofilter pressed with inert material. They watched hindrance of p-xylene in proximity of toluene whereas the proximity of pxylene upgraded the toluene evacuation. Li et al. [37] outlined a coordinate bioreactor system that's a gas division film module introduced after a control biofilter. Due to combining, the expulsion of styrene was upgraded and the general framework productively handled for the fluctuating channel stack. Moreover, within the film module, styrene was condensed and recovered back to biofilter, which in turn amplified the maintenance time. Schiavon et al. [43] recently conducted the try with non-warm plasma (NTP) at the upstream of a biofilter for expulsion of blended VOCs. The NTP utilising the diverse particular vitality densities reduced the VOC concentrations down to the ideal level. Also, plasma treatment converted non-water solvent VOCs to more dissolvable compounds. They utilised NTP effectively for pretreatment some time recently biofiltration.

7.3.5 BIOFILTRATION

Biofiltration for the most part alludes to natural treatment or change of organic or inorganic contaminants into safe compounds whether liquid or gas stage. In spite of the fact that biodegradation in range of wastewater treatment and bio-remediation methods are broadly connected for treatment of soil and ground water, in recent decades biofiltration has been developing as a treatment for VOC expulsion and mechanical application. Most of the biofilters built as an open single bed frameworks® for treating compost gas or odour and continuously developing as VOC expulsion procedures for mechanical application. Within the biofiltration, the sullied stream is dampened and pumped to the bio channel bed. Whereas the stream gradually streams through the filter media, the contaminants within the stream are retained and metabolised. Organic oxidation by microorganisms can be composed as:

$$\text{Organic pollutant} + O_2 + \text{Microorganisms} ® O_2 + H_2O + \text{Heat} + \text{Biomass}$$

Biofiltration is favoured over the ordinary control strategies because of its lower capital and working taken a toll, low chemical utilisation, adaptability in plan, can expel a wide run of contaminants, tall treatment productivity and can be tailor made. Be that as it may, as this can be living it is exceptionally sensitive to the environment and cannot handle great vacillation in stack or extraordinary climate. A few components contribute to the biofiltration handle like channel media, temperature, stickiness, pH, home time, weight drop, supplement, etc. Biofiltration is effective for the evacuation or annihilation of many off-gas poisons, especially natural compounds counting inorganic compounds such as H_2S and NH_3 [44]. Biofiltration is exceptionally compelling and prudent when the volume is tall and contaminant is low.

Essentially, contaminants that can be treated by biofiltration might have the following characteristics [44, 34]:

1. High water solvency – Compounds having great water solvency diffuse to the water layer of the biofilm more effectively, which is required for its biodegradation. Compounds having higher esteem of Henry's steady (H, mol m^{-3} Pa^{-1}) and lower vapor weight display higher water solubility.
2. Ready biodegradability – After the compounds enter the water film (biofilm), it must be biodegraded, something else the concentration of the same will increment in biofilm

which may be inconvenient to microorganisms additionally the encourage dissemination into biofilm may be detrimental to microorganisms additionally the encourage dissemination into biofilm may be diminished. Compounds having lower water solvency and lower biodegradability like halogenated hydrocarbons etc. can moreover be treated in a biofilter. Be that as it may, the treatment proficiency depends on planning specific environments.

Evacuation of contaminants in a biofilter may be a multi-step process beginning with dissemination of contaminants from the gas stage to the fluid phase/biofilm and after that the cell surface of microorganisms and at long last biodegradation by microorganisms (basically microscopic organisms and parasites). Major forms included in biofiltration are outlined as follows [34, 45]:

As the discuss passes through pressing bed, contaminants are exchanged from the gas stream to the water within the biofilm. Once retained within the biofilm layer or broken up within the water layer around the biofilm, the natural compound is accessible as nourishment for the microorganism digestion system, serving as a carbon and vitality source to maintain microbial life and development transported to microbial cells and corrupted. For exceedingly dissolvable compounds, major expulsion happens in water broken down frame, though for hydrophobic contaminants, the major expulsion component may be adsorption on the surface of the medium [34] and

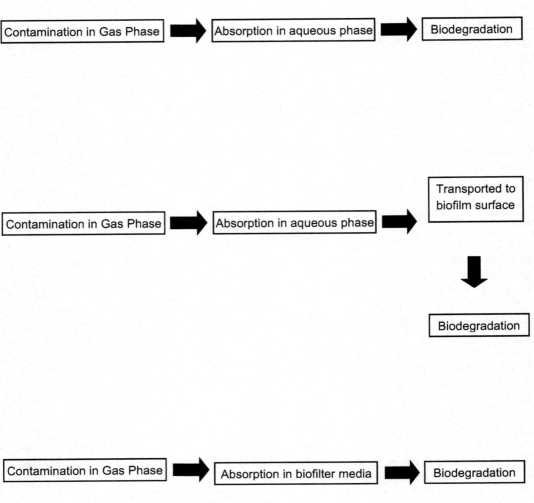

FIGURE 7.4 Biofiltration of contaminants in a biofilter

ensuing biodegradation inside the microorganism cells. The organic deterioration rate depends on concentration of biomass and the particular development rate coefficient. Control parameters in bioreactors are: the nature of pressing fabric, gulf stream rate, temperature, pH, mugginess, and inorganic supplements. In spite of the fact that in biofiltration framework both microscopic organisms and parasites utilised and now and then specialised bacterial or parasitic culture are connected, most of the biofiltration considerations are based on microbial duplication, which is commonly found in biofilters [33].

Biofilters are the best and most seasoned reactor frameworks among the three vapor-phase bioreactors for the treatment of VOCs. VOC-loaded waste gas is passed through a biofilm immobilised back media and are changed over by the microorganisms into carbon dioxide, water and extra biomass. Waste gas is humidified prior to directing it to the bioreactor, which is critical for treatment. The media bed has organic packing fabric like compost, peat, wood or bark chips, etc. or manufactured like plastic, ceramic, etc. Microorganisms create a biofilm onto this pressing media bolster, which (organic media) can moreover be a source for additional supplements to back microbial development. When a synthetic media bed is utilised, supplements ought to be included for microbial development. For organic packing, moreover, supplements are required after a few period as the supplements in media start to dwindle. Supplements are for the most part splashed irregularly on the pressing material. A sprinkling of water over the channel media is additionally connected irregularly to guarantee suitable moisture substance inside the pressing media and aid the wash-out of debasement by-products [33]. In a biofilter, channel-stacking of VOC or odor for the most part is < 1 g m^{-3} [34]. In spite of the fact that biofilters show odor diminishment efficiencies continuously more prominent than 80%, VOC expulsion efficiencies are 20–90% [33]. Subsequently, the operational parameters ought to be kept up securely to guarantee compelling expulsion of VOCs.

7.3.6 NTP AND UV-BTF COORDINATES INNOVATION FOR CL-VOCS REMOVAL

For less-biodegradable and less-water-soluble CleVOCs, NTP oxidation and UV photo-oxidation are included as pretreatment steps to change over CleVOCs to biodegradable intermediates, which may guarantee the intensive treatment of CleVOCs waste gas [46–49]. Jiang et al. [47] inquired about a NTP-BTF framework (RE 97.6%), which had stronger ability to evacuate CB than single-BTF (RE 68.7%, EBRT 60 s) and single-NTP (RE around 59%, EBRT 4.32 s) at low-energy conditions (SIE (particular input vitality) 3,500 J/L). Besides, the NTP pretreatment may be a common-sense strategy for restraining large development of biomass with a small impact on the BTF execution due to the fact that the biomass development in NTP-BTF was slower than within the single-BTF. Jiang et al. and Zhu et al. [50, 51] combined DBD (dielectric barrier discharge, one of the ordinary frameworks for NTP) with BTF for purification of waste discuss containing 1,2-DCA and CB individually. The results showed that the nearness of the catalyst significantly reduced the number of by-products, and the water solvency and biodegradability of the intermediates were altogether improved. In expansion, the test showed that the biodiversity and plenitude of the microbial community within the NTP-BTF system were lower than that in a single BTF framework. This may be because the intermediates shaped by NTP pretreatment influence the development of microorganisms in BTF, or the ozone created by NTP pretreatment kills the microorganisms. Yu et al. [52] coupled UV photooxidation with BTF to expel DCM from squander discuss. Compared with single-BTF, it displayed higher RE of DCM (81.78%) with inlet concentration of 750 mg/m^3, and UV-BTF had higher microbial diversity. UV photo-oxidation might change over hydrophobic VOCs to water-soluble and biodegradable intermediates, which may be carbonyl compounds and unstable greasy acids. These components may have a positive impact on mass exchange of the parent compound and microorganism development, which could greatly decrease the time for biofilm arrangement [46]. The homogeneity of the bacterial community makes a difference in keeping up the high performance of the bioreactor.

7.4 CONCLUSION

Literature indicates that heavy metals and volatile organic compounds being released into the environment is unavoidable. Conversely, their ample treatment is essential to save the environment. The release of these compounds into the environment causes disturbance of aquatic flora and fauna, a risk to human health and also air pollution. Published investigation revealed that biological methods in comparison to other processes could be an efficient treatment process for their removal. Based on the present study, bioreactors (biofilter, biotrickling filter, and bioscrubber) can be used as potential technology for removal of VOCs as compared to the other available elimination techniques. However, use of biological systems to remove VOCs still has limitations and challenges. Additionally biofilters have emerging applications for the treatment of heavy-metals contaminated wastewater whereas BES for heavy metal eradication is still in the development stage, and as a consequence to understand the microbe metal interactions more research is required. Electron transfer mechanisms, and electrode materials to enhance the efficiency of bioelectrochemical processes. Hence advancement in the field of research is required for the development of optimised eco-friendly as well as economically feasible technology to protect the planet for future generations.

REFERENCES

[1] Barthe P, Chaugny M, Roudier S, Delgado Sancho L. Best available techniques (BAT) reference document for the refining of mineral oil and gas. *European Commission*. 2015;754.

[2] Version–Corrected FI. *Emission Estimation Protocol for Petroleum Refineries*, 2011.

[3] Malakar S, Saha PD. Estimation of VOC emission in petroleum refinery ETP and comparative analysis with measured VOC emission rate. *The IJES*. 2015;4(10):20–29.

[4] Dermentzis K, Christoforidis A, Valsamidou E, Lazaridou A, Kokkinos N. Removal of hexavalent chromium from electroplating wastewater by electrocoagulation with iron electrodes. *Global NEST Journal*. 2011;13(4):412–418.

[5] van der Maas P, Peng S, Klapwijk B, Lens P. Enzymatic versus nonenzymatic conversions during the reduction of EDTA-chelated Fe (III) in BioDeNO x reactors. *Environmental Science & Technology*. 2005;39(8):2616–2623.

[6] Francis AJ, Nancharaiah YV. In situ and ex situ bioremediation of radionuclide-contaminated soils at nuclear and norm sites. In *Environmental Remediation and Restoration of Contaminated Nuclear and Norm Sites* (pp. 185–236). Woodhead Publishing, 2015.

[7] Patil SA, Arends JB, Vanwonterghem I, Van Meerbergen J, Guo K, Tyson GW, Rabaey K. Selective enrichment establishes a stable performing community for microbial electrosynthesis of acetate from $CO2$. *Environmental Science & Technology*. 2015;49(14):8833–8843.

[8] Ter Heijne A, Hamelers HV, De Wilde V, Rozendal RA, Buisman CJ. A bipolar membrane combined with ferric iron reduction as an efficient cathode system in microbial fuel cells. *Environmental Science & Technology*. 2006;40(17):5200–5205.

[9] Strycharz-Glaven SM, Glaven RH, Wang Z, Zhou J, Vora GJ, Tender LM. Electrochemical investigation of a microbial solar cell reveals a nonphotosynthetic biocathode catalyst. *Applied and Environmental Microbiology*. 2013;79(13):3933–3942.

[10] Malyan SK, Kumar SS, Singh L, Singh R, Jadhav DA, Kumar V. Bioelectrochemical systems for removal and recovery of heavy metals. In *Bioremediation, Nutrients, and Other Valuable Product Recovery* (pp. 185–203). Elsevier.

[11] Nies DH. Microbial heavy-metal resistance. *Applied Microbiology and Biotechnology*. 1999;51(6): 730–750.

[12] Javanbakht V, Alavi SA, Zilouei H. Mechanisms of heavy metal removal using microorganisms as biosorbent. *Water Science and Technology*. 2014;69(9):1775–1787.

[13] Qin H, Hu T, Zhai Y, Lu N, Aliyeva J. The improved methods of heavy metals removal by biosorbents: A review. *Environmental Pollution*. 2020;258:113777.

[14] Arıca MY, Kacar Y, Genç Ö. Entrapment of white-rot fungus Trametes versicolor in Ca-alginate beads: Preparation and biosorption kinetic analysis for cadmium removal from an aqueous solution. *Bioresource Technology*. 2001;80(2):121–129.

Biological-Based Methods for Removal of VOCs and Heavy Metals

[15] Arıca MY, Arpa C, Ergene A, Bayramoğlu G, Genç Ö. Ca-alginate as a support for Pb (II) and Zn (II) biosorption with immobilized Phanerochaetechrysosporium. *Carbohydrate Polymers*. 2003;52(2):167–174.

[16] Xu P, Zeng GM, Huang DL, Feng CL, Hu S, Zhao MH, Lai C, Wei Z, Huang C, Xie GX, Liu ZF. Use of iron oxide nanomaterials in wastewater treatment: A review. *Science of the Total Environment*. 2012;424:1–10.

[17] Xu P, Zeng GM, Huang DL, Lai C, Zhao MH, Wei Z, Li NJ, Huang C, Xie GX. Adsorption of Pb (II) by iron oxide nanoparticles immobilized Phanerochaetechrysosporium: Equilibrium, kinetic, thermodynamic and mechanisms analysis. *Chemical Engineering Journal*. 2012;203:423–431.

[18] Lebrero R, Estrada JM, Muñoz R, Quijano G. Toluene mass transfer characterization in a biotrickling filter. *Biochemical Engineering Journal*. 2012;60:44–49.

[19] Wu H, Yan H, Quan Y, ZHao H, JiaNg N, Yin C. Recent progress and perspectives in biotrickling filters for VOCs and odorous gases treatment. *Journal of Environmental Management*. 2018;222:409–419.

[20] Wu H, Yin Z, Quan Y, Fang Y, Yin C. Removal of methyl acrylate by ceramic-packed biotrickling filter and their response to bacterial community. *Bioresource Technology*. 2016;209:237–245.

[21] Purswani J, Juárez B, Rodelas B, Gónzalez-López J, Pozo C. Biofilm formation and microbial activity in a biofilter system in the presence of MTBE, ETBE and TAME. *Chemosphere*. 2011;85(4):616–624.

[22] Skiba A, Hecht V, Pieper DH. Formation of protoanemonin from 2-chloro-cis, cis-muconate by the combined action of muconatecycloisomerase and muconolactone isomerase. *Journal of Bacteriology*. 2002;184(19):5402–5409.

[23] Zhao L, Huang S, Wei Z. A demonstration of biofiltration for VOC removal in petrochemical industries. *Environmental Science: Processes & Impacts*. 2014;16(5):1001–1007.

[24] Giri BS, Kim KH, Pandey RA, Cho J, Song H, Kim YS. Review of biotreatment techniques for volatile sulfur compounds with an emphasis on dimethyl sulfide. *Process Biochemistry*. 2014;49(9):1543–1554.

[25] Li Y, Zhang W, Xu J. Siloxanes removal from biogas by a lab-scale biotrickling filter inoculated with Pseudomonas aeruginosa S240. *Journal of Hazardous Materials*. 2014;275:175–184.

[26] Zheng M, Li C, Liu S, Gui M, Ni J. Potential application of aerobic denitrifying bacterium Pseudomonas aeruginosa PCN-2 in nitrogen oxides (NOx) removal from flue gas. *Journal of Hazardous Materials*. 2016;318:571–578.

[27] Cheng CY, Mei HC, Tsao CF, Liao YR, Huang HH, Chung YC. Diversity of the bacterial community in a bioreactor during ammonia gas removal. *Bioresource Technology*. 2010;101(1):434–437.

[28] Repečkienė J, Švedienė J, Paškevičius A, Tekorienė R, Raudonienė V, Gudeliūnaitė E, Baltrėnas P, Misevičius A. Succession of microorganisms in a plate-type air treatment biofilter during filtration of various volatile compounds. *Environmental Technology*. 2015;36(7):881–889.

[29] Jin Y, Guo L, Veiga MC, Kennes C. Fungal biofiltration of α-pinene: Effects of temperature, relative humidity, and transient loads. *Biotechnology and Bioengineering*. 2007;96(3):433–443.

[30] Cheng Z, Lu L, Kennes C, Yu J, Chen J. Treatment of gaseous toluene in three biofilters inoculated with fungi/bacteria: Microbial analysis, performance and starvation response. *Journal of Hazardous Materials*. 2016;303:83–93.

[31] Quan Y, Wu H, Yin Z, Fang Y, Yin C. Effect of static magnetic field on trichloroethylene removal in a biotrickling filter. *Bioresource Technology*. 2017;239:7–16.

[32] Quan Y, Wu H, Guo C, Han Y, Yin C. Enhancement of TCE removal by a static magnetic field in a fungal biotrickling filter. *Bioresource Technology*. 2018;259:365–372.

[33] Omil F. *Biological Technologies for the Removal of VOCs, Odours and Greenhouse Gases*. Departmente of Chemical Engineering, University of Santiago de Compostela, 2014.

[34] Frederickson J, Boardman CP, Gladding TL, Simpson AE, Howell G, Sgouridis F. *Evidence: Biofilter Performance and Operation as Related to Commercial Composting* Environment Agency, Bristol. 2013.

[35] Rene ER, Mohammad BT, Veiga MC, Kennes C. Biodegradation of BTEX in a fungal biofilter: Influence of operational parameters, effect of shock-loads and substrate stratification. *Bioresource Technology*. 2012;116:204–213.

[36] Zamir SM, Halladj R, Nasernejad B. Removal of toluene vapors using a fungal biofilter under intermittent loading. *Process Safety and Environmental Protection*. 2011;89(1):8–14.

[37] Li L, Lian J, Han Y, Liu J. A biofilter integrated with gas membrane separation unit for the treatment of fluctuating styrene loads. *Bioresource Technology*. 2012;111:76–83.

[38] Chen X, Qian W, Kong L, Xiong Y, Tian S. Performance of a suspended biofilter as a new bioreactor for removal of toluene. *Biochemical Engineering Journal*. 2015;98:56–62.

[39] Rene ER, Kar S, Krishnan J, PaKshirajan K, López ME, Murthy DV, Swaminathan T. Start-up, performance and optimization of a compost biofilter treating gas-phase mixture of benzene and toluene. *Bioresource Technology.* 2015;190:529–535.

[40] Natarajan R, Al-Sinani J, ViSwanathan S, Manivasagan R. Biodegradation of ethyl benzene and xylene contaminated air in an up flow mixed culture biofilter. *International Biodeterioration & Biodegradation.* 2017;119:309–315.

[41] Mathur AK, Balomajumder C. Biological treatment and modeling aspect of BTEX abatement process in a biofilter. *Bioresource Technology.* 2013;142:9–17.

[42] Gallastegui G, Ramirez AÁ, ElíAs A, Jones JP, Heitz M. Performance and macrokinetic analysis of biofiltration of toluene and p-xylene mixtures in a conventional biofilter packed with inert material. *Bioresource Technology.* 2011;102(17):7657–7665.

[43] Schiavon M, Schiorlin M, Torretta V, Brandenburg R, Ragazzi M. Non-thermal plasma assisting the biofiltration of volatile organic compounds. *Journal of Cleaner Production.* 2017;148:498–508.

[44] Adler SF. Biofiltration – a primer. *Chemical Engineering Progress.* 2001;97(4):33–41.

[45] Berenjian A, Chan N, Malmiri HJ. Volatile organic compounds removal methods: A review. *American Journal of Biochemistry and Biotechnology.* 2012;8(4):220–229.

[46] Den W, Ravindran V, Pirbazari M. Photooxidation and biotrickling filtration for controlling industrial emissions of trichloroethylene and perchloroethylene. *Chemical Engineering Science.* 2006;61(24): 7909–7923.

[47] Jiang L, Li H, Chen J, Zhang D, Cao S, Ye J. Combination of non-thermal plasma and biotrickling filter for chlorobenzene removal. *Journal of Chemical Technology & Biotechnology.* 2016;91(12):3079–3087.

[48] Shah MP. *Removal of Emerging Contaminants through Microbial Processes.* Springer, 2021.

[49] Shah MP. *Advanced Oxidation Processes for Effluent Treatment Plants.* Elsevier, 2020.

[50] Jiang L, Li S, Cheng Z, Chen J, Nie G. Treatment of 1, 2-dichloroethane and n-hexane in a combined system of non-thermal plasma catalysis reactor coupled with a biotrickling filter. *Journal of Chemical Technology & Biotechnology.* 2018;93(1):127–137.

[51] Zhu R, Mao Y, Jiang L, Chen J. Performance of chlorobenzene removal in a nonthermal plasma catalysis reactor and evaluation of its byproducts. *Chemical Engineering Journal.* 2015;279:463–471.

[52] Jianming Y, Wei L, Zhuowei C, Yifeng J, Wenji C, Jianmeng C. Dichloromethane removal and microbial variations in a combination of UV pretreatment and biotrickling filtration. *Journal of Hazardous Materials.* 2014;268:14–22.

8 Bacterial Biofilters for Arsenic Removal

Rahul Nitnavare, Joorie Bhattacharya, and Sougata Ghosh

CONTENTS

8.1 Introduction 123
8.2 Arsenic Toxicity 124
8.3 Removal of Arsenic by Bacteria and Mechanism Involved 125
    8.3.1 Future Prospects 132
8.4 Conclusion 133
References 134

8.1 INTRODUCTION

Arsenic (As) is a metalloid and is derived from both natural as well as anthropogenic sources. It is widely found in the earth's crust and can be recovered from several minerals out of which arseno-pyrite, orpiment, and realger are the most prominent sources. In these minerals, arsenic occurs in the oxidation state As (III) formed to the reduced conditions. When oxidised arsenic converts to As (V) found in iron arsenate and calcareous arsenolite. Arsenic contamination and toxicity are global problems and several regions have been affected due to this. The primary accountable source of arsenic contamination is due to groundwater pollution by minerals. Apart from this a substantial amount of arsenic also occurs from anthropogenic sources such as mining and industrial wastes. Arsenic has also been introduced into the ecosystem via its usage in agricultural products in the form of pesticides as well as preservatives. The speciation of arsenic is greatly dependent on the physio-chemical condition of the environment it exists in, relying on factors such as pH and redox conditions (Garelick et al., 2008). The inorganic forms, i.e., As (III) and As (V) are comparatively more toxic than the organic forms of arsenic. As mentioned earlier, groundwater contamination of arsenic is the major source causing chronic health problems. Also, the usage of the same groundwater for agricultural purposes has led to its introduction into the biological system. Further, the continuous usage of polluted groundwater for the same purpose has led to the increase in arsenic concentration in the soil strata as well making them unsuitable for agriculture. Long term exposure to arsenic can cause severe skin diseases along with affecting the lungs, and kidneys as well as being a causal factor for cancer of various organs. Arsenic toxicity is therefore a serious issue and needs to be mitigated efficiently before it penetrates the ecosystem (Srivastava et al., 2011).

There are various techniques through which the contamination of arsenic can be addressed. Traditional techniques include finding water sources free of arsenic such as switching water sources (wells). The usage of surface water is yet another alternative that has been taken up. Surface water usually contains lesser concentrations of arsenic and therefore can be used as safe drinking water. Utilisation of rainwater has also been practiced in several regions around the world for arsenic-affected areas. These methods, however, come with the added expense of durability, feasibility, and cost-effectiveness, due to which reforms for arsenic contamination remediation are required. Therefore, physical methods of arsenic removal are believed to provide a more permanent solution to arsenic contamination. Oxidation of As (III) to As (V) and then further precipitation of the arsenate can be achieved via various oxidative agents such as ozone, oxygen, and manganese

DOI: 10.1201/9781003165149-8

123

dioxide polished sand as well as photochemical means of oxidation using ultraviolet (UV) or solar light. Another method of physical means of arsenic removal is coagulation and flocculation. This includes using iron- (Fe) and aluminium- (Al) based coagulants. Due to the copious amounts of contaminated sludge and waste produced in this method, adsorption of arsenic on activated surfaces was adopted as a method. For this purpose, Fe- and Al-based oxides and hydroxides are used, which are cost effective in nature and efficient as well (Shankar et al., 2014). However, with the evolving technologies, a technique for arsenic removal was required that was environmentally friendly along with being efficient in nature. Taking this into context, microbial removal of arsenic came into perspective wherein various kinds of microorganisms were explored for bioremediation properties. Microbial transformation of arsenic includes a series of processes of assimilatory and dissimilatory mechanisms wherein the metal is taken up by the microbial cell followed by detoxification through oxidation or reduction. Bacterial oxidation is an alternative to chemical oxidation and the process allows the bacteria to tolerate high concentrations of arsenic. The As(V) that is then formed can be removed from wastewater using conventional and physical methods. Such processes are temperature- and pH-dependent and show maximum bioremediation capacities under optimum conditions. A wide range of bacterial species have been reported that are capable of arsenic removal including autotrophic and heterotrophic bacteria. These bacteria are known to possess certain oxidases that help in the oxidation process. Several alpha, beta, and gamma-proteobacteria have been studied that contain these oxidases. Further sequencing of the genomes of these bacteria would provide a more extensive fundamental and applied perspective to the mechanism of arsenic removal by these species. It would also provide an insight into their behaviour under various environmental conditions (Ike et al., 2008; Cavalca et al., 2013).

8.2 ARSENIC TOXICITY

Groundwater contamination due to arsenic is one of the worst health and environmental hazards globally. According to the World Health Organization (WHO), the maximum permissible limit for arsenic is $<10\mu g/L$ (WHO, 2006). Ingestion of the inorganic form of arsenic leads to serious chronic health issues such as skin lesions and cancer of the skin, lungs, kidneys, and liver. It also causes other health problems such as respiratory, cardiovascular, and neurological anomalies as well as hypertension. After ingestion into the biological system, arsenate is reduced to arsenite through enzymes and then into mono-methylarsenate by the addition of a methyl group. This is then subsequently converted into mono-methylarsonous acid (MMA[III]). Further methylation into dimethylarsinic acid (DMA[V]) may also occur followed by reduction into dimethylarsinous acid (DMA[III]). Both of these components have been found in the urine of populations exposed to arsenic-contaminated waters (Bjørklund et al., 2018). Continuous ingestion of arsenic is known to have long-term effects, which may also lead to death. Based on the time of exposure and concentration of arsenic consumed, the toxicity of arsenic is defined as acute and sub-acute.

As previously mentioned, the trivalent and pentavalent states of arsenic oxidation are the most toxic. The elevated toxicity of As(III) is due to its ability to bind to sulfhydryl groups of cysteine residues and subsequently inactivating them. The toxicity of an element is dependent upon its ability to inhibit and impairment of the enzyme activity. Arsenite also hinders biological metabolic processes such as gluconeogenesis, fatty acid oxidation, uptake of glucose into the cell, and the production of acetyl coenzyme A (CoA; Jomova et al., 2011). While arsenate is not able to directly bind to enzyme active sites, its reduced form, arsenite, can make the latter more toxic. In itself, the toxicity of arsenate comes from being converted to arsenite. On the other hand, while not being highly toxic, it hinders biological processes such as phosphate metabolisms. This in turn can affect DNA repair mechanisms. In the environment, however, due to the abundance of phosphates, the cumulative effect of arsenate is diminished. Arsenate also has a similar structure to phosphate, and it often replaces it in several essential biochemical pathways such as anion exchange in sodium pump. Arsenate also hinders the formation of ATP (adenosine-5'-triphosphate) by a process called

Bacterial Biofilters for Arsenic Removal 125

arsenolysis at both substrate and mitochondrial levels causing serious damage at the cellular levels (Hughes, 2002).

In the human body, the effects of arsenic toxicity are observed in diseases such as bronchitis, non-malignant pulmonary diseases, cardiovascular abnormalities, gastritis and colitis, anaemia and leukopaenia, chronic hepatic and renal diseases, encephalopathy, congenital malformations, and dermal diseases including cancer. The localisation of arsenic is seen in high concentrations in the skin due to the presence of keratin. Keratin is rich in sulfhydryl groups, which get bound by arsenite, causing their accumulation in these tissues (Hare et al., 2018). The metabolism of inorganic arsenic is defined by the reduction of pentavalent arsenic to its trivalent state aided by glutathione and then to the organic pentavalent arsenic by oxidative methylation. The methylation of inorganic arsenic is a means of detoxification; however, the intermediate components formed during conversion, such as MMA^{III} and DMA^{III}, have been found in the urine of individuals exposed to arsenic (Jomova et al., 2011). The trivalent radicals MMA^{III} and DMA^{III} bind with thiol proteins such as cysteine and are known to show greater binding affinity to dithiols as compared to monothiols. In addition, binding to these essential proteins causes hindrance of cellular pathways. Binding to non-essential proteins may be a means of detoxification of trivalent arsenic (Hughes, 2002).

Arsenic is an extremely lethal element and causes serious damage to the human body as mentioned earlier. It has various modes of action to impart toxicity such as genotoxicity, altered DNA repair and methylation, enhanced oxidative stress, uncontrolled cell proliferation, impaired cell signalling, and oncogene amplification (Hughes, 2002). Even though several studies have been conducted describing the effects of arsenic on human health, the effect is relative to the dose and risk to the individual. Therefore, an elaborate understanding of the mechanism will eliminate risk factors associated with it.

8.3 REMOVAL OF ARSENIC BY BACTERIA AND MECHANISM INVOLVED

The conventional methods that have been adopted, such as chemical precipitation, reverse osmosis, chemically induced oxidation and reduction, and filtration, even though they can prove to be effective to a certain extent, have their share of disadvantages. Such methods of arsenic removal also are not able to remove lower concentrations of the metal. Due to this, bioremediation of heavy metals using microorganisms has emerged as a highly sought after technique owing to its potential. Additionally, microorganisms such as bacteria possess the unique ability to convert arsenic to its other oxidation states and also play an important role in governing various aspects of arsenic cycle in the environment. Under metal stress conditions, bacteria have evolved several mechanisms to tolerate high concentrations of heavy metal such as oxidation/reduction of the metal to a less toxic state, biosorption and bioaccumulation into the cell, and efflux of metal. Certain arsenic-resistant bacteria exist in the environment that use arsenic as a part of an energy-generating metabolism (Banerjee et al., 2011). Heterotrophic bacteria directly oxidise As (III; HAOs) while certain other bacteria such as chemolithoautotrophic As(III) oxidisers (CAOs) use arsenite as an electron donor. Further, chemotrophic conversion of As(III) occurs via aerobic respiration, phototrophy, and nitrate and selenate-dependent anaerobic respiration. Such conversion of toxic arsenite to arsenate allows gain of energy of the bacteria leading to an ecological advantage of the bacteria. In both HAOs and CAOs, the oxidase enzyme consists of two subunits, a larger subunit that is the molybdopterin centre with a [3Fe-S]cluster and a smaller subunit containing a Rieske [2Fe-S] cluster. Bacteria isolated from arsenic-rich environments have been found to contain an As(III) oxidising gene, *aioA*, which codes for arsenic oxidase enzyme. The larger subunit is denoted as *aioA* and the smaller subunit as *aioB*. This enzyme is found in both autotrophic and heterotrophic bacteria across several genera. An As(V) reducing gene was also identified, *ArrA* gene, which encodes a respiratory reductase. The enzymology and characteristics of *aioA* and *ArrA* have been found to be similar. A novel oxidase gene identified from *Alkalilimnicola ehrlichii* MLHE-1, as *arxA* has also shown sequence similarity to the *aioA* and *ArrA* genes. Similarly, the detoxification of Ar(V) through anaerobic respiration or through reduction has been observed in certain microorganisms, dissimilatory As[V]-respiring prokaryotes [DARPs] and As[V]-resistant microbes[ARMs] respectively. In the case of DARPS, the enzyme ArrA, which is a reductase, is the key enzyme. DARPS can exhibit reduction abilities as heterotrophs that gain energy from oxidation or as chemolithoautotrophs, which gain energy from hydrogen and sulphide. For ARMS, the reduction of A(V) is mediated by the *ArsC* encoded enzyme in the presence of glutaredoxin, glutathione, and thioredoxin. Such mechanisms were observed in a wide range of bacterial phyla. The first DARP strain bacterium was *Sulfurospirillum* spp. Bacteria belonging to this species are known to utilise As(V) as the electron acceptor and to reduce As(V). Some of the bacteria belonging to this species are S. *barnesii,S. multivorans*, and *S. halorespirans* (Cavalca et al., 2013; Yamamura and Amachi, 2014). The *ars* operon consists of three genes, viz, *arsR, arsB*, and *arsC*, which transcript at the same time and are responsible for reductase activity and efflux of arsenic. Apart from these, two other genes are also found, which are *ArsA and ArsD*. Therefore, the whole cluster is known as *arsRDABC* (Sher and Rehman, 2019).

Microorganisms such as bacteria adopt various physiological and biochemical means for the bioremediation of arsenic. In biosorption, the metal ions interact with the cellular wall of the bacteria, which is an ATP-independent process. Since the intake of heavy metals is only on the surface and it does not enter the cell, it protects the cell from any toxic effect. Biosorption is achieved by species such as *Bacillus* and *Rhodococcus*, which was found to be as high as 77.33 mg/g of As(III) within a contact time of 30 min in *Rhodococcus* sp. WB12. Further, the respiratory-reducing bacteria, *Bacillus selenatarsenatis* SF-1, in the presence of anthraquinone-2,6-disulfonate, saw an improved efficiency of As(V) reduction.

Pseudomonas aeruginosa AT-01 was able to remove arsenic with 98% efficiency at 37°C for 24 h at a pH of 7.0 through biosorption. Methylation is known to be a detoxification process; however, extensive studies have not been performed in bacterial species. Methylation of arsenic requires the enzyme *S*-adenosylmethionine (SAM) and a methyltransferase. Methyltransferase such as ArsM is encoded by the *arsM* gene. The conversion of arsenic into methyl arsenicals allows the transport of arsenic out of the bacterial cell. Bacterial species such as *Pseudomonas putida* have been known to express the SAM gene and have the potential for methylation of arsenic. *P. putida* also contains the ArsC protein, which reduces As(V) and ArsB proteins, which facilitates the efflux of As(III) from the bacterial cell. **Bioaccumulation** is an energy-intensive mechanism and regulates the transport of heavy metals outside the cellular membrane using ion exchange pumps, channels, carriers, endocytosis, and lipid permeation. **Bioleaching** is a mechanism that is significantly effective wherein the insoluble forms of toxic metals are converted to the soluble form for the easy removal by microorganisms. **Biomineralisation** can be used as an accessory mechanism to enhance the effectiveness of other mechanisms. It is a precipitation reaction for bioleaching of toxic heavy metals. The Mn oxides generated by *Marinobacter* oxidised As(III) to the less toxic As(V) and decreased the concentration of As(III) from 55 μM to 5.55 μM (Cavalca et al., 2013; Satyapal and Rani, 2016; Hare et al., 2018; Sher and Rehman, 2019; Tariq et al., 2019). Biofilms of bacteria produce extracellular polymeric substance (EPS), which contributes in bioremediation of heavy metals. Arsenic-reducing bacteria, which produce EPS, were isolated from industrial wastewater. The identified bacteria *Exiguobacterium profundum* PT2 and *Ochrobactrum ciceri* SW1 saw an increase in proteins and carbohydrates in EPS under arsenic stress. SEM images demonstrated the presence of absorbed arsenic in the interstitial space as well as alteration in the structure of EPS. Therefore, it was concluded that the bacterial production of EPS aids in absorption of arsenic by serving as a source of biosorbent. The sequestration of toxic metals as such is known to be due to covalent to electrostatic interactions. EPS is thus an eco-friendly and cost-effective alternative for the bioremediation of heavy metals. Such bacterial strains need to be explored further for applications of bioremoval techniques (Saba et al., 2019). Siderophore mediated arsenic tolerance also confer As(V) mobilisation. In a study by Das and Barooah (2018), *Staphylococcus* sp. TA6 showed siderophore production and a subsequent biotransformation

Bacterial Biofilters for Arsenic Removal

capacity of As(V) into As(III) within 72 h at 55°C and pH of 5.5. TA6 had a siderophore activity of 78.7 ± 0.004 μmol and exhibited a transformation efficiency of 88.2%. The study demonstrated the role of siderophore-producing bacteria in arsenic bioremediation and also provided insight into its role in arsenic mobilisation.

In a more recent discovery, the green synthesis of iron-nanoadsorbents by bacteria and other microorganisms was reported. In this, FeOOH nanoparticles generated by bacterial strains such as *Klebsiella oxytoca* BAS-10 showed bioremediation capacity due to 95% removal of As(V) in FeEPS hydrogel and dried powder FeEPS in a ratio of 1:5 (Casentini et al., 2015; Fazi et al., 2016).

Several strains of bacteria have hence been isolated from arsenic contaminated sites and identified. Most of the bacteria belong to genera such as *Bacillus, Pseudomonas, Escherichia, Acinetobacter, Desulfitobacterium, Shewanella, Agrobacterium, Stenotrophomonas, Sulfurospirillum,* and *Aeromonas* (Banerjee et al., 2011; Ghodsi et al., 2011).

16srRNA analysis of bacterial species isolated from arsenic-rich environments has led to the identification of several bacterial species that possess the property of arsenic bioremediation. Around ten bacterial isolates collected from arsenic-contaminated soil were found to belong to *γ—proteobacterium, Firmicutes* and *Kocuria* genera. All the bacterial species showed bioaccumulation properties after altering the oxidation states of arsenate and arsenite. The bacteria exhibited 60% reduction in cellular growth in a media containing As(III). In media containing As(III), the pH was found to range between 7.0–8.8, while in As(V) media, the pH was between 7.0–9.2. It was also observed that the bacteria showing maximum arsenic resistance also produced siderophores. These allow the uptake of ferric/ferrous from minerals during which arsenic is mobilised into the aqueous phase and subsequently taken up easily by the bacterial cell. This enhances the intracellular uptake of arsenic by bacteria. This mechanism can be used to study bioremediation capacities of microorganisms (Banerjee et al., 2011). In a similar study, arsenite-resistant bacteria isolated from arsenic-contaminated soil were identified as *Bacillus macerans, Bacillus megaterium,* and *Corynebacterium vitarumen,* which had arsenic removal efficiencies of 60%, 43%, and 38% after 48 h, respectively. After 144 h, *Bacillus macerans* and *Corynebacterium vitarumen* showed 92% and 80% arsenic removal, while after 120 h, *Bacillus megaterium* showed 73% arsenic removal. The bacterial species also exhibited arsenite bioaccumulation properties up to 36%, 24% and 12% in *Bacillus macerans, Bacillus megaterium,* and *Corynebacterium vitarumen,* respectively (Ghodsi et al., 2011). Bacteria isolated from low arsenic-containing soil was found to be closely related to *Stenotrophomonas panacihumi* and the strain has been identified as MM-7. The strain was able to completely oxidise 50 μM of arsenite within 12 h at a pH range of 5–7. It was also able to tolerate arsenite concentrations of up to 60 m. MM-7 showed exceptional oxidation capacity due to the presence of the arsenic oxidase gene (Bahar et al., 2012).

Bacteria isolated from arsenic-contaminated water revealed a novel strain of As(III) oxidising bacteria, As7325. The bacteria was able to oxidise 2,300 μg/L of arsenite within 24 h under aerobic conditions. After oxidisation of arsenite, arsenate was taken up by the cell pellets with an efficiency up to 99% and 100% at As(V) concentrations of 500 and 100 μg/L, respectively after 6 days. This study, however, was preliminary and therefore would require further elaboration on factors such as temperature and pH. Also, the stability of converted As(V) has not been well defined (Kao et al., 2013). Another bacterial strain identified from contaminated arsenic soil is *Brevibacillus* sp. KUMAs2. The strain could resist 265 mM of As(V) and 17 mM of As(III) while removing 40% of arsenic under aerobic conditions. The optimum temperature and pH for growth of the bacteria in an arsenic-containing environment were 37°C and 7.0, respectively (Mallick et al., 2014).

In another study, 50 bacteria were isolated from arsenic-contaminated soil out of which two strains were arsenic tolerant. These strains initially named BC1 and BC2 were identified as *Enterobacter asburiae* and *Enterobacter cloacae* respectively after 16srRNA analysis. The bacterial strains also contained arsenite oxidising genes, *aoxA,* and the reducing gene, *arsC.* The optimum pH observed for growth of the bacteria was 6 and the temperature was seen to be 37°C. The two strains exhibited a minimum inhibitory concentration (MIC) of 40 mM for arsenite and 400 mM for arsenate in 24 h. The significantly high MIC values imply the potential of these bacterial strains for arsenic bioremediation (Selvi et al., 2014). Bacterial isolates from groundwater contaminated by arsenic were identified as Gram-positive *Bacillus* sp. KM02 and *Aneurinibacillus aneurinilyticus* BS-1. The bacterial species were able to oxidise arsenite to arsenate. *Bacillus* sp. was able to remove 51.45% of As(III) and 53.29% of As(V), while *Aneurinibacillus aneurinilyticus* removed 51.99% and 50.37% of As(III) and As(V) respectively after 72 h of incubation. With increasing concentrations of arsenic the resistance of the two strains was also found to increase. This tolerance can be attributed to the development of microbial biofilms.

Arsenic-resistant bacteria contain operons that govern resistance through specific genes. In this scenario, the increasing tolerance of bacteria might imply the generation of additional energy when oxidation occurs. Further, both the strains were Gram-positive in nature, which means that they would possess a thicker cell wall, thus protecting the bacteria from toxic arsenic entering the cell. Scanning electron microscopy (SEM) exhibited change in morphology of the two strains such as chain formation and reduction in size, which could mean the adoption of bioaccumulation as a method for bioremediation by the two strains (Dey et al., 2016).

Arsenic-resistant bacteria isolated from agricultural soil showed As(III) oxidising capabilities. Out of the eight strains identified, four demonstrated oxidase enzyme activity (*Pseudomonas, Acinetobacter, Klebsiella* and *Comamonas*) while four (*Geobacillus, Bacillus, Paenibacillus,* and *Enterobacter*) failed to do so. All of the bacterial strains were able to grow in a wide range of pH (5–9) and temperature (20–40°C) with the optimum pH being 7 and temperature being 30°C. Additionally, bacteria possessing plant growth-promoting (PGP) traits have been reported to enhance metal uptake and translocation and increase heavy-metal tolerance. Out of all of the identified strains, *Pseudomonas* sp. ASR1 was comparatively more tolerant to As(III) and had greater oxidising capacity along with having phosphate-solubilisation potential and the ability to produce siderophores. These properties make *Pseudomonas* a better candidate for arsenic bioremediation. However, the other strains also offer tremendous potential in bioremediation of heavy metals due to their high resistance to arsenic as well as PGP traits (Das et al., 2014).

Among the several *Bacillus* sp. that have been identified to date, another addition is *B. aryabhattai* (NBRI014). The strain was able to remove significant concentrations of arsenic by the expression of *ars* genes and upregulation of seven proteins that potentially increase the tolerance of the bacteria to arsenic. Multiple *ars* genes were studied for their activity in arsenic tolerance. The gene *arsB* was found to play a role in movement of arsenic outside the bacterial cell through a transport membrane. Likewise, *arsB* encodes for a membrane protein that pumps As(III) out of the bacterial cell by the use of proton motive force. The *arsC* gene encodes for a reductase that reduces As(V) to As(III). Further, *arsH* confers resistance to arsenite. The genes in *ars* operon have also been characterised in bacterial strains such as *Escherichia, Pseudomonas, and Streptococcus*. The capacity of bioaccumulation of arsenic by NBRI014 was also determined and it was observed that maximum uptake of arsenic occurred at 48 h of incubation at 32°C at about approximately 10%. The strain also underwent biovolatilisation wherein it exudes the bioaccumulated arsenic from the cell to the medium. This has been characterised as a detoxification mechanism. At 48 h 40% of the arsenic was detoxified from the bacterial cell. Fourier transform infrared (FTIR) spectroscopic techniques demonstrated the involvement of functional groups in arsenic binding within the bacterial cell such as free hydroxyl alcohols, alkenes, phenols, carbonyls, amines, alkanes, aromatics, and alkyl halides (Singh et al., 2016).

Bacteria isolated from shallow aquifers with high arsenic concentrations was identified as Gram-positive bacteria that was able to tolerate arsenic concentrations of 70 mM (As^{3+}) and 1,000 mM (As^{5+}). Two strains were able to effectively convert toxic As(III) to less toxic As(V). The optimum temperature for As(III) oxidation was found to be 33.5°C at 100 µM concentration of arsenic in 8.5 h. Under these conditions, the two strains oxidised 88% of As(III). 16srRNA analysis showed that both the strains were similar to *Bacillus* sp. (Biswas and Sarkar, 2019).

Bacterial Biofilters for Arsenic Removal

Halophilic bacteria also exhibit resistance to arsenic. Strains isolated from mangrove rhizosphere, *Kocuria flava* AB402 and *Bacillus vietnamensis* AB403, were able to tolerate 20–35 mM of arsenite. In rhizospheric soil contaminated with arsenic, the microbes play an essential role in defining the bioavailability and arsenic phytotoxicity. In soil possessing high salinity and arsenic, plant growth promoting bacteria (PGPB) and arsenic-tolerant halophilic bacteria will aid sustainable agricultural practices. AB402 and AB403 were also able to resist As(V) at a concentration of 350–450 mM after 48 h at 37°C with a pH of 7.4. The bacteria inhabited well on the root system as compared to the shoot. Additionally, the accumulation of arsenic was observed to be greater at the root as compared to the stem establishing an association with root and arsenic uptake by microorganisms as seen in Figure 8.1. Transmission electron microscopy (TEM) showed that there was significant accumulation of arsenic, as dark, electron-dense regions were observed. The strains reduced the uptake of arsenic by the plant by bioaccumulation under salt stress, implying their potential for bioremediation of arsenic in saline rhizosphere (Mallick et al., 2018).

FIGURE 8.1 (i) SEM of the cells of isolate AB402 (Bar 200 nm) (a) and AB403 (b) in biofilm (Bar 1 μm), formed in presence of 2 mM As(III) after 96 h and 24 h of incubation respectively; (ii) phenotypic expression of the rice plants under salt and arsenic stress: (a) plant with salt treatment (b) plant with arsenic treatment (c) plant with salt and arsenic treatment.

Source: Reprinted with permission from Mallick, I., Bhattacharyya, C., Mukherji, S., Dey, D., Sarkar, S. C., Mukhopadhyay, U. K., Ghosh, A. (2018). Effective rhizoinoculation and biofilm formation by arsenic immobilising halophilic plant growth promoting bacteria (PGPB) isolated from mangrove rhizosphere: A step towards arsenic rhizoremediation. *Sci. Total Environ*. 610–611, 1239–1250. doi:10.1016/j.scitotenv.2017.07.234.

Copyright © 2017 Elsevier B.V.

Rhizobacteria are being used as plant growth promoters under stress conditions such as in heavy-metal environments. Such bacteria also have the ability to produce siderophores. Indigenous hypertolerant strains identified as *Bacillus flexus* (NM02) and *Acinetobacter junii* (NM03) were studied for their ability to resist arsenic. Analysis of the metalloregulatory ars operon revealed the presence of *arsC* gene. The *arsC* gene codes a reductase that employs glutathione (GSH) for the reduction of As(V) to As(III). FTIR spectroscopy demonstrated that both the strains showed absorbance greater than the control and NM02 had a higher absorbance as compared to NM03. This was associated with the presence of functional groups such as amines, aromatics, nitro compounds alcohols, carboxylic acids, and esters. The biovolatilisation potential of the strains were also studied, which showed that at 72 h of incubation, NM03 was able to biovolatilise 14% while NM02 biovolatilised only 8%. Further, energy dispersive spectroscopy (EDS) analysis determined the arsenic bioaccumulation capacity of NM02 (Marwa et al., 2019).

Thermal-power stations that are fuelled by coal often emit copious amounts of fly-ash (FA), which leaches arsenic. Contamination caused by arsenic from ash ponds can cause pollution of drinking water above the permissible limit. Ten bacterial isolates were characterised from ash ponds belonging to *Bacillus, Brevibacillus, Micrococcus, Kytococcus,* and *Staphylococcus.* The optimum pH and temperature for growth and arsenic transformation was observed as 7.0–8.0 and 30–37°C, respectively. The strains were found to be halotolerant as well, tolerating salt concentrations of as high as 12%. A few of the strains, *Micrococcus* sp. strain HMR9 and *B. subtilis* strain HMR5, were seen to produce exopolysaccharide, which absorbed arsenic into the bacterial cell. The study demonstrated the role of indigenous bacteria inhabiting ash ponds in arsenic bioremediation and subsequently preventing its leaching into drinking-water sources (Roychowdhury et al., 2018).

A study performed on native bacterial population resistant to arsenate and arsenite revealed 14 such strains. Out of these, only two strains showed the ability to convert arsenic, which were characterised as As-11 and As-12 and showed phylogenetic similarity to *Pseudomonas* and *Bacillus* sp. Both the strains were determined to be mesophilic and the optimum temperature for growth under arsenic stress are 25°C and 38°C. As-11 has the conversion capacity of As(V) to As(III) at an efficiency of 78% and vice versa of 48%. Similarly, As-12 had the transformation efficiencies of 28% and 45% respectively. Studies like these emphasise the need for elaborate and extended research on native species of bacteria for bioremediation as they are highly compatible with the environment (Jebelli et al., 2018).

Characterisation of bacteria from high-arsenic-content water led to identification of two strains of *Pseudomonas* viz, AK1 and AK9, which had significantly high MIC. Both the strains possessed *aox* operon with genes pertaining to oxidase activity. Notably, *aoxR, aoxB,* and *aoxC* genes were present that are associated with oxidation of arsenic and further arsenic bioremediation. The optimal growth of the strains under arsenic stress conditions was found to be pH of 7.0 and temperature of 30°C. AK1 and AK9 showed resistance to As(III) at approximate concentrations of 13 mM and 15 mM also exhibiting complete aerobic reduction of arsenate after 48h of incubation. The strains oxidised 25% of As(III) after 72 h of incubation (Satyapal et al., 2018).

Dam sludge at gold mining tailings is usually highly contaminated with arsenic. Bacterial strains isolated from such a site were identified as *Bacillus thuringiensis* strain WS3, *Pseudomonas stutzeri* strain WS9, and *Micrococcus yunnanensis* strain WS11. Mixed dried biomass (MDB) of the three strains were used to analyse their efficacy in removing As(III) and As(V) under different conditions. Combinations of two were taken in a total set of three at 37°C and pH 7.0 at a contact time of 10 h against arsenite and arsenate. The absorption percentage of the three individual strains was found to be 77%, 71%, and 74% for As(III) and 78%, 75%, and 76% for As(V), for WS3, WS9, and WS11 respectively. For the MDB, adsorption capacity of WS3 and WS9 was 80%, WS3 and WS11 was 83%, and WS9 and WS11 was 81% for As(III), while it was seen to be 81%, 85%, and 82% for As(V). The highest absorption was observed for a MDB of all the strains at 86% for As(III) and 88% for As(V). Field emission scanning microscopy/energy dispersive x-ray analysis

Bacterial Biofilters for Arsenic Removal

(FESEM–EDX) exhibited that there was significant change in the morphology of the bacterial cell upon absorption. Further, FTIR spectroscopy demonstrated the involvement of functional groups such as thiols, amines, amides, and hydroxyl in the binding and removal of arsenic. The high absorption and bioremoval capacity of the MDBs implied the potential of such methods for bioremediation of toxic arsenic from contaminated water as well (Altowayti et al., 2020a).

The bioremediation techniques adopted usually depend on isolation of arsenic-resistant bacteria from arsenic-contaminated samples. However, it is essential to study the similarity of the cultured isolates compared to that of the native strains found in arsenic-rich environments. For this purpose, bacteria isolated from gold mine soil and tailings were identified as belonging to eight genera, viz *Staphylococcus* (89.8%), *Pseudomonas* (1.25%), *Corynebacterium* (0.82%), *Prevotella* (0.54%), *Megamonas* (0.38%), *Sphingomonas* (0.36%), *Pseudonocardia* (0.39%), and *Prevotellaceae* (0.33%). The resistance of the strains against As(III) and As(V) at 100–1,000 ppm was studied at 37°C for 24 h. The MIC value observed for As(III) was 600–800 ppm and that of As(V) was 800–1,000 ppm. In corroboration with existing data, the strains were found to contain genes of the *ars* system. This can be further utilised for identification and characterisation of native bacterial species in arsenic-rich environments (Altowayti et al., 2020b).

In mine tailings, arsenic is found in forms such as arsenopyrite, arsenian pyrites, and arsenates. Gold mine tailings are one of the major sources of arsenic contamination. Biological leaching as a means of heavy-metal removal has gained popularity due to the ability of microorganisms to convert solid compounds into soluble components. In acid-contaminated soils with high concentrations of arsenic, leaching is observed, which is attributed to the acidophilic iron and sulphur oxidising bacteria. Such methods have an advantage of requiring less energy, lacking emission of gaseous pollutants, and producing a high concentration of leaching agents. Temperature plays a major role in defining the leaching speed. Several studies on bioleaching use mesophilic, thermophilic, and extremely thermophilic bacteria. At lower temperatures, however, the leaching of metals is reduced significantly. Two acidophilic strains isolated from lower temperatures, *Acidithiobacillus ferrivorans* 535 and *Acidithiobacillus ferrooxidans* 377, were used to study their efficiency in removal of arsenic. At a pH of 1.6, 377 showed highest leaching properties and removal of arsenic (up to 68%) at 28°C, while 535 showed maximum removal at 8°C (up to 61%). Studies on all the pure cultures implied that at a higher temperature the arsenic bioleaching efficiency was greater. Additionally, the study opened avenues for the impact of psychrotolerant and mesophilic bacteria prevalent in mines against removal of arsenic under low temperatures (Seitkamal et al., 2020). Other bacteria isolated from gold mines are *Rhizobium* sp., *Pseudomonas xanthomarina* S11, and *Halomonas* A3H3, all of which exhibit processes such as biosorption, bioaccumulation, and biovolatilisation (Plewniak et al., 2018).

Microbial biomethylation of arsenic forms volatile arsines such as mono-, di- and tri-methyl arsine. As mentioned earlier, the conversion of inorganic arsenic into the volatile trimethylarsine allows for easy transport of the arsenic exterior to the cell. Genetic engineering of the bacteria for the expression of *arsM* genes that encode methyltransferase has been done to study the biovolatilisation capacity.

Rhizoremediation, utilisation of plant roots and associated microbes in the rhizosphere, in bioremediation of heavy metals is also a promising method used for soil remediation. However, often the microbes fail to colonise, thus causing hinderance in the bioremediation. Therefore, a successful approach is the introduction of functional genes that make these organism efficient colonisers. *Bacillus idriensis* and *Sphingomonas desiccabilis* were isolated from arsenic-contaminated sites. The *Rhodopseudomonas palustris* strain consists of the *arsM* gene, which aids in the methylation of arsenic. The *arsM* gene from *R. palustris* was introduced into *B. idriensis* and *S. desiccabilis*. The wild type and recombinant bacteria were grown at 37°C for 24 h in a medium supplemented with inorganic arsenic. The recombinant bacterial cells exhibited a greater growth curve at all times compared to the wild time implying that the *arsM* expression conferred it with resistance. Further, when incubated in a medium containing As(III), the concentration of arsenite was also found to decrease in the solution. The disappearance of arsenite was correlated with the amount of volatile arsenic present, and it was found that the amount was equivalent. Additionally, there was the production of DMA(V), As(V), and arsenic, which was generated due to the conversion of As(III). The biovolatilisation of arsenic into methylated arsenic observed was almost 10-fold that of the wild type. In a soil system, the biovolatilisation was observed to be 2.2–4.5%, which suggests that these bacteria may be suitable for arsenic bioremoval from contaminated soil (Liu et al., 2011).

In a similar study, the *arsM* gene from *R. palustris* was genetically engineered into a strain of *Pseudomonas putida*, KT2440 (GE *P. putida*). The strain showed elevated levels of arsenic methylation and subsequent volatilisation activity. After 12 h of incubation at 30°C, GE *P. putida* demonstrated resistance to arsenic and biomethylation of As(III) to DMA(V) and trimethylarsine, TMA(V)O. The toxicity of TMA(V)O is extremely low. Wild type *P. putida* showed resistance to arsenic but no traces of volatilised arsenicals. Of the total arsenic, the methylated species were about $62 \pm 2.2\%$ DMA(V), $25 \pm 1.4\%$MA(V), and $10 \pm 1.2\%$ TMA(V) in GE *P. putida*. In soil biosphere, the biovolatilised arsenic radicals were identified as dimethylarsine (Me_2AsH) and TMA(III). GE *P. putida* showed a biovolatilisation capacity about nine-fold greater than the wild type. Therefore, genetically engineered *P. putida* can show significant ability in reducing the inorganic content of arsenic in soil (Chen et al., 2014).

The gene encoding for As(III) S-adenosylmethionine methyltransferase in *Chlamydomonas reinhardtii*, *CrarsM*, was introduced into the genome of the soil symbiont bacteria *Rhizobium leguminosarum* and studied for methylation ability. It was also tested in symbiosis with red clover plant for methylation capacity. Recombinant *R. leguminosarum* had gained the ability to methylate inorganic arsenic to arsenical forms such as DMA(V), TMA(V)O, and MA(V). Additionally, red clover plant in symbiosis with *CrarsM* recombinant bacteria showed volatilisation of arsenic along with presence of all three methylated species in the root and stem system of the plant. Out of the total As(III), $78.8 \pm 1.3\%$ was converted to DMA(V) and $8.6 \pm 0.5\%$ was converted to TMA(V)O, after 72 h of incubation at 28°C. The species also produced minute concentrations of DMA(III) and TMA(III). Additionally, the generation of TMA(III) was found to be due to gradual methylation of DMA(V). As previously mentioned, the trivalent radicals of methylated arsenic bind to essential proteins such as thiols causing hindrance of vital cellular pathways and, thus, are not desirable. Due to this, in spite of showing significant bioremediation potential, the enzymatic ability of the species to convert DMA(V) to TMA(III) might be a limiting factor for achieving efficient arsenic removal. The volatilisation ability of soil bacteria, therefore, needs to be explored further. *Arsenicibacter rosenii*, a soil bacterium, is known to encode for an *arsM* gene, which shows high efficiency of arsenic methylation and volatilisation (Zhang et al., 2017).

8.4 FUTURE PROSPECTS

There is a need for developing sustainable and effective arsenic bioremediation strategies. Microbial bioremoval, thus, is an exceptional alternative to the conventional methods used. The existing microbe-assisted bioremediation of arsenic is being supported with biological materials such as granular-activated carbon, iron-oxide, manganese-oxide, combination of activated charcoal and alumina, and zero-valent iron. Several studies were performed using arsenic-resistant bacteria along with bioreactors of the previously mentioned material, and significant enhancement was observed in removal of As(V) and As(III). Laboratory-based experiments were carried out with species such as *Ralstonia eutropha* MTCC 2487, *Gallionella ferruginea*, and *Leptothrix ochracea*. However, only a single technology out of all of them showed promising effects under field conditions. An integrated arsenic removal system with oxidation-charcoal/alumina was developed with the bacterial strain *Mycobacterium lacticum*. Oxidation of As(III) to As(V) was achieved followed by sorption into the activated charcoal/alumina. Such bio-physiochemical techniques also aid in the removal of multiple heavy metals. Technologies like such, even though they have tremendous potential, come with the added disadvantage of usage of an external carbon source. Further, after every breakthrough, the bioreactors would need to be changed along with fresh bacterial inoculations. Therefore, alternatives with cost-effectiveness and that are environmentally friendly need to be developed. Further research on the sustainable use of biosorbents should be explored to achieve efficient arsenic bioremediation. For example, EPS may serve as a potential alternative to chemical biosorbents used for arsenic removal (Bahar et al., 2013; Saba et al., 2019).

Another aspect that needs to be taken into consideration is comprehending the pathway of arsenic metabolism and its origin. Modern-day biotechnological techniques such as next-generation sequencing and assembly software allow the sequencing of a complete genome of native arsenic-resistant bacteria and give a peek into the genes associated with tolerance. The presence of genomic data makes it possible to understand the behaviour of such microbial species under varying metal stress conditions and accordingly allows the manipulation of the desirable genes. Metagenome projects are underway whose goal is to generate a database of microbial communities encompassing a large quantity of genomes across various microbial species. Other branches of meta-approaches, namely metatrancriptomics, metaproteomics, and metabolomics, help in understanding the functionality of an environment and biosphere. In-depth research using these techniques would provide an insight into rarer species of undiscovered biospheres. This would also ensure the continuous evolution of a database comprising information associated with environment-specific arsenic-resistant microbes.

Additionally, exploring native bacterial cultures that grow in natural open systems can be used to enhance the performance of biosorption as they are adaptive in nature. Even though pure cultures have also demonstrated remarkable bioremoval capacity, they need to be maintained in specific conditions with substrates and sterile bioreactors, making this a limiting factor. Evaluation of natural consortia of microbes would help in designing and optimising large-scale bioremediation systems, which often prove to be expensive in comparison with pure cultures.

Understanding the genes and the regulatory mechanism exhibited by microorganisms like bacteria are needed to be elaborated. Deciphering the genetic makeup of arsenic-tolerant bacteria would help in generating genetically engineered microbes, which would possess desirable traits, such as enhance resistance and biovolatilisation capability. These can also be utilised to improve crop resistance and regulate phytoremediation as with rhizobium bacteria as a part of sustainable approaches (Cavalca et al., 2013; Plewniak et al., 2018; Shukla et al., 2020).

8.5 CONCLUSION

Microbes are an excellent source of heavy metal bioremediation and are a more cost-effective and environmentally friendly alternative in comparison with conventional methods. The appropriate selection of microbial species exhibiting maximum efficiency in a specific environment is critical for effective bioremediation systems. A wide variety of such bacterial species have been isolated and studied for their bioremoval properties in arsenic-contaminated ecosystems. These bacterial species adopt various inherent mechanisms such as bioaccumulation, biosorption, biovolatilisation, biotransformation, and production of proteins such as siderophores and EPS, which individually or in an integrated manner contribute to the effective bioremoval of arsenic. All of these are bacterial mechanisms for detoxification. However, they possess certain shortcomings. For example, bioaccumulation as a method is known to significantly reduce arsenic, but it still persists in the ecosystem due to the geochemical cycle. Also, biovolatilisation is a more effective means as the arsenic is released in its volatile form from soil and water. However, the arsenate has a good chance of being demethylated into its oxidised form. Therefore, the usability of microorganisms for arsenic bioremediation is debatable until proven otherwise. Biosorption as a technology is more promising as it is cheap and renewable. The arsenic is adsorbed on the surface of the biomass and thus is reversible. It also allows the procurement of arsenic appropriately and ensures that it is not disposed into the environment. Researchers have also treated the biological adsorbents with chemical material to enhance the adsorption capacity. These treatments increase the surface charge or enhance/expose the functional groups present on the bacterial cells. Biochemical pathways and genetic makeup of these bacteria were explored elaborately and led to the identification of genes that produce enzymes such as oxidases and reductases. The *ars* operon is one such factor that is found in arsenic-tolerant species of bacteria. The operon encodes for various kinds of proteins each contributing to the biotransformation of arsenic. At optimised environmental conditions, bacterial species can remove toxic arsenite species up to 98%. Most of the studies carried out to date are based on a laboratory scale. Thus, large-scale bioremediation systems need to be optimised and regulated with microbial consortia in order to accelerate the applicability in natural environments. Additionally, information on genomics, microbial diversity, and the biotransformation of arsenic is still fragmentary and therefore needs to be extended in depth. The extensive distribution of arsenic in soil, minerals, water, and the ecosystem in general makes its bioremediation an immensely vital process. Hence, arsenic-removing microbes can serve as potential biofilters for generation of clean and safe water.

REFERENCES

Altowayti, W. A. H., Haris, S. A., Almoalemi, H., Shahir, S., Zakaria, Z., and Ibrahim, S. (2020a). The removal of arsenic species from aqueous solution by indigenous microbes: Batch bioadsorption and artificial neural network model. *Environ. Technol. Innov.* 19, 100830. doi:10.1016/j.eti.2020.100830.

Altowayti, W. A. H., Almoalemi, H., Shahir, S., and Othman, N. (2020b). Comparison of culture-independent and dependent approaches for identification of native arsenic-resistant bacteria and their potential use for arsenic bioremediation. *Ecotoxicol. Environ. Saf.* 205, 111267. doi:10.1016/j.ecoenv.2020.111267.

Bahar, M. M., Megharaj, M., and Naidu, R. (2012). Arsenic bioremediation potential of a new arsenite-oxidizing bacterium Stenotrophomonas sp. MM-7 isolated from soil. *Biodegradation.* 23, 803–812. doi:10.1007/s10532-012-9567-4.

Bahar, M. M., Megharaj, M., and Naidu, R. (2013). Bioremediation of arsenic-contaminated water: Recent advances and future prospects. *Water. Air. Soil Pollut.* 224, 1–20. doi:10.1007/s11270-013-1722-y.

Banerjee, S., Datta, S., Chattyopadhyay, D., and Sarkar, P. (2011). Arsenic accumulating and transforming bacteria isolated from contaminated soil for potential use in bioremediation. *J. Environ. Sci. Heal.—Part A Toxic/Hazardous Subst. Environ. Eng.* 46, 1736–1747. doi:10.1080/10934529.2011.623995.

Biswas, R., and Sarkar, A. (2019). Characterization of arsenite-oxidizing bacteria to decipher their role in arsenic bioremediation. *Prep. Biochem. Biotechnol.* 49, 30–37. doi:10.1080/10826068.2018.1476883.

Bjørklund, G., Aaseth, J., Chirumbolo, S., Urbina, M. A., and Uddin, R. (2018). Effects of arsenic toxicity beyond epigenetic modifications. *Environ. Geochem. Health.* 40, 955–965. doi:10.1007/s10653-017-9967-9.

Casentini, B., Rossetti, S., Gallo, M., and Baldi, F. (2015). Potentialities of biogenerated iron hydroxides nanoparticles in arsenic water treatment. In *Conference Proceedings 6th European Bioremediation Conference*, Chania, Crete, Greece.

Cavalca, L., Corsini, A., Zaccheo, P., Andreoni, V., and Muyzer, G. (2013). Microbial transformations of arsenic: Perspectives for biological removal of arsenic from water. *Future Microbiol.* 8, 753–768. doi:10.2217/FMB.13.38.

Chen, J., Sun, G. X., Wang, X. X., Lorenzo, V. De, Rosen, B. P., and Zhu, Y. G. (2014). Volatilization of arsenic from polluted soil by Pseudomonas putida engineered for expression of the arsM arsenic(III) S-adenosine methyltransferase gene. *Environ. Sci. Technol.* 48, 10337–10344. doi:10.1021/es502230b.

Das, S., and Barooah, M. (2018). Characterization of siderophore producing arsenic-resistant Staphylococcus sp. Strain TA6 isolated from contaminated groundwater of Jorhat, Assam and its possible role in arsenic geocycle. *BMC Microbiol.* 18, 1–11. doi:10.1186/s12866-018-1240-6.

Das, S., Jean, J. S., Kar, S., Chou, M. L., and Chen, C. Y. (2014). Screening of plant growth-promoting traits in arsenic-resistant bacteria isolated from agricultural soil and their potential implication for arsenic bioremediation. *J. Hazard. Mater.* 272, 112–120. doi:10.1016/j.jhazmat.2014.03.012.

Dey, U., Chatterjee, S., and Mondal, N. K. (2016). Isolation and characterization of arsenic-resistant bacteria and possible application in bioremediation. *Biotechnol. Rep.* 10, 1–7. doi:10.1016/j.btre.2016.02.002.

Bacterial Biofilters for Arsenic Removal

Fazi, S., Amalfitano, S., Casentini, B., Davolos, D., Pietrangeli, B., Crognale, S., Lotti, F., and Rossetti, S. (2016). Arsenic removal from naturally contaminated waters: A review of methods combining chemical and biological treatments. *Rend. Fis. Acc. Lincei.* 27, 51–58. doi: 10.1007/s12210-015-0461-y

Garelick, H., Jones, H., Dybowska, A., and Valsami-jones, E. (2008). Arsenic pollution sources. *Rev. Environ. Contam.* 197, 17–60. doi:10.1007/978-0-387-79284-2.

Ghodsi, H., Hoodaji, M., Tahmourespour, A., and Gheisari, M. M. (2011). Investigation of bioremediation of arsenic by bacteria isolated from contaminated soil. *African J. Microbiol. Res.* 5, 5889–5895. doi:10.5897/ajmr11.837.

Hare, V., Chowdhary, P., Kumar, B., Sharma, D. C., and Baghel, V. S. (2018). Arsenic toxicity and its remediation strategies for fighting the environmental threat. *Emerg. Eco-Friendly Approaches Waste Manag.*, 143–170. doi:10.1007/978-981-10-8669-4_8.

Hughes, M. F. (2002). Arsenic toxicity and potential mechanisms of action. *Toxicol. Lett.* 133, 1–16. doi:10.1016/S0378-4274(02)00084-X.

Ike, M., Miyazaki, T., Yamamoto, N., Sei, K., and Soda, S. (2008). Removal of arsenic from groundwater by arsenite-oxidizing bacteria. *Water Sci. Technol.* 58, 1095–1100. doi:10.2166/wst.2008.462.

Jebelli, M. A., Maleki, A., Amoozegar, M. A., Kalantar, E., Gharibi, F., Darvish, N., et al. (2018). Isolation and identification of the native population bacteria for bioremediation of high levels of arsenic from water resources. *J. Environ. Manage.* 212, 39–45. doi:10.1016/j.jenvman.2018.01.075.

Jomova, K., Jenisova, Z., Feszterova, M., Baros, S., Liska, J., Hudecova, D., et al. (2011). Arsenic: Toxicity, oxidative stress and human disease. *J. Appl. Toxicol.* 31, 95–107. doi:10.1002/jat.1649.

Kao, A. C., Chu, Y. J., Hsu, F. L., and Liao, V. H. C. (2013). Removal of arsenic from groundwater by using a native isolated arsenite-oxidizing bacterium. *J. Contam. Hydrol.* 155, 1–8. doi:10.1016/j.jconhyd.2013.09.001.

Liu, S., Zhang, F., Chen, J., and Sun, G. (2011). Arsenic removal from contaminated soil via biovolatilization by genetically engineered bacteria under laboratory conditions. *J. Environ. Sci.* 23, 1544–1550. doi: 10.1016/S1001-0742(10)60570-0.

Mallick, I., Bhattacharyya, C., Mukherji, S., Dey, D., Sarkar, S. C., Mukhopadhyay, U. K., et al. (2018). Effective rhizoinoculation and biofilm formation by arsenic immobilizing halophilic plant growth promoting bacteria (PGPB) isolated from mangrove rhizosphere: A step towards arsenic rhizoremediation. *Sci. Total Environ.* 610–611, 1239–1250. doi:10.1016/j.scitotenv.2017.07.234.

Mallick, I., Hossain, S. T., Sinha, S., and Mukherjee, S. K. (2014). Brevibacillus sp. KUMAs2, a bacterial isolate for possible bioremediation of arsenic in rhizosphere. *Ecotoxicol. Environ. Saf.* 107, 236–244. doi:10.1016/j.ecoenv.2014.06.007.

Marwa, N., Singh, N., Srivastava, S., Saxena, G., Pandey, V., and Singh, N. (2019). Characterizing the hypertolerance potential of two indigenous bacterial strains (Bacillus flexus and Acinetobacter junii) and their efficacy in arsenic bioremediation. *Appl. Microbiol. Int.* 126(4), 1117–1127. doi:10.1111/jam.14179.

Plewniak, F., Crognale, S., Rossetti, S., and Bertin, P. N. (2018). A genomic outlook on bioremediation: The case of arsenic removal. *Front. Microbiol.* 9, 1–8. doi:10.3389/fmicb.2018.00820.

Roychowdhury, R., Roy, M., Rakshit, A., Sarkar, S., and Mukherjee, P. (2018). Arsenic bioremediation by indigenous heavy metal resistant bacteria of fly ash pond. *Bull. Environ. Contam. Toxicol.* 101, 527–535. doi:10.1007/s00128-018-2428-z.

Saba, Rehman, Y., Ahmed, M., and Sabri, A. N. (2019). Potential role of bacterial extracellular polymeric substances as biosorbent material for arsenic bioremediation. *Bioremediat. J.* 23, 72–81. doi:10.1080/10 889868.2019.1602107.

Satyapal, G. K., Mishra, S. K., Srivastava, A., Ranjan, R. K., Prakash, K., Haque, R., et al. (2018). Possible bioremediation of arsenic toxicity by isolating indigenous bacteria from the middle Gangetic plain of Bihar, India. *Biotechnol. Rep.* 17, 117–125. doi:10.1016/j.btre.2018.02.002.

Satyapal, G. K., and Rani, S. (2016). Potential role of arsenic resistant bacteria in bioremediation: Current status and future prospects. *J. Microb. Biochem. Technol.* 8. doi:10.4172/1948-5948.1000294.

Seitkamal, K. N., Zhappar, N. K., Shaikhutdinov, V. M., Shibayeva, A. K., Ilyas, S., Korolkov, I. V., et al. (2020). Bioleaching for the removal of arsenic from mine tailings by psychrotolerant and mesophilic microbes at markedly continental climate temperatures. *Minerals.* 10, 1–13. doi:10.3390/min10110972.

Selvi, M. S., Sasikumar, S., Gomathi, S., Rajkumar, P., Sasikumar, P., and Sadasivam, S. G. (2014). Isolation and characterization of arsenic resistant bacteria from agricultural soil, and their potential for arsenic bioremediation. *Int. J. Agric. Policy Res.* 2, 393–405. Available at: www.journalissues.org/IJAPR/%0Ahttp://dx.doi.org/10.15739/IJAPR.012.

Shankar, S., Shanker, U., and Shikha. (2014). Arsenic contamination of groundwater: A review of sources, prevalence, health risks, and strategies for mitigation. *Sci. World J.* 2014. doi:10.1155/2014/304524.

Sher, S., and Rehman, A. (2019). Use of heavy metals resistant bacteria—a strategy for arsenic bioremediation. *Appl. Microbiol. Biotechnol.* 103, 6007–6021. doi:10.1007/s00253-019-09933-6.

Shukla, R., Sarim, K. M., and Singh, D. P. (2020). Microbe-mediated management of arsenic contamination: Current status and future prospects. *Environ. Sustain.* 3, 83–90. doi:10.1007/s42398-019-00090-0.

Singh, N., Gupta, S., Marwa, N., Pandey, V., Verma, P. C., Rathaur, S., et al. (2016). Arsenic mediated modifications in Bacillus aryabhattai and their biotechnological applications for arsenic bioremediation. *Chemosphere.* 164, 524–534. doi:10.1016/j.chemosphere.2016.08.119.

Srivastava, P. K., Vaish, A., Dwivedi, S., Chakrabarty, D., Singh, N., and Tripathi, R. D. (2011). Biological removal of arsenic pollution by soil fungi. *Sci. Total Environ.* 409, 2430–2442. doi:10.1016/j.scitotenv.2011.03.002.

Tariq, A., Ullah, U., Asif, M., and Sadiq, I. (2019). Biosorption of arsenic through bacteria isolated from Pakistan. *Int. Microbiol.* 22, 59–68. doi:10.1007/s10123-018-0028-8.

WHO—World Health Organization. (2006). *Guidelines for Drinking Water Quality.* First addendum to 3rd addition, Volume 1. Geneva: WHO Press. Available at: www.who.int/water_sanitation_health/dwq/gdwq0506.pdf. Accessed 28 March 2016.

Yamamura, S., and Amachi, S. (2014). Microbiology of inorganic arsenic: From metabolism to bioremediation. *J. Biosci. Bioeng.* 118, 1–9. doi:10.1016/j.jbiosc.2013.12.011.

Zhang, J., Xu, Y., Cao, T., Chen, J., Rosen, B. P., and Zhao, F. J. (2017). Arsenic methylation by a genetically engineered Rhizobium-legume symbiont. *Plant Soil.* 416, 259–269. doi:10.1007/s11104-017-3207-z.

9 Bio-nanoparticle
Synthesis and Application in Wastewater Treatment

Swatilekha Pati, Somok Banerjee, and Shaon Ray Chaudhuri

CONTENTS

9.1 Introduction 137
9.2 Types of NP Preparation 138
    9.2.1 Physical Synthesis 138
    9.2.2 Chemical Synthesis 138
    9.2.3 Biological Synthesis 139
9.3 Biosynthesis of NPs 139
    9.3.1 Synthesis by Microorganisms 139
        9.3.1.1 Bacteria 140
        9.3.1.2 Fungi 141
        9.3.1.3 Algae 141
        9.3.1.4 Yeast 142
        9.3.1.5 Virus 142
        9.3.1.6 Human Cell Line 142
    9.3.2 Synthesis by Plants (Phytofabrication) 142
9.4 Factors Affecting the Biosynthesis of NPs 143
9.5 Different Sources and Types of Pollutants in Wastewater 145
    9.5.1 Hydrocarbon Pollutants 145
    9.5.2 Metal Pollutants 148
    9.5.3 Pollutants from Textile Industries 148
    9.5.4 Radionuclide Contamination 149
    9.5.5 Pesticides 149
    9.5.6 Pharmaceuticals 149
9.6 Applications of BNPs in Remediating the Pollutants of Wastewater 150
9.7 Future Perspective 151
9.8 Acknowledgement 152
References 152

9.1 INTRODUCTION

High-quality potable water is an important resource for survival of humans. However, in the past few decades a number of natural and anthropogenic activities have led to the pollution of fresh water, with urban and industrial wastewater being the major source of organic and inorganic pollutants. According to the World Health Organization in 2012, a global population of 780 million people still lacks access to basic clean drinking-water supply (Qu et al., 2013a).

In developing countries, the major source of water pollution is wastewater. The organic and inorganic constituents of wastewater are toxic to both humans and the environment (Gautam et al., 2017; Saxena et al., 2016). It is estimated that in India approximately 22,900 million litres per day

DOI: 10.1201/9781003165149-9

(MLD) and 13,500 MLD of wastewater are generated from urban and industrial centres respectively. However, the treatment capacity available for domestic wastewater is 5,900 MLD while that of small-scale industries is 8,000 MLD (www.scribd.com/document/421191172/sewagepollution-pdf). Thus, there is a big gap in treatment of wastewater. Therefore, now there is an urgent need for water management and wastewater treatment in order to supply safe and clean potable water to all. Currently, there are various conventional methods used in water purification and wastewater treatment such as reverse osmosis, electrostatic precipitation, etc. but these technologies are less efficient and are expensive. Hence, in this context, the nanoparticles (NPs) can be used as a very good alternative for wastewater treatment in terms of effectiveness and cost.

The NPs are used in wastewater treatment because of their small size (1–100 nm) and large surface area for adsorption (due to high surface to volume ratio) (Qu et al., 2013b), high chemical reactivity, and photocatalysis (Yang et al., 2015). A number of NPs including TiO_2 (Titanium oxide), ZnO (Zinc oxide), AgNP (Silver), and FeNP (Iron) are reported to be used in wastewater treatment. The NPs can be produced by physical, chemical, or biological methods. But the synthesis of nanoparticles by physical and chemical processes generates toxic substances, is highly energy intensive, and is very costly (Parveen et al., 2016; Arshad, 2017). However, bio synthesis of NPs is an environmentally friendly, single-step process that makes use of safe reagents.

Bio nanoparticles (BNPs) are utilised in the treatment of wastewater (generated from industries or household chores) by removing toxic textile dyes, pesticides, heavy metals, pathogenic microorganisms, organic micro-pollutants, etc. (Ali et al., 2019) that can be reused further for some industrial applications or irrigational purposes. It is also used in the purification of drinking water that is contaminated with pesticides and heavy metals. Thus, conclusively, biosynthesis and application of nanoparticles is an emerging approach for treating wastewater.

This chapter attempts to provide descriptive information on the biosynthesis of NPs as an eco-friendly, cost-effective, and single-step process; various factors governing their biosynthesis, different pollutants present in wastewater; and subsequent application of BNPs in wastewater treatment.

9.2 TYPES OF NP PREPARATION

9.2.1 PHYSICAL SYNTHESIS

The physical route of metal NP (MNP) synthesis is the top-down approach. The concept of the top-down approach is to restructure (decompose) large material into a nano level by applying some external forces like degradation, disruption, crushing, cutting, grinding, cryo-grinding, processing, and homogenisation (Mandava et al., 2017; Lade and Shanware, 2020). This includes methods like ball milling, chemical vapour deposition (CVD), laser sputtering, spray pyrolysis, plasma arcing, etc. (Singh et al., 2020; Parveen et al., 2016). The NPs produced using these methods are generally larger in size and volume and vary in size (Mandava et al., 2017). However, the physical processes use sophisticated machineries whose maintenance and assemblance is not easy and that are highly energy-intensive, time consuming, and exorbitant (Parveen et al., 2016; Arshad, 2017). For example, fabrication of AgNPs by physical means can be done at atmospheric pressure by using a tube furnace where the furnace requires a large area, a high amount of energy, and time to achieve thermal stability. Though the NP generation of this process is very stable because the temperature does not fluctuate with time, it still has certain drawbacks (Iravani, 2014).

9.2.2 CHEMICAL SYNTHESIS

The chemical route of NP synthesis takes place through a bottom-up approach, which is used very frequently for synthesising AgNPs. The NPs are constructed from the very basic building blocks (atoms or molecules) in the case of the bottom-up approach, which is a better approach and is well suited for synthesising MNPs of a definite shape, size, and structure (Mandava et al., 2017). The

Bio-Nanoparticle

various chemical processes include plasma arcing, sol–gel process, pyrolysis, hydrothermal processes, and chemical vapour deposition (CVD), which also requires the use of excessive radiation, highly toxic reductants, and other stabilising agents that are environmentally unfriendly (Parveen et al., 2016; Arshad, 2017). The reductants that are mainly used in chemical synthesis are citrate, borohydrate, elemental hydrogen, and ascorbate (Lade et al., 2019).

9.2.3 Biological Synthesis

Biological synthesis uses living organisms (bacteria, fungi, yeast, and plants) for generating the NPs. This is also known as 'Green Synthesis'. Bio-synthesis of NPs is an environmentally friendly, single-step process that makes use of safe reagents. The BNPs also have higher stability with desired dimension as they are synthesised through a one-step process. Microbes have several biomolecules (e.g., organic acids, enzymes, polysaccharides, reductases, quinones, etc.) that play a major role in synthesising these NPs (Parveen et al., 2016). For instance, an extracellular cell free supernatant is known to generate nanocrystals of sulphate (Nasipuri et al., 2011) due to presence of extracellular reductases. Similarly, plant extracts are also known for their ability to synthesise NPs from a salt solution due to presence of polyphenols, proteins, and terpenoids (as reducing agent) as well as phenolics, proteins, and alkaloids (as capping and stabilising agent; Makarov et al., 2014). The efficiency of NP synthesis varies among the plant varieties and depends on their phytochemicals. These NPs are efficient in terms of their antimicrobial efficacy both on bacterial planktonic cells and biofilms (Ray Chaudhuri et al., 2021) and hence can be actively involved in adsorption during wastewater treatment. NP synthesis was possible using water extract of spent coffee grounds as the aforementioned reducing agent, and stabilisers are present in coffee grounds. Based on the synthesis parameters like temperature, time, and pH, the particle size can be varied and is reflected in the colour of the solution (Sarkar et al., 2021).

Unlike chemical synthesis, BNP synthesis poses a lesser threat to the environment (Ijaz et al., 2020). Moreover, a biologically synthesised NP with greater surface area can absorb higher amounts of pollutants, have higher catalytic reactivity, and remain in an unaggregated fashion because the previously stated biomolecules that encapsulate the NPs, thereby acting as a capping agent (Pandey et al., 2015). Chemical synthesis requires high-purity feed stocks (Yu et al., 2009), whereas BNPs can be synthesised from impure feed stocks (e.g. metal-contaminated water) and can also be recovered and re-utilised post-reaction. Additionally, BNPs don't require high energy like that in physical synthesis. The biosynthesis process is conducted in mild temperature ranging from 20–30°C under ambient conditions and does not require any sophisticated machineries, thus reducing the operational cost (Capeness et al., 2019).

Because of these advantages of biologically synthesised NPs over the physico-chemical ones, BNPs are preferred as an efficient alternative for treating wastewater.

9.3 BIOSYNTHESIS OF NPS

Each and every biological entity possesses some active molecules and compounds that act as natural reducing and stabilising agents that are used for the fabrication of NPs having various shapes, sizes, compositions, and different physical and chemical properties (Fawcett et al., 2017).

9.3.1 Synthesis by Microorganisms

Microbes have this unique potential to change the metal(loid)'s oxidation state, which provides them access to synthesise MNPs for novel applications. The major biochemical agents that are responsible for reducing the metal ions and forming NPs involves proteins, enzymes, carbonyl groups, polysaccharides, peptides, and sulfur. Though the biosynthesis of NPs takes place due to the microbial enzymes generated by the cellular activities that grab the targeted metal ions from the environment and convert them into the element metal, the exact mechanism behind this is yet to be discovered (Ng et al., 2015).

Microorganisms (bacteria, fungi, algae) are capable of synthesising all types of NPs, for example, single metallic NPs (silver, gold, copper, iron), metal oxides (zinc oxide, titanium oxide, iron oxide, aluminium oxide), or metal sulphides (CdS, PbS, PbSe, ZnS, CdSe; Yadav et al., 2020).

The synthesis mechanism can be classified into the following types based on the location of NP formation:

1. **Intracellular synthesis**: In this case, the metal ions are transferred into the microbial cell for the formation of NPs in the presence of cellular enzymes (Ng et al., 2015).
2. **Extracellular synthesis**: Here, the metal ions are trapped and reduced on the microbial cell surface in the presence of enzymes. It is a more desirable process than intracellular synthesis because the cell culture can be reused (no need for cell lysis) again for biosynthesis i.e. an economical process (Grasso et al., 2020).

Examples of a few biologically synthesised NPs are listed in Table 9.1.

9.3.1.1 Bacteria

The rapid growth rate and the various advanced processes of genetic manipulation make bacteria a suitable candidate for NP synthesis (Vithiya and Sen, 2011). They are also capable of

TABLE 9.1
Biologically Synthesised NPs and Their Uses.

Microorganism	Form of Growth	Strain	Nanoparticle	Use	Reference
Bacteria	Planktonic	*S. putrefaciens CN-32*	As-S	Metal remediation	Jiang et al. (2009)
Bacteria	Planktonic	*Pseudomonas putida*	Pd	Dechlorination	Bunge et al. (2010)
Bacteria	Planktonic	*S. oneidensis*	Au/Pd	Dehalogenation	De Corte et al. (2012)
Bacteria	Biofilm	Bacterial sludge	Ag	Ag NPs cluster in wastewater treatment	Tanzil et al. (2016)
Fungi	Planktonic	*Aspergillus flavus*	ZnS	Optical detection of Pb, Cd	Uddandarao et al. (2019)
Fungi	Planktonic	*Aspergillus flavus*	PbS	Optical detection of As in water	Priyanka et al. (2017)
	Planktonic	*Fusarium acuminatum*	Ag	Shows antimicrobial activity against human pathogens	Ingle et al. (2008)
Microalgae	Planktonic	*Amphora*-46	Polycrystalline Ag	Antimicrobial activity against *E. coli, S. mutans*	Jena et al. (2014)
Yeast	Planktonic	*Saccharomyces cerevisiae* BU-MBT-CY1	Ag	Arsenate removal	Selvakumar et al. (2011)
Plants	Planktonic (leaf extracts)	*Magnusiomyces ingens* LH-F1	Au	Catalytic reduction of nitrophenols	Zhang et al. (2016)
	(leaf extracts)	*Ocimum sanctum*	Ni	Adsorption of dyes and pollutant	Pandian et al. (2015)
	(leaf extracts)	*Withania coagulans*	Fe2O3/Pd/RGO	Acts as an catalyst in the reduction of 4-nitrophenol	Atarod et al. (2016)
		Morinda morindoides	Fe2O3/Cu	Acts as a catalyst for reducing organic dyes	Nasrollahzadeh et al. (2016)

Bio-Nanoparticle

reducing and absorbing heavy-metal ions. In addition, they are capable of interacting with the surrounding environment, which arises because of the polarity of their lipid bilayer, which helps in catalysing various oxidation-reduction processes (Singh et al., 2018). NP formation by bacteria also depends on the species type because each species of bacteria possesses its particular genetic constitution, which results in different functional expression for the same redox enzyme. Bacteria can synthesise MNPs (e.g., AgNPs) by both intracellular and extracellular routes (Hulkoti et al., 2014).

Some strains of bacteria used prolifically for synthesising silver NPs are *Bacillus cereus, E. coli, Pseodomonas proteolytica, Geobacter spp.* and *Shewanella oneidensis*

Similarly, the strains that are used for synthesising gold NPs are: *Desulfovibrio desulfuricans, E. coli* DH5α, *Bacillus megaterium* D01, *Bacillus subtilis* 168, *Shewanella alga, etc.* (Singh et al., 2018).

Though bacteria-mediated synthesis of NPs is very advantageous as it produces NPs at a large scale without using any harmful chemicals, there lie certain disadvantages such as the fact that the bacterial cell culturing process is laborious and, moreover, its size, shape and also the distribution can rarely be controlled. (Yadav et al., 2020).

9.3.1.2 Fungi

The biosynthesis of MNPs by fungi is a very efficient approach for generating monodispersed NPs with defined morphologies. Fungi are considered suitable biological agents for generating MNPs because they possesses several intracellular enzymes. As compared to bacteria, fungi can produce NPs in larger amounts (Mohanpuria et al., 2008). The expected mechanism behind fungi-mediated NP synthesis inside the cell or on the cell wall is the enzymatic reduction (Singh et al., 2018). The major drawback is in the difficulty of gene manipulation in the eukaryotic system rather than in the prokaryotes (Vithiya and Sen, 2011). Various fungal species are used for synthesising metal/metal oxide NPs such as Au, Ag, ZnO, and TiO_2.

AgNPs were obtained intracellularly when the fungal biomass of *Verticillium* sp. was mixed with the aqueous solution of silver nitrate ($AgNO_3$), whereas, extracellular AgNPs were obtained by using *Fusarium oxyporum* biomass (Senapati et al., 2004). Similarly, the technically useful semiconductor, CdS NPs, were produced outside the cell by an enzymatic procedure when *F. oxyporum* was added to the aqueous solution of SO_4^{2-} and Cd^{2+} ions (Ahmad et al., 2002).

9.3.1.3 Algae

Algae have the capability of hyper-accumulating ions of heavy metals and are also capable of remodelling the metal ions into a form that is more malleable. These alluring properties make algae a model organism for synthesising MNPs (Ijaz et al., 2020).

The steps that are involved in the synthesis of NPs by algae are:

1. Algal extract preparation in an organic solvent or in water by boiling/heating.
2. Molar solution preparation of metallic compounds (ionic).
3. Incubation of the aforementioned solution (with/without stirring) under optimal conditions, followed by a colour change, which indicates the formation of NPs.

NP synthesis can be done using algae either intracellularly or extracellularly. The algal aqueous solution contains reducing agents like reducing sugars, peptides, polysaccharides, pigments, proteins, etc. The activity of these reducing agents may lead to the precipitation of the metallic ions, which infers with the extracellular formation of NP, whereas, for intracellular synthesis of algal NPs, metabolic processes of the algae (i.e., respiration and photosynthesis) are responsible for reducing the metal ions.

Arockiya et al. (2012) demonstrated the formation of gold (Au) NPs of size 18.7 nm to 93.7 nm by using a brown alga (*Stoechospermum marginatum*). A change in colour of the reaction mixture from pale brown to ruby red within 10 mins confirmed the formation of NPs. Another study by Romero-González et al. (2003) showed the formation of AuNPs by using dealginated seaweed waste that reduces Au ion formation to AuNPs of size ranging up to 6 μm. This study demonstrated that the functional groups present in the seaweed are responsible for the formation of stable NPs of different shapes. Other MNPs like silver, cadmium, palladium, and metal oxide (e.g. ZnO, CuO, Cu_2O, Fe_3O_4) NPs can also be synthesised using algal extracts (Fawcett et al., 2017).

9.3.1.4 Yeast
Kowshik et al. (2003) in his study demonstrated that a silver-tolerant species of yeast, MKY3 in its log phase, produces MNPs extracellularly when added to silver solution.

9.3.1.5 Virus
Viruses are being used for synthesising NPs in a template dependent manner. For example, MNPs and micro structured elements are generated in a template dependent manner by using viroid capsules (Douglas and Young, 1998). Iron oxides are synthesised by oxidative hydrolysis using Tobacco Mosaic Virus (TMV) as a template (Bansal et al., 2014).

9.3.1.6 Human Cell Line
Various human cancerous and non-cancerous cell lines including SiHa (malignant cervical epithelial cell), SKNSH (human neuroblastoma), HEK-293 (non-malignant human embryonic kidney cells), and HeLa (malignant cervical epithelial cell) are also capable of synthesising AuNPs of size ranging from 20–100 nm. These NPs are present either in the cytoplasm or in the nucleus of the cells (Anshup et al., 2005; Larios-Rodriguez et al., 2011).

9.3.2 SYNTHESIS BY PLANTS (PHYTOFABRICATION)

Apart from the microorganisms, plants are also a very good option for NP synthesis. Plant extracts are very useful in NP synthesis because of its reducing and stabilising availability, safe handling nature, and the fact that it contains a wide range of metabolites (Taheriniya et al., 2016). Moreover, the use of plant extracts (leaf, stem, seeds etc.) for synthesising BNPs can be advantageous over other biogenic process of synthesis to some extent because in this case there is no need to maintain cell cultures, which requires a longer duration for reduction than water soluble phytochemicals and it can be scaled-up easily for mass production of NPs (Razavi et al., 2015; Lee et al., 2011).

Various plants such as aloe vera, tulsi, mustard, lemon grass, coriander, ramie, lemon, asian pigeonwings, asiatic pennywort, night-flowering jasmine, and drumsticks are reported to be capable of synthesising NPs including Ag, Au, Ni (Nickel), Cu (Copper), ZnO, etc. (Marchiol, 2012; Anastas and Warner, 1998; Ray Chaudhuri et al., 2021). Plant extract consists of amino acid, proteins, phytochemicals, enzymes, and other components, which are responsible for synthesis of the NPs. The phytochemical components of the plant extracts act as capping agents in synthesising metal NPs (Jameel et al., 2020). Different metal precursors are added with plant extracts and under suitable reaction conditions (pH, temperature, concentration of metal ions, etc.) the phytochemicals reduce the metal ions to metal NPs (Mittal et al., 2013; Dwivedi and Gopal, 2010). The basic phytochemicals that are responsible for NPs synthesis are ketones, flavones, aldehyde amides, terpenoids, etc. (Taheriniya and Behboodi, 2016). Plants can successfully synthesise various NPs, for instance, Cu, Au, Ag, Co (Cobalt), ZnO, magnetite, palladium, and platinum (Parveen et al., 2016; Ray Chaudhuri et al., 2021). A study by Shankar et al. (2003) demonstrated the reduction of silver ions into NPs by terpenoids present in the leaf extract of Geranium that acts as an antioxidant and stabilising agent. Another report stated that enol to keto transformation of flavonoids (including luteolin and apigenin) is responsible for the production of biogenic silver NPs from *Ocimum basilicum* plant extract (Ahmad et al., 2010). Amino acids like cysteine, methionine, and lysine bind to metal ions and reduce them to form metal NPs (Gruen, 1975).

Bio-Nanoparticle 143

Ray Chaudhuri et al. (2021) showed Ramie (4.55 gm/gm of fresh leaf) plant to be a better candidate for AgNP synthesis when compared to lemon (0.48 gm/gm of fresh leaf), Asian pigeon-wings (0.92 gm/gm of fresh leaf), Asiatic pennywort (0.91 gm/gm of fresh leaf), night-flowering jasmine (0.67 gm/gm of fresh leaf) and drumsticks (0.38 gm/gm of fresh leaf). Representative images of AgNPs synthesised using these plant extracts are shown in Figure 9.1.

Some other examples of plant mediated NP synthesis include the synthesis of AgNPs, gold nano-triangles using aloe vera plant extracts (Chandran et al., 2006), conversion of gold ions into NPs by *Pelargonium graveolens* plant extract (Shankar et al., 2003), and production of 15–25 nm AuNPs from extract of *Triginella foenum-graecum* (Aromal and Philip, 2012). A pictorial representation of the NP synthesis process from plant extract is provided in Figure 9.2.

9.4 FACTORS AFFECTING THE BIOSYNTHESIS OF NPS

The synthesis of BNPs is common but the size of the particle and its morphology depend upon different physico-chemical parameters, which are as follows:

- **pH** – One of the most important factor is the reaction medium's pH, which can affect the size and morphology of biosynthesized NPs. A study shows that there is an alteration in

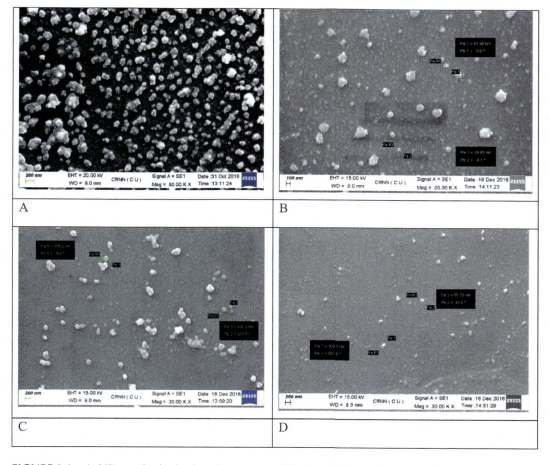

FIGURE 9.1 AgNPs synthesised using plant extract of Ramie (A), Lemon (B), night-flowering jasmine (C), and drumstick (D) as visualised under scanning electron microscope (SEM). The confirmation of AgNP was done in each case using Energy Dispersive X-Ray (EDX) analysis during SEM analysis (E). Atomic force micrograph of AgNPs synthesised using Ramie (F) and night-flowering jasmine (G) extract.

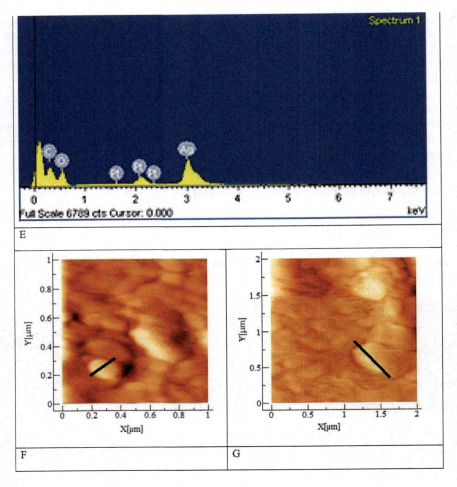

FIGURE 9.1 (Continued)

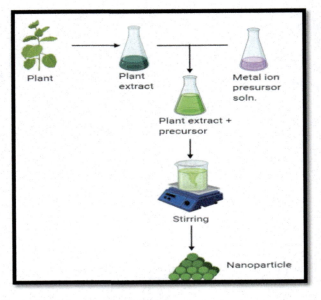

FIGURE 9.2 Work flow of green synthesis of NPs from plant extract.

Bio-Nanoparticle

size and number of AuNPs formed from *Avena sativa* based on the pH of the reaction medium. The NPs formed at pH 2, 5, and 6 were found to be larger in size in comparison to pH 3 and 4. However, there was a significant increase in the amount of NP formed at pH 3 and 4 (Armendariz et al., 2004). The change in pH of extracted spent coffee grounds shows distinct change in AgNP synthesis with variation in time as shown in Figure 9.3. Hence pH has a distinct role in metal nanoparticle formation.

- **Temperature** – Another important factor is temperature, which affects the production of NPs and determines its nature. Generally, the physical and chemical methods of NP synthesis require a temperature of more than 350°C and less than 330°C respectively whereas biological synthesis requires much less temperature (< 100°C). Rai et al. (2006) in an experiment on synthesis of AuNPs from lemongrass leaf extract showed that by increasing the reaction temperature, there is a decrease in both percentages of Au nanotriangles in comparison to the spherical particles as well as their size. Sarkar et al. (2021) have also shown the effect of temperature on AgNP synthesis from spent coffee ground extract.
- **Time** – The incubation time for the reaction is another important factor that affects the type and quality of the NP to be synthesised by biological means. Variation in time may cause aggregation of NPs because of long storage time, shrinkage of NPs, and so on. Soni and Prakash (2011) showed an increased production of AgNPs from *C. tropicum* and *F. oxysporium* with increasing time. Sarkar et al. (2021) reiterated the finding that period of incubation decides the quantity as well as the size of the MNPs. The scan data in Figure 9.4A shows the AgNP synthesis under optimum pH and temperature from spent coffee ground water extract following different times of incubation. The NP synthesis during incubation is reflected through colour change as depicted in Figure 9.4B.
- **Pressure** – The shape and size of NPs are also affected by the pressure, which is applied on the reaction medium. Tran et al. (2013) showed that there is an increase in reduction of metal ions by biological agents under ambient pressure.
- **Environment** – The nature of NPs to be synthesised also depends on the surrounding environment. Sometimes, a single NP transforms quickly into a core-shell NP by absorption or reaction of other materials from the surrounding environment whereas BNPs forms a thick coat and thus increases in size (Sarathy et al., 2008; Lynch et al., 2007).

OTHER FACTORS

Several other factors affect the synthesis of NPs such as concentration of reaction medium, proximity, etc. Soni and Prakash (2011) showed there is an increase in the production of AgNPs by *C. tropicum* and *F. oxysporium* with an increase in concentration of the reaction medium. Moreover, when one NP comes within proximity of another NP, there is an alteration in one of their properties (Baer et al., 2008) such as particle charge, substrate interaction, and magnetic property.

9.5   DIFFERENT SOURCES AND TYPES OF POLLUTANTS IN WASTEWATER

Pollutants are the elements or particles or components introduced into the environment and they have adverse effects on both the environment and living things, including animals, plants, and humans. They can be easily characterised into two types based on their source of generation, as shown in Figure 9.5. These are broadly pollution due to natural activities and anthropogenic activities. However, the majority of wastewater pollution arises directly from human activities such as from the industrial sector (including food, pharmaceutical, chemical, fertilisers, textile, agricultural fields, refineries, municipal sewage sludge, etc.).

9.5.1   HYDROCARBON POLLUTANTS

Hydrocarbon pollutants, including petroleum and its derivatives that are released into the environment through oil spills, industrial effluents, and agricultural wastes (Kennish, 2017) are found in a

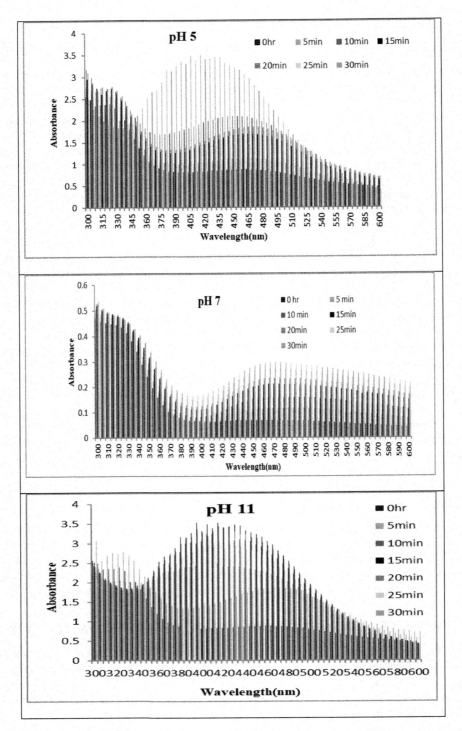

FIGURE 9.3 Experimental data of AgNP production from silver nitrate solution using spent coffee ground water extract at different pH following incubation for different time internal. The graph represents the absorption spectra of the solution using spectrophotometer at different wavelength from 300 nm to 600 nm. The peak around 340–350 nm represents small and spherical AgNPs while that around 400–440 nm represents large-size NP (Khodashenas and Ghorbani, 2015).

Bio-Nanoparticle

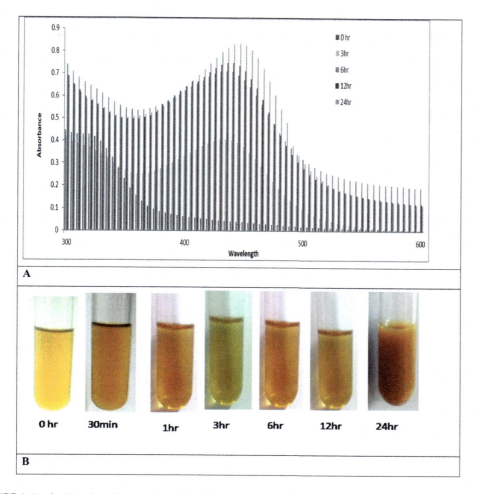

FIGURE 9.4 **A.** Wavelength scan data of AgNP synthesis at optimum pH and temperature at different time points using silver nitrate solution and spent coffee ground water extract. **B.** Pictorial representation of the change in colour of the silver nitrate solution and spent coffee ground water extract solution following different periods of incubation. This is an indication of the particle size synthesised in the solution.

variety of habitats and pose a potential threat to various organisms, thereby disrupting the balance of the ecosystem.

Among all these hydrocarbon pollutants, the polyaromatic hydrocarbons (chemical compounds having two or more benzene rings that are put together in a linear, angular, or cluster arrangement) are considered the highest priority pollutants worldwide due to their toxicity and carcinogenicity. They are generally discharged into the environment through insufficient combustion of fossil fuels and accidental discharge, as well as through incineration of some other waste materials.

The toxic aromatic hydrocarbons, when released into the environment, affect the different trophic levels ranging from microbes to humans. They may be mutagenic, carcinogenic, immunotoxic, or teratogenic. According to Varjani et al. (2017) the polyaromatic hydrocarbons possess a high affinity towards macromolecules like DNA, RNA, and protein; as a result, they induce mutation and tumour development, thereby affecting the skin as well as other organs. Long term exposure to these contaminants can also cause nausea, skin irritation, immune suppression, etc.

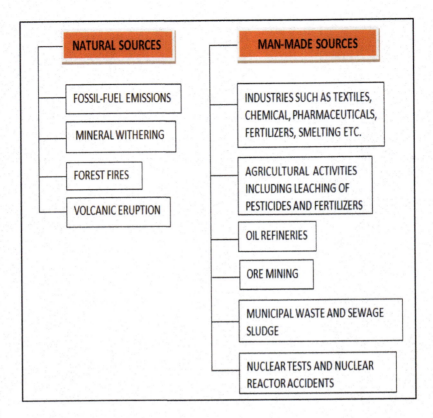

FIGURE 9.5 Different sources of pollutants.

9.5.2 Metal Pollutants

The recent growth in the industrial sector has led to the gradual increase of metal contamination in the world. Metal pollutants are a global concern because of their toxic nature, long term presence in the ecosystem, and subsequent bioaccumulation in the trophic levels (Douay et al., 2013). Moreover, the living beings are continuously exposed to these xenobiotics through occupational and environmental routes.

Continuous exposure to heavy metal pollutants causes neurological, circulatory, endocrine, and immune system disorders, mutagenicity, diabetes, and cancer through changes at the molecular and cellular levels. For instance, the attachment of cadmium (Cd) and mercury (Hg) with the thiol groups of the proteins leads to the formation of stable complexes and results in the dysfunctioning of these biomolecules (Farina et al., 2011; Kasprzak, 2002; Shah, 2020, 2021). Other metals like Cd, Pb, Hg, and Al through their metabolic interaction with the nutritionally vital elements like Ca and Fe (Goyer, 1997) interfere with the major physiological functions in humans.

9.5.3 Pollutants from Textile Industries

Dyes are the primary substances present in the waste effluent of textile industries. In general, the dyeing process requires a lot of water, and not all the places have effective ways of treating the wastewater before releasing it into the environment. Nearly 15% of the dyes used are discarded through wastewater and are one of the significant pollutants because of their recalcitrant nature. Some dyes (mainly the synthetic ones) don't degrade in the water, while others that degrade produce harmful substances as they decompose.

Bio-Nanoparticle

In general, most of the classes of synthetic dyes are toxic and carcinogenic in nature that causes kidney, bladder, and liver cancer. They generate undesirable turbidity in the water and reduce sunlight penetration, which in turn resists the photochemical synthesis of aquatic plants such as algae and also harms the aquatic environment (Saini, 2017).

9.5.4 Radionuclide Contamination

Large areas of land or ecosystems all over the world have been contaminated with radionuclides as a result of nuclear bomb explosions, above-ground nuclear experiments, nuclear reactor accidents, etc. Rapid industrialisation and natural biogeochemical cycle disturbance have also led to an increase in radionuclide pollution.

Radionuclide isotopes including ^3H, ^{14}C, ^{90}Sr, ^{99}Tc, ^{129}I, ^{137}Cs, ^{237}N, and several U and Th series isotopes originating from nuclear-related activities are of environmental importance due to high mobility, toxicity, and abundance. Mining activities and phosphate fertiliser factories also produce wastes containing ^{230}Th, ^{226}Ra, and ^{210}Pb radionuclides. Besides these, commercial fuel reprocessing plants also release liquid and gaseous wastes of ^{99}Tc and ^{129}I into the soil and atmosphere, thereby polluting it. However, nuclear weapons testing and nuclear facilities are the primary public concerns for radionuclide contamination (Jagetiya et al., 2014).

The radionuclides from the soil are taken up by the plants, redistributed throughout the ecosystem, and eventually passed through the food chain to human beings, having a wide-ranging impact on human, animal, and plant populations. For instance, human depositions of large quantities of the radionuclides, including ^{238}U, ^{232}Th, and ^{40}K in various organs weaken the immune system and cause various diseases, while radionuclide toxicity in plants causes stunted growth, nutrient deficiency, and leaf scorch.

9.5.5 Pesticides

In recent times because of technological development, global populations have been exposed to a plethora of chemical substances with a broad spectrum, including pesticides. This group of chemicals plays an essential role in protecting the crops and livestock from pests and also enhances yield. Despite their obvious advantages, the bioaccumulation through the food chain has adverse effects on the environment as well as animal and human population (Ozkara et al., 2016).

The wastewater from fertiliser industries contains toxic chemical compounds that are often discarded into water bodies without proper treatment. Moreover, once pesticides and herbicides are applied to agricultural fields, some portion of them remains in the fields, while a major portion enters into the surrounding environments as contaminants after precipitation and irrigation. Some of these chemical compounds, including organochlorine insecticides, reside in the environment for an extended period of time, which makes it even more dangerous (Larson et al., 1997).

According to several studies, these compounds are toxic and mutagenic in nature and can damage the immune system and lead to spontaneous abortions in female, non-Hodgkin's lymphoma, leukaemia, and other types of tumours in humans.

9.5.6 Pharmaceuticals

Process wastewater from pharmaceutical industries contains large number of active pharmaceutical ingredients, which are hazardous if discharged into various water bodies. According to a few studies, half of the global pharmaceutical wastewaters are discharged without proper treatment (Enick and Moore, 2007; Lange et al., 2006). Such wastewater generally contains endocrine-disrupting compounds that harm the endocrine system of humans (Anderson, 2005) and other end products reduce nitrite oxidation, causing remasculinisation of male fish and defeminisation in female fish (Orlando et al., 2004).

9.6 APPLICATIONS OF BNPS IN REMEDIATING THE POLLUTANTS OF WASTEWATER

Removal of petroleum hydrocarbons: Murgueitio et al. (2018) exhibited the capability of BNPs in remediating hydrocarbon from contaminated soil and water. Iron NPs were synthesised using *Vaccinum floribundum* extract as the reducing and stabiliser agent. Hydrocarbon-contaminated water and soil samples were subjected to treatment with these NPs and showed 85.94% and 88.34% removal of hydrocarbon from 9.32 mg/L and 94.20 mg/L hydrocarbon-containing water samples and 81.90% from a 5,000 mg/Kg hydrocarbon-containing soil sample.

Removal of pollutant dyes: Lately, with the increased amount of pollution on a daily basis, the urgent need for safe drinking water is also increasing at par. In this context, the metal and metal oxide NPs such as AgNPs, ZnO NPs, CuO NPs, and TiO2 NPs are of great use in the removal of toxic pollutants like various synthetic dyes due to their enhanced photocatalytic activity and high surface area to volume ratio in comparison to the bulk nanomaterials. Moreover, this increased rate of pollutant removal even at low concentrations is due to the presence of a huge number of surface reactive sites on the NP surfaces. Thereby, only a small amount of these nanoparticles is required for the removal of the contaminants. One such study of MNP in dye removal was performed by Jyoti and Singh (2016) in which AgNPs produced from *Z. armatum* leaf extract were capable of degrading Safranine O, Methyl Blue, and Methylene Orange.

Catalytic activity: 4-Nitrophenol is an organic compound that is used in manufacturing various pesticides, insecticides, and synthetic dyes and is a potential organic contaminant of wastewater because of its toxic and inhibitory characteristics. Thus, its remediation is crucial. One of the simplest ways of reducing 4-nitrophenol is by using the green synthesised MNPs like AgNPs, AuNPs, PdNPs, and CuONPs as a catalyst along with sodium borohydride ($NaBH_4$) as a reductant. MNPs show an excellent catalytic potential due to a high rate of surface adsorption and high surface area to volume ratio. In general, the viability of the reaction decreases because of the considerable potential difference between donor (H_3BO_3/$NaBH_4$) and acceptor molecules (nitrophenolate ion), which is responsible for the higher activation energy barrier. In this context, the reaction rate is promoted by the metallic NPs by increasing the adsorption of reactants on their surfaces thereby declining the activation energy barriers (Gangula et al., 2011). According to Yuan et al. (2017) the UV–visible spectrum of 4-Nitrophenol was characterised by a sharp band at 400 nm as a nitrophenolate ion was produced in the presence of NaOH. However, the addition of AgNPs (synthesised from *Chenopodium aristatum L.* stem extract) to the reaction medium led to a fast decay in the absorption intensity at 400 nm, concomitantly, followed by the emergence of a comparatively wide band at 313 nm, which demonstrated the formation of 4-aminophenol. Moreover, 4-Aminophenol which is the reduction product of 4-Nitrophenol, has applications in various fields, i.e., as an intermediate for paracetamol, rubber antioxidants, sulphur dyes, preparing black/white film developers, inhibitors of corrosion, and also as an antipyretic drug precursor and pain killers (Panigrahi et al., 2007; Woo and Lai, 2012).

Heavy-metal sensor: Heavy metals such as Ni, Pb (lead), Cu, Fe (iron), Zn, Hg (mercury), Mn (manganese), etc. are common pollutants of water. Few metals like Pb, Cu, Cd (cadmium), and Hg+ (mercury) ions even at trace (ppm/mg L^{-1}) levels shows high toxicity. Detection of such hazardous metals in the marine environment is one of the major requirements for their proper remediation. Normally, conventional approaches, based upon instrumentation, offer very good sensitivity in multi-element analyses but the whole set up that is required for performing such analyses is very expensive, tedious, non-portable, and requires skilled persons.

The metallic NPs can be used as an effective alternative for detecting heavy-metal ions in metal-contaminated water bodies because of their tunable size, cost effectiveness, and great sensitivity even at sub-ppm levels (Maiti et al., 2016). According to a study, AgNPs synthesised from extracts of various plants exhibited colorimetric sensing of zinc and mercury ions (Zn^{2+} and Hg^{2+}). Likewise, AgNPs fabricated using fresh mango leaves and dried leaves showed selective sensing for Hg^{2+} and Pb^{2+} ions. AgNPs synthesised from pepper seed extracts and green tea extracts also exhibited selective sensing properties for Hg^{2+}, Pb^{2+}, and Zn^{2+} ions (Karthiga and Anthony, 2013).

As supporting materials for embedding NPs:

Cellulose nanofiber: Bendi and Imae (2013) built a CuNPs – cellulose nanofibre catalytic system that showed high levels of catalytic activity towards the reduction of 4-Nitrophenol, in contrast to a Cu-free TEMPO (2,2,6,6-tetramethylpiperidine-1-oxyl radical)-oxidised cellulose nanofibre (TOCNF) film having no catalytic activity. The cellulose film could be recycled easily for ten cycles without any notable decrease in catalytic activity suggesting that the CuNPs embedded in the film efficiently promoted the catalytic activity.

Carbon nanofibers supporting AgNPs: Composite nanofibers, such as carbon nanofiber (CNF) supporting AgNP, were fabricated by the electrospinning and hydrothermal methods, respectively. The CNF were functionalised by treating them with HNO_3 that enhances the dispersion of AgNPs on the nanofibers whose amount can be controlled by changing the concentrations of the total reactants. A study showed that the catalytic activity of CNF/AgNP composite nanofibers in reducing 4-Nitrophenol can be elevated by increasing the content of Ag on the CNF/AgNPs composite nanofibers. Additionally, these composite nanofibers because of their one-dimensional nanostructural property can be recycled easily (Zhang et al., 2011).

9.7 FUTURE PERSPECTIVE

BNPs have become an emerging research trend in today's world because of their extensive use in almost every field. The synthesis of NPs by biological means (e.g. bacteria, fungi, algae, plant) are stable, economical, sustainable, less toxic, and eco-friendly as compared to the NPs synthesised by the physico-chemical means. Their area of usage is not only limited to treating wastewater, but further their applications also find a place in various research areas and industries, such as electronics and in the medical field where purity is the major concern. Current treatment procedures are also capable of removing organic and inorganic pollutants from wastewater, but those are uneconomical and energy intensive. Moreover, these methods do have the ability to reuse the retentates, whereas both can be achieved by using BNPs in treating the impure water.

As seen in this chapter, it can be concluded that biologically synthesised NPs could be a very promising approach in terms of handling water quality issues because they are highly proficient in recycling and removal of various pollutants (heavy metals, pesticides, dyes, etc.). More focused study should be done to dive deep into the mechanisms of bio-synthesis and the strategies of bio-remediation in order to unveil the unknown mechanisms so that we can customise the NPs accordingly and use them not only for a selective pollutant but to enhance multi-pollutant removal from wastewater. To date most of the experiments are carried out on a lab scales, and with the positive progress of this treatment procedures using BNPs, researchers are planning to explore more of this in terms of pilot scale trials followed by industrial scale usage, so as to fulfil the future requirements of the growing world and thus to provide the best service for human welfare.

9.8 ACKNOWLEDGEMENT

Authors acknowledge Tripura University for the laboratory and computational facility, and project students of the Department of Microbiology for the nanoparticle data generated during MSc work. Financial assistance from Ministry of Education under the Frontier Area of Science and Technology Scheme as well as Department of Biotechnology, Government of India under the Twinning scheme is acknowledged.

REFERENCES

Ahmad A., Mukherjee P., Mandal D., Senapati S., Khan M. I., Kumar R., Sastry M. (2002). Enzyme mediated extracellular synthesis of CdS nanoparticles by the fungus Fusarium oxyporum. *Journal of the American Chemical Society* 124, 12108–12109pp.

Ahmad N., Sharma S., Alam M. K. (2010). Rapid synthesis of silver nanoparticles using dried medicinal plant of basil. *Colloids Surfaces B Biointerfaces* 81, 81–86pp.

Ali I., Peng C., Khan Z. M., Naz I., Sultan M., Ali M., Abbasi I. A., Islam T., Ye T. (2019). Overview of microbes based fabricated biogenic nanoparticles for water and wastewater treatment. *Journal of Environmental Management* 15, 128–150pp.

Anastas P. T., Warner J. C. (Eds.). (1998). *Green Chemistry: Theory and Practice.* Oxford: Oxford University Press, 148pp. ISBN: 9780198506980.

Anderson P. D. (Ed.). (2005). *Technical Brief: Endocrine Disrupting Compounds and Implications for Wastewater Treatment,* WERF Report: Surface Water Quality, 04-WEM-6. London: IWA, 4, 48pp. ISBN: 9781780404400.

Anshup A., Venkataraman J. S., Subramaniam C., Kumar R. R., Priya S., Kumar T. R., Omkumar R. V., John A., Pradeep T. (2005). Growth of gold nanoparticles in human cells. *Langmuir* 21, 11562–11567pp.

Armendariz V., Herrera I., Peralta-Videa J. R. (2004). Size controlled gold nanoparticle formation by *Avena sativa* biomass: Use of plants in nanobiotechnology. *Journal of Nanoparticle Research* 6, 377–382pp.

Arockiya A. R. F., Parthiban C., Ganesh Kumar V., Anantharaman P. (2012). Biosynthesis of antibacterial gold nanoparticle using brown alga, Stoechospermum marginatum (kützing). *Spectrochim seta. Molecular and Biomolecular Spectroscopy* 99, 166–173pp.

Aromal S. A., Philip D. (2012). Green synthesis of gold nanoparticles using Trigonella foenum-graecum and its size dependent catalytic activity. *Spectrochim Acta A Mol Biol Spectrosc* 97, 1–5pp.

Arshad A. (2017). Bacterial synthesis and applications of nanoparticles. *Nano Science & Nano Technology* 11, 119pp.

Atarod M., Nasrollahzadeh M., Sajadi S. M. (2016). Green synthesis of Pd/RGO/Fe3O4 nanocomposite using Withania coagulans leaf extract and its application as magnetically separable and reusable catalyst for the reduction of 4-nitrophenol. *Journal of Colloid and Interface Science* 1, 249–258pp.

Baer D. R., Amonette J. E., Engelhard M. H. (2008). Characterization challenges for nanomaterials. *Surface and Interface Analysis* 40, 529–537pp.

Bansal P., Duhan J. S., Gahlawat S. K. (2014). Biogenesis of nanoparticles: A review. *African Journal of Biotechnology* 13, 2778–2785pp.

Bendi R., Imae T. (2013). Renewable catalyst with Cu nanoparticles embedded into cellulose nanofiber film. *RSC Advances* 3, 16279–16282pp.

Bunge M., Søbjerg L. S., Rotaru A. E., Gauthier D., Lindhardt A. T., Hause G. (2010). Formation of palladium (0) nanoparticles at microbial surfaces. *Biotechnology and Bioengineering* 107, 206–215pp.

Capeness M. J., Echavarri-Bravo V., Horsfall L. E. (2019). Production of biogenic nanoparticles for the reduction of 4-nitrophenol and oxidative laccase-like reactions. *Frontiers in Microbiology* 10, 997pp.

Chandran S. P., Chaudhary M., Pasricha R., Ahmad A., Sastry M. (2006). Synthesis of gold nanotriangles and silver nanoparticles using *Aloe vera* plant extract. *Biotechnology Progress* 22, 577–583pp.

De Corte S., Sabbe T., Hennebel T., Vanhaecke L., De Gusseme B., Verstraete W., Boon N. (2012). Doping of biogenic Pd catalysts with Au enables dechlorination of diclofenac at environmental conditions. *Water Research* 46, 2718–2726pp.

Douay F., Pelfrêne A., Planque J., Fourrier H., Richard A., Roussel H., Girondelot B. (2013). Assessment of potential health risk for inhabitants living near a former lead smelter. Part 1: Metal concentrations in soils, agricultural crops, and home-grown vegetables. *Environmental Monitoring and Assessment* 185, 3665–3680pp.

Bio-Nanoparticle

Douglas T., Young M. (1998). Host-guest encapsulation of materials by assembled virus protein cages. *Nature* 393, 152–155pp.

Dwivedi A. D., Gopal K. (2010). Biosynthesis of silver and gold nanoparticles using *Chenopodium album* leaf extract. *Colloids and Surfaces A: Physicochemical and Engineering Aspects* 369, 27–33pp.

Enick O. V., Moore M. M. (2007). Assessing the assessments: Pharmaceuticals in the environment. *Environmental Impact Assessment Review* 27, 707–729pp.

Farina M., Aschner M., Rocha J. B. T. (2011). Oxidative stress in MeHg-induced neurotoxicity. *Toxicology and Applied Pharmacology* 256, 405–417pp.

Fawcett D., Verduin J. J., Shah M., Sharma S. B., Poinern G. E. J. (2017). A review of current research into the biogenic synthesis of metal and metal oxide nanoparticles via marine algae and seagrasses. *Journal of Nanoscience* 2017, 1–15pp.

Gangula A., Podila R., Ramakrishna M., Karanam L., Janardhana C., Rao A. M. (2011). Catalytic reduction of 4-nitrophenol using biogenic gold and silver nanoparticles derived from *breyniarhamnoides*. *Langmuir* 27, 15268–15274pp.

Gautam S., Kaithwas G., Bharagava R. N., Saxena G. (2017). Pollutants in tannery wastewater, pharmacological effects and bioremediation approaches for human health protection and environmental safety. In: Bharagava R. N. (Ed.), *Environmental Pollutants and Their Bioremediation Approaches*, 1st edition. Boca Raton, FL: CRC Press, Taylor & Francis Group, 369–396pp. ISBN: 9781315173351.

Goyer R. A. (1997). Toxic and essential metal interactions. *Annual Review of Nutrition* 17, 37–50pp.

Grasso G., Zane D., Dragone R. (2020). Microbial nanotechnology: Challenges and prospects for green biocatalytic synthesis of nanoscale materials for sensoristic and biomedical applications. *Nanomaterials* (Basel) 10, 11pp.

Gruen L. C. (1975). Interaction of amino acids with silver (I) ions. *Biochim Biophys Acta* 386, 270–274pp.

Hulkoti N. I., Taranath T. C. (2014). Biosynthesis of nanoparticles using microbes: A review. *Colloids Surfaces B Biointerfaces* 121, 474–483pp.

Ijaz I., Gilani E., Nazir A., Bukhari A. (2020). Detail review on chemical, physical and green synthesis, classification, characterizations and applications of nanoparticles. *Green Chemistry Letters and Reviews* 13, 223–245pp.

Ingle A., Gade A., Pierrat S., Soennichsen C., Rai M. (2008). Mycosynthesis of silver nanoparticles using the fungus fusarium acuminatum and its activity against some human pathogenic bacteria. *Current Nanoscience* 4, 141–144pp.

Iravani S. (2014). Bacteria in nanoparticle synthesis: Current status and future prospects. *International Scholarly Research Notices* 2014, 18pp.

Jagetiya B., Sharma A., Soni A., Khatik U. K. (2014). Phytoremediation of radionuclide: A report on the state of the art. *Radionuclide Contamination and Remediation through Plants*, 1–31pp.

Jameel M. S., Aziz A. A., Dheyab M. A. (2020). Green synthesis: Proposed mechanism and factors influencing the synthesis of platinum nanoparticles. *Green Processing and Synthesis* 9, 386–398pp.

Jena J., Pradhan N., Dash B. P., Panda P. K., Mishra B. K. (2014). Pigment mediated biogenic synthesis of silver nanoparticles using diatom Amphora sp. and its antimicrobial activity. *Journal of Saudi Chemical Society* 19, 661–666pp.

Jiang S., Lee J. H., Kim M. G., Myung N. V., Fredrickson J. K., Sadowsky M. J., Hur H. G. (2009). Biogenic formation of As-S nanotubes by diverse Shewanella strains. *Applied and Environmental Microbiology* 75, 6896–6899pp.

Jyoti K., Singh A. (2016). Green synthesis of nanostructured silver particles and their catalytic application in dye degradation. *Journal of Genetic Engineering and Biotechnology* 14, 311–317pp.

Karthiga D., Anthony S. P. (2013). Selective colorimetric sensing of toxic metal cations by green synthesized silver nanoparticles over a wide pH range. *RSC Advances* 3, 16765–16774pp.

Kasprzak K. S. (2002). Oxidative DNA and protein damage in metal-induced toxicity and carcinogenesis. *Free Radical Biology and Medicine* 32, 958–967pp.

Kennish M. J. (2017). Polyaromatic hydrocarbon. In: Petralia P., Careli C., Fortener S., Jaffe G. (Eds.), *Practical Handbook of Estuarine and Marine Pollution*. 1st edition. Boca Raton, FL: CRC Press. Taylor & Francis, 544pp. ISBN: 0849384249.

Khodashenas B., Ghorbani H. R. (2015). Synthesis of silver nanoparticles with different shapes. *Arabian Journal of Chemistry* 12, 1823–1838pp.

Kowshik M., Ashtaputre S., Kharrazi S., Vogel W., Urban J., Kulkarni S., Paknikar K. (2003). Extracellular synthesis of silver nanoparticles by a silver-tolerant yeast strain MKY3. *Nanotechnology* 14, 95–100pp.

Lade B. D., Gogle D. P., Lade D. B., Moon G. M., Nandeshwar S. B., Kumbhare S. D. (2019). *Nano-Biopesticides Today and Future Perspectives CHAPTER 7: Nanobiopesticide Formulations: Application Strategies Today and Future Perspectives*. Academic Press, 179–206pp.

Lade B., Shanware A. (2020). Phytonanofabrication: Methodology and factors affecting biosynthesis of nanoparticles. In: Shabatina Y., Bochenkov V. (Eds.), *Smart Nanosystems for Biomedicine, Optoelectronics and Catalysis*. London: Intech, 1–17pp. ISBN: 9781838802547

Lange F., Cornelissen S., Kubac D., Sein M. M., Von Sonntag J., Hannich C. B. (2006). Degradation of macrolide antibiotics by ozone: A mechanistic case study with clarithromycin. *Chemosphere* 65, 17–23pp.

Larios-Rodriguez E., Rangel-Ayon C., Castillo S. J., Zavala G., Herrera-Urbina R. (2011). Bio-synthesis of gold nanoparticles by human epithelial cells, *in vivo*. *Nanotechnology* 22, 355601 (8pp).

Larson S. J., Capel P. D., Majewski M. S. (Eds.). (1997). *Pesticides in surface waters—distribution, trends and governing factors*. 1st edition. Boca Raton, FL: CRC Press Taylor & Francis, 400pp. ISBN 9780367455828

Lee J., Kim H. Y., Zhou H., Hwang S., Koh K., Han D. W., Lee J. (2011). Green synthesis of phytochemical-stabilized Au nanoparticles under ambient conditions and their biocompatibility and antioxidative activity. *Journal of Materials Chemistry* 21, 13316–13326pp.

Lynch I., Cedervall T., Lundqvist M., Cabaleiro-Lago C., Linse S., Dawson K. A. (2007). The nanoparticle-protein complex as a biological entity; a complex fluids and surface science challenge for the 21st century. *Advances in Colloid and Interface Science* 134–135, 167–174pp.

Maiti S., Gadadhar B., Laha J. K. (2016). Detection of heavy metals (Cu^{+2}, Hg^{+2}) by biosynthesized silver nanoparticles. *ApplNanosci* 6, 529–538pp.

Makarov V. V., Love A. J., Sinitsyna O. V., Makarova S. S., Yaminsky I. V., Taliansky M. E., Kalinina N. O. (2014). Green nanotechnologies: Synthesis of metal nanoparticles using plants. *Acta Naturae* 6, 35–44pp.

Mandava K. (2017). Biological and non-biological synthesis of metallic nanoparticles: Scope for current pharmaceutical research. *Indian Journal of Pharmaceutical Sciences* 79, 501–512pp.

Marchiol L. (2012). Synthesis of metal nanoparticles in living plants. *Italian Journal of Agronomy* 7, 274–282pp.

Mittal A. K., Chisti Y., Banerjee U. C. (2013). Synthesis of metallic nanoparticles using plant extracts. *Biotechnology Advances* 31, 346–356pp.

Mohanpuria P., Rana N. K., Yadav S. K. (2008). Biosynthesis of nanoparticles: Technological concepts and future applications. *Journal of Nanoparticle Research* 10, 507–517pp.

Murgueitio E., Cumbal L., Abril M., Izquierdo A., Debut A., Tinoco O. (2018). Green synthesis of iron nanoparticles: Application on the removal of petroleum oil from contaminated water and soils. *Journal of Nanotechnology* 2018, 1–8pp.

Nasipuri P., Alex L. E., Mukherjee I., Pandit G. G., Thakur A. R., Ray Chaudhuri S. (2011). Enhancement of crystal size by Microbes. *Research Journal of Nanoscience and Nanotechnology* 1, 42–47pp.

Nasrollahzadeh M., Atarod M., Sajadi S. M. (2016). Green synthesis of the Cu/Fe3O4 nanoparticles using Morinda morindoides leaf aqueous extract: a highly efficient magnetically separable catalyst for the reduction of organic dyes in aqueous medium at room temperature. *Applied Surface Science* 364, 636–644pp.

Ng K. C., Cao B., Mohanty A. (2015). Biofilms in Bio-nanotechnology: Opportunities and Challenges. In: Singh O. V. (Ed.), *Bio-nanoparticles: Biosynthesis and Sustainable Biotechnological Implications*, 1st edition. Hoboken, NJ: John Wiley & Sons, Inc., 83–100pp. ISBN: 9781118677681.

Orlando E. F., Kolok A. S., Binzcik G. A., Gates J. L., Horton M. K., Lambrigth C. S. Gray L. E., Soto A. M., Guillette L. J. (2004). Endocrine-disrupting effects of cattle feedlot effluent on an aquatic sentinel species, the fathead minnow. *Environmental Health Perspectives* 112, 353–358pp.

Ozkara A., Akyil D., Konuk M. (2016). Pesticides, environmental pollution, and health. In: Larramendy M. L., Solonesky S. (Eds.), *Environmental Health Risk: Hazardous Factors to Living Species*. Croatia: Intech. ISBN: 978-953-51-2402-3.

Pandey S., Kumari M., Singh S., Bhattacharya A., Mishra S., Chauhan P., Mishra A. (2015). Bioremediation via nanoparticles: An innovative microbial approach. In: Singh S., Srivastava K. (Eds.), *Handbook of Research on Uncovering New Methods for Ecosystem Management through Bioremediation*. Hershey, PA: IGI Global, 491–515pp. ISBN: 9781466686823.

Pandian C. J., Palanivel R., Solairaj D. (2015). Green synthesis of nickel nanoparticles using Ocimum sanctum and their application in dye and pollutant adsorption. *Chinese Journal of Chemical Engineering* 23, 1307–1315pp.

Panigrahi S., Basu S., Praharaj S. (2007). Synthesis and size-selective catalysis by supported gold nanoparticles: Study on heterogeneous and homogeneous catalytic process. *The Journal of Physical Chemistry C* 111, 4596–4605pp.

Parveen K., Bansi V., Ledwani L. (2016). Green synthesis of nanoparticle: Their advantages and disadvantages. *American Institute of Physics* 1724, 020048-1–020048-7pp.

Priyanka U., Akshay Gowda K. M., Elisha M. G., Nitish N. (2017). Biologically synthesized PbS nanoparticles for the detection of arsenic in water. *International Biodeterioration & Biodegradation* 119, 78–86pp.

Qu X., Alvarez P. J. J., Li Q. (2013a). Applications of nanotechnology in water and wastewater treatment. *Water Research* 47, 3931–3946pp.

Qu X., Brame J., Li Q., Pedro J., Alvarez J. (2013b). Nanotechnology for a safe and sustainable water supply: Enabling integrated water treatment and reuse. *Accounts of Chemical Research* 46, 834–843pp.

Rai A., Singh A., Ahmad A., Sastry M. (2006). Role of Halide ions and temperature on the morphology of Biologically synthesized Gold nanotriangles. *Langmuir* 22, 736–741pp.

Ray Chaudhuri S., Agarwala B. K., Sett S. K., Chaudhuri P., Paul P., Bhattacharjee G., Deb S., Chowdhury S., Devi P., Barman S., Gogoi M., Biswas T., Baidya P., Bora A., Chakraborty A., Chanda C., Saha S., Modak A., Das G., Sarkar P., Jamatia R., Mukherjee A., Kumar A., Thakur A. R., Sudarshan M., Nath R., Mishra L., Mukherjee I., Bose G., Singh A., Naik R. K. (2021). Self-sustained ramie cultivation: An alternative livelihood option. In: Thatoi H., Das S. K., Mahapatra S. (Eds.), *Bioresource Utilization and ManagementApplications in Therapeutics, Biofuels, Agriculture, and Environmental Science*. Burlington: Apple Academic Press (AAP), Inc., Taylor & Francis, 365–382pp. ISBN: 9781771889339.

Razavi M., Salahinejad E., Fahmy M., Yazdimamaghani M., Vashaee D., Tayebi L. (2015). Green chemical and biological synthesis of nanoparticles and their biomedical applications. In: Basiul V. A., Basiul E. V. (Eds.), *Green Processes for Nanotechnology: From Inorganic to Bioinspired Nanomaterials*. Cham: Springer, 207–235pp. ISBN: 9783319154602.

Romero-González M. E., Williams C. K., Gardiner P. H. E., Gurman S. J., Habesh S. (2003). Spectroscopic studies of the biosorption of gold (III) by dealginated seaweed waste. *Environmental Science & Technology* 37, 4163–4169pp.

Saini R. D. (2017). Textile Organic Dyes (TOD): Polluting effects and elimination methods from textile waste water. *International Journal of Chemical Engineering Research* 9, 121–136pp.

Sarathy V., Tratnyek P. G., Nurmi J. T. (2008). Aging of iron nanoparticles in aqueous solution: Effects on structure and reactivity. *The Journal of Physical Chemistry C* 112, 2286–2293pp.

Sarkar P., Biswas T., Chanda C., Saha A., Sudarshan M., Majumder C., Ray Chaudhuri S. (2021). Spent coffee waste conversion to value added products for pharmaceutical industry. In: Thatoi H., Das S. K., Mahapatra S. (Eds.), *Bioresource Utilization and ManagementApplications in Therapeutics, Biofuels, Agriculture, and Environmental Science*. Burlington: A Taylor & Francis Group. Apple Academic Press (AAP), Inc., 471–487pp. ISBN: 9781771889339.

Saxena G., Chandra R., Bharagava R. N. (2016). Environmental pollution, toxicity profile and treatment approaches for tannery wastewater and its chemical pollutants. *Reviews of Environmental Contamination and Toxicology* 240, 31–69pp.

Selvakumar R., Jothi N. A., Jayavignesh V., Karthikaiselvi K., Antony G. I., Sharmila P. R., Kavitha S., Swaminathan K. (2011). As (V) removal using carbonized yeast cells containing silver nanoparticles. *Water Research* 45, 583–592pp.

Senapati S., Mandal D., Ahmad A., Khan M. I., Sastry M., Kumar R. (2004). Fungus mediated synthesis of silver nanoparticles: A novel biological approach. *Indian Journal of Physics* 78A, 101–105pp.

Shah M. P. (2020). *Advanced Oxidation Processes for Effluent Treatment Plants*. Elsevier.

Shah M. P. (2021). *Removal of Emerging Contaminants through Microbial Processes*. Springer.

Shankar S. S., Ahmad A., Pasricha R., Sastry M. (2003). Bioreduction of chloroaurate ions by geranium leaves and its endophytic fungus yields gold nanoparticles of different shapes. *Journal of Materials Chemistry* 13, 1822pp.

Singh J., Dutta T., Kim K. H. (2018). 'Green' synthesis of metals and their oxide nanoparticles: Applications for environmental remediation. *Journal of Nanobiotechnology* 16, 84pp.

Singh J., Vyas A., Wang S., Prasad R. (Eds.). (2020). Microbial biotechnology: Basic research and applications. *Environmental and Microbial Biotechnology* 14, 370pp.

Soni N., Prakash S. (2011). Factors affecting the geometry of silver nanoparticles synthesis in *Chrysosporium tropicum* and *Fusarium oxysporum*. *Current Research in Nanotechnology* 2, 112–121pp.

Taheriniya S., Behboodi Z. (2016). Comparing green chemical methods and chemical methods for the synthesis of titanium dioxide nanoparticles. *International Journal of Pharmaceutical Sciences and Research* 7, 4927–4932pp.

Tanzil A., Sultsna S. T., Saunders S. R., Shi L., Marsili E., Beyenal H. (2016). Biological synthesis of nanoparticles in biofilms. *Enzyme and Microbial Technology* 95, 4–12pp.

Tran Q. H., Nguyen V. Q., and Le A. T. (2013). Silver nanoparticles: Synthesis, properties, toxicology, applications and perspectives. *Advances in Natural Sciences: Nanoscience and Nanotechnology* 4, 20pp.

Uddandarao P., Balakrishnan R. M., Ashok A., Swarup S., Sinha P. (2019). Bioinspired ZnS: Gd Nanoparticles Synthesized from an Endophytic Fungi Aspergillus flavus for fluorescence-based metal detection. *Biomimetics* 4, 11pp.

Varjani S. J., Gnansounou E., Pandey A. (2017). Comprehensive review on toxicity of persistent organic pollutants from petroleum refinery waste and their degradation by microorganisms. *Chemosphere* 188, 280–291pp.

Vithiya K., Sen S. (2011). Biosynthesis of nanoparticles. *IJPSR* 2, 2781–2785pp.

Woo Y., Lai D. Y. (2012). Aromatic amino and nitro–amino compounds and their halogenated derivatives. In: Bingham E., Cohrssen B., Powell C. H. (Eds.), *Patty's Toxicology*. Hoboken, NJ: John Wiley & Sons, Inc., 6, 1–96pp. ISBN: 9780470410813.

Yadav V. K. (2020). Microbial synthesis of nanoparticles and their applications for wastewater treatment. In: Singh J., Vyas A., Wang S., Prasad R. (Eds.), *Microbial Biotechnology: Basic Research and Applications. Environmental and Microbial Biotechnology*. Singapore: Springer, 147–187pp. ISBN: 9789811528170.

Yang D., Zhang X., Zhou Y., Xiu Z. (2015). The principle and method of wastewater treatment in biofilm technology. *Journal of Computational and Theoretical Nanoscience* 12, 2630–2638pp.

Yu J., Yang J., Liu B., Ma X. (2009). Preparation and characterization of glycerol plasticized-pea starch/ZnO–carboxymethylcellulose sodium nanocomposites. *Bioresource Technology* 100, 2832–2841pp.

Yuan C. G., Huo C., Gui B. (2017). Green synthesis of silver nanoparticles using *Chenopodium aristatum* L. stem extract and their catalytic/antibacterial activities. *Journal of Cluster Science* 28, 1319–1333pp.

Zhang P., Shao C., Zhang Z., Zhang M., Mu J., Guo Z., Liu Y. (2011). In situ assembly of well-dispersed Ag nanoparticles (AgNPs) on electrospun carbon nanofibers (CNFs) for catalytic reduction of 4-nitrophenol. *Nanoscale* 3, 3357–3363pp.

Zhang X., Qu Y., S., Wang J., Li H., Zhang Z., Li S., Zhou J. (2016). Biogenic synthesis of gold nanoparticles by yeast Magnusiomyces ingens LH-F1 for catalytic reduction of nitrophenols. *Colloids and Surfaces A: Physicochemical and Engineering Aspects* 497, 280–285pp.

10 Applications of Nanofiltration for Wastewater Treatment

Charles Oluwaseun Adetunji, Olugbemi T. Olaniyan, Kshitij RB Singh, Ruth Ebunoluwa Bodunrinde, Abel Inobeme, John Tsado Mathew, Ogundolie Abimbola Frank, Olalekan Akinbo, Jay Singh, Vanya Nayak, and Ravindra Pratap Singh

CONTENTS

10.1 Introduction 157
10.2 Principle and the Mechanism of Action Involved in the Process of Biofiltration 158
10.3 Relevance of the Process of Filtration for Suspended Solid, Biodegradation of Biological Component, and Adsorption of Micro-Pollutants 160
10.4 Process of Maintenance and Operation of the Biofilter 162
10.5 Several Immobilisation Process Using Artificial Entrapment of Microorganisms within Polymer Beads and the Self-Attachment of Microorganisms to the Bedding Material 162
10.6 The Application of Nanofiltration in Water Purification and Wastewater Treatment 164
10.7 The Mechanism of Action Using Nanofiltration for Treatment of Wastewater 164
10.8 Methods Involved in the Biofabrication of Nanofiltration 165
10.9 Conclusion and Future Recommendation 165
References 165
Review Article 165

10.1 INTRODUCTION

Membrane processes have been identified as a sustainable approach that could be applied as a permanent substitute to traditional unit operations, most especially in the management of wastewater and in the process of desalination. This might be linked to their eco-friendly attribute, proficient separation competencies, and compact size as a result of their minimal chemical nature (Dąbrowski and Robens, 2004; Drioli et al., 2017).

Nanofiltration (NF) has been identified as a pressure-driven membrane process that involves the separation effectiveness that exists between the process of ultrafiltration and reverse osmosis. Nanofiltration is normally performed by the application of asymmetric polymeric membranes, which entails a functionally active porous top layer that has a lower resistance supporting layer. The pore size around the active layers varies within 1 nm and most of them possess a fixed charge. Moreover, together with their mode of action that involves sieving, the surface charge of the support layer enables declining of charged ions with a minimal size that are greater than that of the membrane pores. These features enable nanofiltration to maintain multivalent ions while permitting the movement of smaller, uncharged monovalent ions. This process entails minimal energy utilisation in comparison with reverse osmosis when enable nanofiltration an effective techniques for denunciation of heavy metals from wastewater (Fane et al., 1992; Andrade et al., 2017).

Numerous scientists have documented the application of nanofiltration of the management of complex wastewater that represents a standalone process (Liu et al., 2008; Muthukrishnan and Guha,

DOI: 10.1201/9781003165149-10

2008; Mnif et al., 2017; Mohammad et al., 2004; Lin et al., 2005) together with some other process such as ultrafiltration or coagulation (Ates and Uzal, 2018). Therefore, this chapter intends to provide detailed information on the application of nanofiltration for effective purification of wastewater.

10.2 PRINCIPLE AND THE MECHANISM OF ACTION INVOLVED IN THE PROCESS OF BIOFILTRATION

Adsorption has become increasingly important in environmental protection and industry over the last few decades. Adsorption processes are extensively used for purification and separation because of energy efficiency, high reliability, technological maturity, design flexibility, and the ability to redevelop the depleted adsorbent. Biofiltration is one of the tools for expanding the adsorption treatment methods. The biological filter is based on the activities of a population of microbes that bind themselves to the filter media.

Microorganisms generate energy by oxidising organic matter in water, so available nutrients in feed water are critical for their growth. Biofiltration can effectively extract organic matter that cannot be separated from water or chemically treated sewage effluent in traditional sewage treatment. In this study, microbial attachment procedure, the kinetics of microbial growth, the factors that influence biological filtration, and descriptions of the microbial community in the biofilter are investigated. There are numerous biofilters, including trickling filters, submerged filters, fluidised beds, and beds (Chen and Hoff, 2012).

Different membranes are required to separate various types of particles and microorganisms; therefore, the membrane techniques are also categorised based on the pressure gradient as membrane filter (MF), ultra-filtration (UF), nanofiltration (NF), and reverse osmosis (RO), as illustrated in Figure 10.1. The figure also shows the filtration spectrum of each membrane and its applicable range. However, RO and NF are classified under the main membrane separation in which water is pressurised and forced at a semi-permeable membrane.

Biofiltration has long been a hopeful biotechnology in the treatment of contaminated air's volatile organic contaminates.

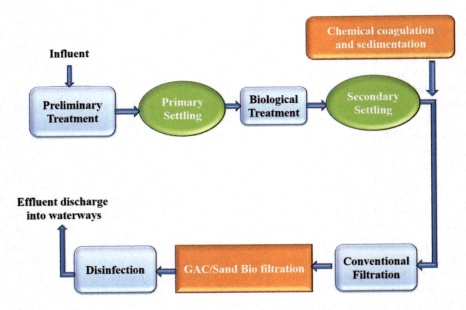

FIGURE 10.1 The filtration spectrum classified based on the pressure gradient (reprinted with permission from Abdel-Fatah (2018) [CC BY-NC-ND 4.0]).

Figure 10.2 is a schematic diagram of the laboratory scale biofilter system designed to treat air-containing toluene (reprinted with permission from Mohamed et al. (2016)).

Biofiltration removes biodegradable volatile organic compounds, odours, and other harmful compounds from contaminated air using biologically active media. Contaminants are removed in a phase involving several steps in which the contaminated air is passed into one or more beds containing biologically active groups, where the pollutants are biologically oxidised into simple inorganic materials like water and carbon dioxide by these microorganisms.

Diffusion and biodegradation are the two primary treatment routes. As polluted gas passes through the reactor, contaminants are transferred from the gaseous state to the liquid or solid phase and then to the media, where microorganisms biodegrade the contaminants (Shareefdeen et al., 2011). The several vital stages in the process of biofiltration are dissolution of odorous contaminants in water within the media, movement from bulk waste gas to media through the process of diffusion and then within the particles in the media.

Bio-oxidation (biodegradation) is a process through which microbes in media coverts metals in the insoluble form to a soluble form (Sakunthala et al., 2013). The most important element of biofiltration has an organic media that can sustain certain microorganisms that could biologically react to the pollutants. Once absorbed in the water layer or biofilm layer surrounding the biofilm, the contaminants, usually organic molecules, become available as food, acting as a carbon and energy source for microorganism growth and metabolic activities. The end products of the biodegradation process, especially water, carbon dioxide, and all treated air, are frequently exhausted from the biofilter. The biochemical reactions at work are incredibly complicated. Several groups of microorganisms collaborate in a network of metabolic processes, where a particular compound can be broken down into less complex compounds at each point (Irfana et al., 2016). Any filter with

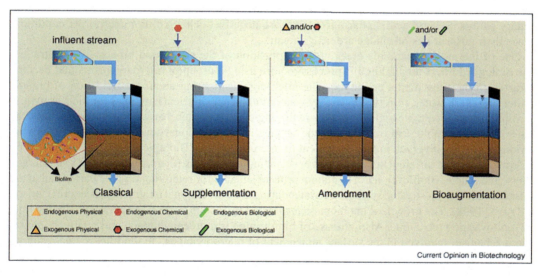

FIGURE 10.2 Schematic illustration of a bench-scale biofilter system that was designed for this study. A cylinder-type stainless steel column biofilter of 1.4 cm internal diameter and 24 cm height is built to remove toluene. Biofilter bed is packed with *S. griseus* DSM-40759, immobilised filter material (12 g of activated carbon). To establish continuous biofiltration, it was composed of nutrient reservoirs and nutrient pump for the supply of nutrients to the bacteria and also to prevent the packing material from drying; a liquid toluene storage bottle; a pump of air to evaporate toluene through a volatilisation chamber before entering the biofilter; an airflow meter; and finally inlet and outlet samples ports. The toluene vapor concentration was adjusted with the help of flow rate. Air was blown via a pump that was connected by a waste-gas generating unit containing liquid toluene operated at a fixed flow rate (2 L/min) at room temperature (25°C) with empty bed retention times (EBRTs) in the 80s.

attached biomass on the filter media could be referred to as a biological filter. It is one of the most recent processes for the separation of organic and micropollutants from wastewater.

Biological filtration involves the biodegradation of the contaminants, which is aided by the activities of microorganisms found attached to the filter media surface. The most remarkable operation parameter that determines the process is the attachment of the biomass to the medium. The mass of microorganisms such as bacteria found attached to the filter media as the biofilm aid in the oxidation of the organic matter through the use of energy using carbon as their energy sources. Few models exist in the literature that have reported to aid the prediction of the performance of the biofilters. But, the majority of the models focus on steady-state condition assumption. The biofiltration process consists of basically aerobic microorganisms, which implies the need for oxygen for their metabolic processes. The oxygen can be obtained from the biofilm either simultaneously or together with the flow of water. The process of aeration takes place passively through the natural flow of the air from the blowers. Biofiltration is suitable for the removal of industrial pollutants as well as organic matter such as surfactants.

10.3 RELEVANCE OF THE PROCESS OF FILTRATION FOR SUSPENDED SOLID, BIODEGRADATION OF BIOLOGICAL COMPONENT, AND ADSORPTION OF MICRO-POLLUTANTS

One of the remarkable fates of micropollutants in wastewater is the adsorption on the surfaces of particulate solid matter in the secondary and primary treatment phases. The occurrence of adsorption could result from hydrophobic associations within the aromatic and aliphatic groups of molecules with the different fractions of lipids and fats present in the primary sludge and the cell membranes of the microorganisms that are lipophilic. There are also electrostatic interactions between the positively charged groups in the micropollutants and the microbes present in the secondary sludge. Several acidic pharmaceuticals are negatively charged in a neutral pH condition; hence their adsorptions on sludge are insignificant.

One of the most remarkable areas in the application of the adsorption process is biofiltration. The biological filters applied in this process require using a group of microorganisms that become attracted to the filter media employed. The microorganisms bring about the oxidation of organic matter present in water, resulting in energy generation; hence, nutrients are vital for their growth.

Biofiltration is suitable and reliable for removing organic matter that was not removed from the water during the biological treatment of the sewage in the traditional sewage treatment. There are various processes involved in the attachment of microorganisms. Biological filtration is affected by several factors, and different kinds of filters exist, including trickling filters, fluidised beds, submerged filters, and others. Biofiltration is a treatment process suitable for removing organic particles that are not removed from bio-treated sewage and water during the conventional treatment procedure (Carlson and Amy, 1998). The growth of the biomass is affected by factors such as temperature, loading rate, pH, and washing technique. Also, the use of biofiltration is a green approach and relatively cheap. The mechanism involved in the surface attachment of microorganisms to the filter media resulting in biofilm formation is complex. Several methods have been employed in the studies, such as high-resolution microscopy, microbalance applications, laser microscopy, and scanning electron microscopy (Percival et al., 2000).

Numerous parameters apply during the attachment of microorganisms to the surface, during which the intensity of the attachment depends on the conditions within the environment, fluid properties, type of microbes, and surface characteristics. There are five steps involved during the attachment of microbes to the filter media surface: Formation of surface conditioning layer of film, cell transport to the surface, adhesion process, colonisation within the surface, and detachment (Percival et al., 2000). The primary essence of biofiltration is the improvement of the effluent quality of wastewater. This approach is relevant, especially in the recirculation process, and it is usually found as a component of the multicomponent system. There are three major mechanisms involved

Applications of Nanofiltration for Wastewater Treatment

in removing contaminants by using biological filters: Biological degradation of the bio component, micropollutant adsorption, and suspended particle filtration.

Biochar as a steady carbon-rich substance indicates the incredible capability to handle wastewater/water contaminants. It has ore potential due to the simplicity of the preparation methods, the enhanced physico-chemical properties, and the availability of feedstock. The efficiency of biochar to eliminate inorganic and organic pollutants differs based on its pore size distribution, surface area, size of the molecules to be removed, and surface functional groups, although the surface properties and physical architecture of biochar vary based on the preparation conditions/method and the nature of the feedstock. The biochar removes the organic and inorganic contaminants mainly through the adsorption technique. The organic contaminants are adsorbed by electrostatic attraction, pore-filling, π-π electron-donor acceptor interaction, hydrogen-bonding, complexes adsorption, and hydrophobic interactions, whereas the inorganic pollutants, like heavy metals, are removed through surface precipitation under alkaline conditions, ion exchange and complexation, and cationic and anionic electrostatic attraction as illustrated in Figure 10.3. For example, high-temperature pyrolysis produces hydrophobic biochars with higher surface area and micropore volume, making them better for sorption of organic contaminants, while low-temperature biochars have a lower surface area, higher oxygen-containing functional groups, and smaller pore size, making them better for eliminating inorganic pollutants. Biochar has a wide range of applications in the wastewater/water treatment sector. Biochar has been commonly used as a filter media to eliminate heavy metals, pathogens, suspended matter, and a support/additive media during anaerobic digestion. Biochar was also put to the test as a support-based catalyst for dye and recalcitrant contaminant degradation. The current analysis examines the various methods for producing biochar and offers an overview of biochar's current applications in wastewater treatment. (Enaime et al., 2020; Rattier et al., 2012).

Ingestion of infected water spreads disease causing microorganisms such as bacteria, protozoa, helminths, and viruses in the human population, posing a significant threat to public health. In both

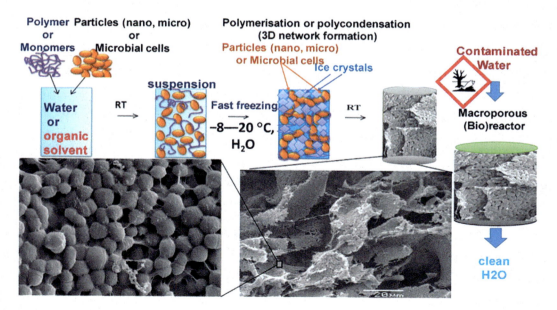

FIGURE 10.3 The different mechanisms for the removal of organic and inorganic contaminants with the help of biochar.

Source: Reprinted with the permission from Enaime, G., Bacaoui, A., Yaacoubi, A., & Lübken, M. (2020). Biochar for wastewater treatment – conversion technologies and applications. *Applied Sciences, 10*(10), 3492 [CC BY 4.0]).

decentralised point-of-use and centralised multistep water treatment systems, the biological filtration approach of treatment is the most important phase. A biofilter, which is a biofilm of localised microorganisms fixed to a solid layer, is an environmentally friendly and cost-effective method for removing contaminants and pathogens. Gravel, stone, soil, sand, mesoporous mineral stone, compost wood, and chips demonstrate a solid bed medium that requires improved surface area meant for water retention and microbial adherence. The slow sand filter, which has a biological surface over the filter bed called schmutzdecke and a slow flow rate (0.1–0.2 m3/h), is reasonably effective at removing coliform microorganisms, including Cryptosporidium, Giardia, Escherichia coli, Salmonella, fecal coliform and total coliform, bacteriophage, fecal streptococci, and MS2 virus from wastewater. The rapid sand filter, which has a higher flow rate and coarser sand than the slow sand filter, effectively eliminates indicator microorganisms. The biological activity and physical retention of the pathogen on filter media in a stormwater biofilter is a common and successful method for pathogen removal. Inoculum size, microbial diversity, moisture of medium and pH, nutrient content, and temperature are the main parameters regulating the optimal performance of the biofilter system (Anurag et al., 2020).

Micropollutants in surface water can jeopardise the production of high-quality, healthy drinking water. In the production of drinking water, the adsorption of micropollutants onto granular activated carbon in fixed-bed filters is frequently applied as a polishing step. Biodegradation can be an additional removal method for micropollutants in granular activated carbon filters because activated carbon can serve biofilm carrier material. It's important to distinguish adsorption from biodegradation as a removal mechanism when evaluating biofilm's ability to biodegrade micropollutants. To test the biodegradation of micropollutants by means of the biofilm grown on the granular activated carbon surface, experiments were conducted at 5°C and 20°C with biologically active and autoclaved granular activated carbon. As model compounds, ten micropollutants were chosen. Three of them including iopamidol metformin and iopromide were biodegraded through the granular activated carbon biofilm.

Furthermore, depending on the micropollutant tested, the temperature has been shown to increase or decrease adsorption. Finally, the adsorption power of granular activated carbon is used for more than 100,000-bed volumes compared to fresh granular activated carbon. When compared to new granular activated carbon, the study found that used granular activated carbon has higher adsorption ability for metformin, hexamethylenetetramine, and guanylurea, with only a minor reduction in adsorption capacity for benzotriazole and diclofenac (Piai et al., 2020).

10.4 PROCESS OF MAINTENANCE AND OPERATION OF THE BIOFILTER

Various groups of microorganisms, both anaerobic and aerobic, including fungi, protozoa, bacteria, and algae, are formed on the filter media's surface, resulting in a slimy layer called the biofilm. The formation of a biofilm considered active could take up to a few days, depending on the concentration of the organic matter. Filter media exhausted, such as granular activated carbon, could be generated again for subsequent usage.

10.5 SEVERAL IMMOBILISATION PROCESS USING ARTIFICIAL ENTRAPMENT OF MICROORGANISMS WITHIN POLYMER BEADS AND THE SELF-ATTACHMENT OF MICROORGANISMS TO THE BEDDING MATERIAL

Bioremediation is a crucial method for using biological agents to recycle contaminated water and soil. Using live microorganisms to neutralise or eliminate harmful contaminants from polluted areas is a popular approach. Bioremediation approaches have traditionally relied on immobilised microorganisms rather than planktonic cells. Inorganic minerals (metal oxides, zeolites, clays), activated carbon, and, although there are limitations with biomass loading and bacteria leaching during the process, agricultural waste products are suitable substrates for the immobilisation of bacteria. Cells and microorganisms have been successfully immobilised using various natural and synthetic polymers with different functional classes. In biotechnology and water treatment applications, macroporous materials such as cryogels through entrapped cells or bacteria have shown promise. A cryogel is a macroporous polymeric gel that is formulated at sub-zero temperatures using the cryogelation process. Supports or scaffolds made of macroporous hydrogels have been used to immobilise viral, bacterial, and other cells. The manufacture of composite materials through immobilised cells possessing appropriate chemical and mechanical stability, elasticity, porosity, and biocompatibility proposes that these materials are likely candidates for a series of uses within applied biotechnology, research, and microbiology (Berillo et al., 2021).

Petroleum hydrocarbons are the world's most important environmental contaminants, and oil spills endanger marine and terrestrial habitats. Oil pollution could arise either operationally or accidentally whenever the oil is transported, produced, stored, and used at sea or processed or on land. Oil spills are a significant environmental threat because they destroy the habitats around them. Bacterial cells must be immobilised to increase the retention and survival of bioremediation agents in polluted areas. Immobilised cells have been extensively studied in a wide range of applications. Depending on the application, there are a variety of immobilisation and assistance techniques to choose from. This study has shown the ability of immobilised microbial cells to degrade petroleum hydrocarbons in this review article. In some studies, immobilised cells for treating oil-contaminated areas have been more effective than free-living bacterial cells. It has been proven that immobilised cells are more efficient, faster, and can occur for longer periods. (Bayat et al., 2015).

The fact that the microorganisms and the processed waste are separated distinguishes biofiltration from other biological waste treatments. Microorganisms are immobilised to the bedding material in biofiltration systems as the processed fluid flows through it. A large volume of literature has been published in recent decades on single experiments involving the treatment of fluids by means of immobilised microorganisms. Several artificial immobilisation methods have been investigated, and promising results in the treatment of fluids have been obtained using one artificial immobilisation method, the entrapment of microorganisms inside polymer beads. Even though it needs to be developed, this process appears to have commercial biofiltration device potential.

In contrast to the handling of fluids through the artificial immobilisation of microorganisms, naturally attached microorganisms within biofiltration systems have many advantages and disadvantages (Mohamed et al., 2016). Understanding the mechanisms and forces that cause microbes to adhere to bedding materials is critical to strengthen this attachment. The artificial entrapment of microorganisms inside polymers can be improved further, allowing the benefits of this approach to be used to treat fluids. This chapter aims to incorporate the key concepts of two immobilisation processes: Microorganism artificial entrapment inside polymer beads and microorganism self-attachment to bedding material. Each immobilisation process is explored in liquid and gas treatments (Cohen, 2001; Suja et al., 2014).

Recent biodegradation investigations have emphasised on the deterioration efficiency while preserving microbial activity stability during the process. Using the adhesive method to immobilise bacterial cells is an easy way to accomplish both goals. This study aimed to see how biofilm immobilisation affects two different bacterial consortia biodegradation efficiencies: High phenol degraders (base deterioration efficiency 90%) and low diesel oil degraders (base deterioration efficiency 40%). Four other carriers were used in the degradation tests (clay pellets, expanded polypropylene, polyvinyl chloride rings, and paperboard). Biofilms can significantly improve the efficiency of low degraders, according to the findings. The best diesel oil oxidation (80%) was achieved when paperboard was used as a carrier. The immobilisation of high degraders, on the other hand, did not affect their biodegradation potential. Only the enlarged clay pellets recorded about a 6% rise (Ławniczak et al., 2011).

10.6 THE APPLICATION OF NANOFILTRATION IN WATER PURIFICATION AND WASTEWATER TREATMENT

The nanofiltration method is widely used in several wastewater treatments and industrially; it works by selectively removing ions and carbon-based substances from water samples and water bodies. This technique is also applied to purifying municipal waste (Bunani et al., 2013). A type of nanofiltration lessened biochemical oxygen demand and conductivity when the process was done in the presence and absence of oxygen (Zulaikha et al., 2014).

Shahmansouri and Bellona (2015) found nanofiltration effective in owning to its reduced organic molecular weight and pore size of 1–5 nm. It was found to have meaningfully greater water permeability capacity at reduced pressure. Some researchers highlighted some rejection mechanisms: Interaction between water and the membrane via hydrogen bonding; capillary rejection; it presents a heterogenous membrane with extremely small pores and occurs from various electrostatic processes (Timmer, 2001; Simpson et al., 1987; Eriksson, 1988). In solution diffusion, the membrane is homogenous and does not permit the transfer of water or air. The dissolution of solute and solvent in the membrane active layer making layer transportation occurs via diffusion for the charged capillary. An ion with a charge similar to that of the membrane is involved, and ions with dissimilar ions exclude streaming potential. Their study discussed the declined influx of molecules that occupy the pores of the surface of the membrane, which was made possible by the hydrophobic nature of the solutes. They also noted that this method could cause low-pressure nanofiltration bioreactors and could be used for a longer period of time in the filtration process without fatal fouling and servicing.

10.7 THE MECHANISM OF ACTION USING NANOFILTRATION FOR TREATMENT OF WASTEWATER

The mode of operation in nanofiltration is dependent on the membrane, not allowing the passage of water, air and ultrafiltration. Choi et al. (2002) reported that the electrostatic features of the nanofiltration sheath were found to be responsible for the rejected ions, usually negatively zeta potential on the surface of the membrane with varying pH and different electrolyte solutions concentrations. Shahmansouri and Bellona (2015) reported that nanofiltration for wastewater through filtering proves effective because it prevents fouling. This method is one of the most important techniques in protecting and improving performance. They also reported that this process protects against mechanical and chemical destruction. When a high solid is loaded, it could cause mechanical damage to the surface of the membrane, thereby hindering free flow during filtration. Nanofiltration could be utilised in tertiary wastewater management, especially in eliminating frequently occurring pollutants from organic sources.

Shahmansouri and Bellona (2015) added a need for pretreatment to achieve effective filtration without experiencing extensive biological fouling. They said that fouling on the surface of the nanofiltration membrane can be minimised to the bare minimum in the presence of high-rate flocculation and exchange resin of magnetic ions. Nanofiltration has been found to meet the quality necessities of effective wastewater treatment.

Studies has shown that the choice of nanofiltration for wastewater purification could be as a result of its techniques. It is relentless and economically viable, affordable and easy to maintain. This method is appropriate for drinking water and reusing wastewater (Schaefer et al., 2005; Wilson-Engineering 2013).

Ghizellaoui et al. (2005) used nanofiltration technology to soften water and also in the removal of colour and some disinfectant dye products. The research of Beardsley and McClellan (1995) reported efficiency in nanofiltration compared to reverse osmosis, producing water with less stabilisation in minimising corrosion in the distribution system. Researchers also reported some hitch in nanoparticle utilisation even though they claimed this process was more effective than others. The effectiveness of this technique could be slowed down owing to the activity

Applications of Nanofiltration for Wastewater Treatment

of microorganisms and bringing about constant fouling. Nanofouling could be used to eliminate arsenic, detergent dye product, heavy metals, fluoride, pesticides, nitrate, oxyanions, and several emerging organic impurities (Saitua et al., 2011; Padilla and Saitua, 2010; Murthy and Choudhary, 2011; Santafe-Moros and Gozalvez-Zafrilla, 2010; Caus et al., 2009; Listiarini et al., 2010; Hajibabania et al., 2011).

Some researchers also reported that nanoparticles singly are not always effective in wastewater treatment and reuse for indirect potable reuse due to inorganic nitrogen and indiscriminate organic pollutants (Bellona and Drewes, 2007; Alzahrani et al., 2013). They concluded that nanofiltration is a newly developed technique in applying membranes in technology. This process has been cooperated into several industries for treating water and separation. The profitability of nanofiltration is dependent on the quality of the influent water capacity of the treatment amenities and the goals to be achieved while treating. Some of the limitations are due to narrow range temperature, membrane fouling, and lack of compatibility with chemical oxidants.

10.8 METHODS INVOLVED IN THE BIOFABRICATION OF NANOFILTRATION

Xing et al. (2010) indicated that nanomaterials obtained from the natural source could be utilised for several purposes such as food industries, pharmaceuticals, etc. In water treating and bioremediation, utilisation of nanotechnology was found to be most efficient, cost-effective, and environmentally friendly (Adetunji et al., 2020, 2021a, 2021b, 2021c, 2021f, 2021g, 2021h, 2021i, 2021j; Adetunji and Ugbenyen, 2019; Adetunji, 2008, 2019; Adetunji and Anani, 2020; Hameed et al., 2022; Bello et al., 2019; Adejumo et al., 2017). Cellulose, being a natural product, was utilised as an adsorbent for contaminants – usually metals – and as a supporting framework for toxic pollutants. Bacterial cellulose formed by *Gluconacetobacter xylinus* from sugar was biofabricated and utilised in engineering tissue, blood vessel fabrication, and other processes (Klemm et al., 2011).

10.9 CONCLUSION AND FUTURE RECOMMENDATION

This chapter has provided detailed information on the management of wastewater. The mode of action involved in the application of nanofiltration of the management of wastewater include largely immobilisation. Several immobilisation process use artificial entrapment of microorganisms within polymer beads and the self-attachment of microorganisms to the bedding material. There is a need to explore the application of biogenic natural materials in the synthesis of biofiltration as an effective sustainable biotechnological tool for management of wastewater.

REFERENCES

Alzahrani, S., Mohammad, A. W., Hilal, N., Abdullah, P., & Jaafar, O. (2013). Comparative study of NF and RO membranes in the treatment of produced water—Part I: Assessing water quality. *Desalination* 315, 18–26.

Beardsley, S. S., & McClellan, S. A. (1995). Membrane softening: An emerging technology helping Florida communities meet the increased regulation for quality potable water. In: *American Waterworks Association Membrane Technology Conference*, Reno, NV.

Bellona, C., & Drewes, J. E. 2007. Viability of a low-pressure nanofilter in treating recycled water for water reuse applications: A pilot-scale study. *Water Research* 41(17), 3948–3958.

REVIEW ARTICLE

Abdel-Fatah, M. A. (2018). Nanofiltration systems and applications in wastewater treatment. *Ain Shams Engineering Journal* 9(4), 3077–3092.

Adejumo, I. O., Adetunji, C. O., Nwonuma, C. O., Alejolowo, O. O., & Maimako, R. (2017). Evaluation of selected agricultural solid wastes on biochemical profile and liver histology of albino rats. *Food and Feed Research* 44(1), 73–79.

Adetunji, C. O. (2008). *The Antibacterial Activities and Preliminary Phytochemical Screening of Vernoni-aamygdalina and Aloe Vera Against Some Selected Bacteria.* M.Sc Thesis, University of Ilorin, pp. 40–43.

Adetunji, C. O. (2019). Environmental impact and ecotoxicological influence of biofabricated and inorganic nanoparticle on soil activity. In: *Nanotechnology for Agriculture* (D. Panpatte & Y. Jhala, eds.). Singapore: Springer. https://doi.org/10.1007/978-981-32-9370-0_12

Adetunji, C. O., & Anani, O. A. (2020). Bio-fertilizer from *Trichoderma*: Boom for agriculture production and management of soil- and root-borne plant pathogens. In: *Innovations in Food Technology* (P. Mishra, R. R. Mishra & C. O. Adetunji, eds.). Singapore: Springer. https://doi.org/10.1007/978-981-15-6121-4_17

Adetunji, C. O., Ajayi, O. O., Akram, M., Olaniyan, O. T., Chishti, M. A., Inobeme, A., Olaniyan, S., Adetunji, J. B., Olaniyan, M., & Awotunde, S. O. (2021c). Medicinal plants used in the treatment of influenza A virus infections. In: *Medicinal Plants for Lung Diseases* (K. Dua, S. Nammi, D. Chang, D. K. Chellappan, G. Gupta & T. Collet, eds). Singapore: Springer. https://doi.org/10.1007/978-981-33-6850-7_19

Adetunji, C. O. Akram, M., Olaniyan, O. T., Ajayi, O. O., Inobeme, A., Olaniyan, S., Hameed, L., & Adetunji, J. B. (2021f). Targeting SARS-CoV-2 novel corona (COVID-19) virus infection using medicinal plants. In: *Medicinal Plants for Lung Diseases* (K. Dua, S. Nammi, D. Chang, D. K. Chellappan, G. Gupta & T. Collet, eds). Singapore: Springer. https://doi.org/10.1007/978-981-33-6850-7_21

Adetunji, C. O., Inobeme, A., Olaniyan, O. T., Ajayi, O. O., Olaniyan, S., & Adetunji, J. B. (2021i). Application of nanodrugs derived from active metabolites of medicinal plants for the treatment of inflammatory and lung diseases: Recent advances. In: *Medicinal Plants for Lung Diseases* (K. Dua, S. Nammi, D. Chang, D. K. Chellappan, G. Gupta & T. Collet, eds). Singapore: Springer. https://doi.org/10.1007/978-981-33-6850-7_26

Adetunji, C. O., Michael, O. S., Kadiri, O., Varma, A., Akram, M., Oloke, J. K., Shafique, H., Adetunji, J. B., Jain, A., Bodunrinde, R. E., Ozolua, P., & Ubi, B. E. (2021a). Quinoa: From farm to traditional healing, food application, and phytopharmacology. In: *Biology and Biotechnology of Quinoa* (A. Varma, ed.). Singapore: Springer. https://doi.org/10.1007/978-981-16-3832-9_20

Adetunji, C. O., Michael, O. S., Varma, A., Oloke, J. K., Kadiri, O., Akram, M., Bodunrinde, R. E., Imtiaz, A., Adetunji, J. B., Shahzad, K., Jain, A., Ubi, B. E., Majeed, N., Ozolua, P., & Olisaka, F. N. (2021g). Recent Advances in the application of biotechnology for improving the production of secondary metabolites from quinoa. In: *Biology and Biotechnology of Quinoa* (A. Varma, ed.). Singapore: Springer. https://doi.org/10.1007/978-981-16-3832-9_17

Adetunji, C. O., Olaniyan, O. T., Adeyomoye, O., Dare, A., Adeniyi, M. J., Alex, E., Rebezov, M., Garipova, L., & Shariati, M. A. (2022). eHealth, mHealth, and telemedicine for COVID-19 pandemic. In: *Assessing COVID-19 and Other Pandemics and Epidemics using Computational Modelling and Data Analysis* (S. K. Pani, S. Dash, W. P. dos Santos, S. A. Chan Bukhari & F. Flammini, eds.). Cham: Springer. https://doi.org/10.1007/978-3-030-79753-9_10

Adetunji, C. O., Olaniyan, O. T., Akram, M., Ajayi, O. O., Inobeme, A., Olaniyan, S., Khan, F. S., & Adetunji, J. B. (2021b). Medicinal plants used in the treatment of pulmonary hypertension. In: *Medicinal Plants for Lung Diseases* (K. Dua, S. Nammi, D. Chang, D. K. Chellappan, G. Gupta & T. Collet, eds). Singapore: Springer. https://doi.org/10.1007/978-981-33-6850-7_14

Adetunji, C. O., Olaniyan, O. T., Anani, O. A., Olisaka, F. N., Inobeme, A., Bodunrinde, R. E., Adetunji, J. B., Singh, K. R. B, Palnam, W. D., & Singh, R. P. (2021h). Current scenario of nanomaterials in the environmental, agricultural, and biomedical fields. In: *Nanomaterials in Bionanotechnology*, 1st edition. Imprint CRC Press, 30pp. eBook ISBN 9781003139744. Ravindra Pratap Singh, Kshitij RB Singh, USA.

Adetunji, C. O., Roli, O. I., & Adetunji, J. B. (2020). Exopolysaccharides derived from beneficial microorganisms: Antimicrobial, food, and health benefits. In: *Innovations in Food Technology* (P. Mishra, R. R. Mishra & C. O. Adetunji, eds.). Singapore: Springer. https://doi.org/10.1007/978-981-15-6121-4_10

Adetunji, C. O., & Ugbenyen, M. A. (2019). Mechanism of action of nanopesticide derived from microorganism for the alleviation of abiotic and biotic stress affecting crop productivity. In: *Nanotechnology for Agriculture: Crop Production & Protection* (D. Panpatte, & Y. Jhala, eds.). Singapore: Springer. https://doi.org/10.1007/978-981-32-9374-8_7

Andrade, L. H., Aguiar, A. O., Pires, W. L., Miranda, G. A., Teixeira, L. P. T., Almeida, G. C. C., & Amaral, M. C. S. (2017). Nanofiltration and reverse osmosis applied to gold mining effluent treatment and reuse. *Brazilian Journal of Chemical Engineering* 34, 93–107. https://doi.org/10.1590/0104-6632.20170341s20150082.

Ates, N., & Uzal, N. (2018). Removal of heavy metals from aluminum anodic oxidation wastewaters by membrane filtration. *Environmental Science and Pollution Research* 25, 22259–22272. http://doi.org/10.1007/s11356-018-2345-z.

Bayat, Z., Hassanshahian, M., & Cappello, S. (2015). Immobilization of microbes for bioremediation of crude oil polluted environments: A mini review. *The Open Microbiology Journal* 9, 48–54.

Bello, O. M., Ibitoye, T., & Adetunji, C. (2019). Assessing antimicrobial agents of Nigeria flora. *Journal of King Saud University-Science* 31(4), 1379–1383.

Berillo, D., Al-Jwaid, A., & Caplin, J. (2021). Polymeric materials used for immobilisation of bacteria for the bioremediation of contaminants in water. *Polymers* 13, 1073. https://doi.org/10.3390/polym13071073.

Bunani, S., Yörükoğlu, E., Sert, G., Yüksel, Ü., Yüksel, M., & Kabay, N. (2013). Application of nanofiltration for reuse of municipal wastewater and quality analysis of product water. *Desalination*, 315, 33–36.

Carlson, K. H., & Amy, G. L. (1998). BOM removal during biofiltration. *Journal-American Water Works Association* 90(12), 42–52.

Caus, A., Vanderhaegen, S., Braeken, L., & Van der Bruggen, B. (2009). Integrated nanofiltration cascades with low salt rejection for complete removal of pesticides in drinking water production. *Desalination* 241(1–3), 111–117.

Chen, L., & Hoff, S. J. (2012). *A Two-Stage Wood Chip-based Biofilter System to Mitigate Odors from a Deep-Pit Swine Building*. Digital Repository @ Iowa State University. Agricultural and Biosystems Engineering (ASBE), No. 9631.

Choi, W., Termin, A., & Hoffmann, M. R. (2002). The role of metal ion dopants in quantum-sized TiO2: Correlation between photoreactivity and charge carrier recombination dynamics. *The Journal of Physical Chemistry* 98(51), 13669–13679.

Cohen, Y. (2001). Biofiltration—the treatment of fluids by microorganisms immobilized into the filter bedding material: a review. *Bioresource Technology* 77(3), 257–274. https://doi.org/10.1016/s0960-8524(00)00074-2.

Dąbrowski, A., & Robens, E. (2004). Selective removal of the heavy metal ions from waters and industrial wastewaters by ion-exchange method. *Chemosphere* 56, 91–106. https://doi.org/10.1016/j.chemosphere.2004.03.006.

Drioli, E., Ali, A., & Macedonio, F. (2017). Membrane operations for process intensification in desalination. *Applied Sciences* 7, 100. https://doi.org/10.3390/app7010100.

Enaime, G., Baçaoui, A., Yaacoubi, A., & Lübken, M. (2020). Biochar for wastewater treatment—conversion technologies and applications. *Applied Sciences* 10(10), 3492. https://doi.org/10.3390/app10103492.

Eriksson, P. (1988). Nanofiltration extends the range of membrane filtration. *Environmental Progress* 7(1), 58–62.

Fane, A. G., Awang, A. R., Bolko, M., Macoun, R., Schofield, R., Shen, Y. R., & Zha, F. (1992). Metal recovery from wastewater using membranes. *Water Science and Technology* 25, 5–18. https://doi.org/10.2166/wst.1992.0233.

Ghizellaoui, S., Taha, S., Dorange, G., Chibani, A., & Gabon, J. (2005). Softening of Hamma drinking water by nanofiltration and by lime in the presence of heavy metals. *Desalination* 171(2), 133–138.

Hajibabania, S., Verliefde, A., McDonald, J. A., Khan, S. J., & Le-Clech, P. (2011). Fate of trace organic compounds during treatment by nanofiltration. *Journal of Membrane Science* 373(1–2), 130–139. https://doi.org/10.3390/microorganisms9010004.

Hameed, A., Condò, C., Tauseef, I., Idrees, M., Ghazanfar, S., Farid, A., Muzammal, M., Al Mohaini, M., Alsalman, A. J., Al Hawaj, M. A., Adetunji, C. O., Dauda, W. P., Hameed, Y, Alhashem, Y. N., & Alanazi, A. A. (2022). Isolation and characterization of a cholesterol-lowering bacteria from *Bubalus bubalis* raw milk. *Fermentation* 8(4), 163. https://doi.org/10.3390/fermentation8040163

Irfana, S., Farooq, A. L., Moieza, A., Mohammad, A. M., & Asmat, R. (2016). Biofilters in mitigation of odour pollution—A review. *Nature Environment and Pollution Technology* 15(4), 1177–1185.

Klemm, D., Kramer, F., Moritz, S., Lindström, T., Ankerfors, M., Gray, D., & Dorris, A. (2011). Nanocelluloses: A new family of nature-based materials. *Angewandte Chemie International Edition* 50(24), 5438–5466.

Ławniczak, Ł., Kaczorek, E., & Olszanowski, A. (2011). The influence of cell immobilization by biofilm forming on the biodegradation capabilities of bacterial consortia. *World Journal of Microbiology and Biotechnology* 27(5), 1183–1188. https://doi.org/10.1007/s11274-010-0566-5.

Lin, S., Wang, T., & Juang, R. (2005). Metal rejection by nanofiltration from diluted solutions in the presence of complexing agents. *Separation Science and Technology* 39, 363–376. https://doi.org/10.1081/SS-120027563.

Listiarini, K., Tor, J. T., Sun, D. D., & Leckie, J. O. (2010). Hybrid coagulation-nanofiltration membrane for removal of bromate and humic acid in water. *Journal of Membrane Science* 365(1–2), 154–159.

Liu, F., Zhang, G., Meng, Q., & Zhang, H. (2008). Performance of nanofiltration and reverse osmosis membranes in metal effluent treatment. *Chinese Journal of Chemical Engineering* 16, 441–445. https://doi.org/10.1016/S1004-9541(08)60102-0.

Maurya, A., Singh, M. K., & Kumar, S. (2020). Biofiltration technique for removal of waterborne pathogens. *Waterborn Pathogen*, 123–141. doi:10.1016/B978-0-12-818783-8.00007-4.

Mnif, A., Bejaoui, I., Mouelhi, M., & Hamrouni, B. (2017). Hexavalent chromium removal from model water and car shock absorber factory effluent by nanofiltration and reverse osmosis membrane. *International Journal of Analytical Chemistry* 2017, 7415708. https://doi.org/10.1155/2017/7415708.

Mohamed, E. F., Awad, G., Andriantsiferana, C., & El-Diwany, A. I. (2016). Biofiltration technology for the removal of toluene from polluted air using *streptomyces griseus*. *Environmental Technology* 37(10), 1197–1207. https://doi.org/10.1080/09593330.2015.1107623.

Mohammad, A. W., Othaman, R., & Hilal, N. (2004). Potential use of nanofiltration membranes in treatment of industrial wastewater from Ni-P electroless plating. *Desalination* 168, 241–252. https://doi.org/10.1016/j.desal.2004.07.004.

Murthy, Z. V. P., & Choudhary, A. (2011). Application of nanofiltration to treat rare earth element (neodymium) containing water. *Journal of Rare Earths* 29(10), 974–978.

Muthukrishnan, M., & Guha, B. K. (2008). Effect of pH on rejection of hexavalent chromium by nanofiltration. *Desalination* 219, 171–178. https://doi.org/10.1016/j.desal.2007.04.054.

Padilla, A. P., & Saitua, H. (2010). Performance of simultaneous arsenic, fluoride and alkalinity (bicarbonate) rejection by pilot-scale nanofiltration. *Desalination* 257(1–3), 16–21.

Percival, S. L., Walker, J. T., & Hunter, P. R. (2000). *Microbiological Aspects of Biofilms and Drinking Water*. CRC Press, Boca Raton, Florida, USA.

Piai, L., Blokland, M., van der Wal, A., & Langenhoff, A. (2020). Biodegradation and adsorption of micropollutants by biological activated carbon from a drinking water production plant. *Journal of Hazardous Materials* 388, 122028. https://doi.org/10.1016/j.jhazmat.2020.122028.

Rattier, M., Reungoat, J., Gernjak, W., Joss, A., & Keller, J. (2012). Investigating the role of adsorption and biodegradation in the removal of organic micropollutants during biological activated carbon filtration of treated wastewater. *Journal of Water Reuse and Desalination* 2(3), 127–139. https://doi.org/10.2166/wrd.2012.012.

Saitua, H., Gil, R., & Padilla, A. P. (2011). Experimental investigation on arsenic removal with a nanofiltration pilot plant from naturally contaminated groundwater. *Desalination* 274, 1–6.

Sakunthala, M., Sridevi, V., Chandana lakshmi, M. V. V., & Vijay Kumar, K. (2013). A review: The description of three different biological filtration processes and economic evaluation. *Journal of Environmental Science, Computer Science and Engineering & Technology* 2(1), 91–99.

Santafe-Moros, A., & Gozalvez-Zafrilla, J. M. (2010). Nanofiltration study of the interaction between bicarbonate and nitrate ions. *Desalination* 250(2), 773–777.

Schaefer, A. I., Andritsos, N., Karabelas, A. J., Hoek, M. V., Schneider, R. P., & Nystrom, M. (2005). Fouling in nanofiltration. In: *Nanofiltration: Principles and Application* (A. I. Schaefer, A. G. Fane & T. D. Waite, eds.). Oxford: Elsevier, pp. 170–240.

Shahmansouri, A., & Bellona, C. (2015). Nanofiltration technology in water treatment and reuse: Applications and costs. *Water Science and Technology* 71(3), 309–319.

Shareefdeen, Z. M., Ahmed, W., & Aidan, A. (2011). Kinetics and modeling of H2 S removal in a novel biofilter. *Advances in Chemical Engineering and Science* 1, 72–76.

Simpson, A. E., Kerr, C. A., & Buckley, C. A. (1987). The effect of Ph on the nanofiltration of the carbonate system in solution. *Desalination* 64, 305–319.

Suja, F., Rahim, F., Taha, M. R., Hambali, N., Rizal Razali, M., Khalid, A., & Hamzah, A. (2014). Effects of local microbial bioaugmentation and biostimulation on the bioremediation of total petroleum hydrocarbons (TPH) in crude oil contaminated soil based on laboratory and field observations. *International Biodeterioration & Biodegradation* 90, 115–122. https://doi.org/10.1016/j.ibiod.20.

Timmer, J. M. K. (2001). *Properties of Nanofiltration Membranes; Model Development and Industrial Application*. PhD thesis, Eindhoven University of Technology, Eindhoven, The Netherlands.

Ukhurebor, K. E., Aigbe, U. O., Onyancha, R. B., & Adetunji, C. O. (2021). Climate change and pesticides: Their consequence on microorganisms. In: *Microbial Rejuvenation of Polluted Environment. Microorganisms for Sustainability*, vol. 27 (C. O. Adetunji, D. G. Panpatte & Y. K. Jhala, eds.). Singapore: Springer. https://doi.org/10.1007/978-981-15-7459-7_5

Wilson-Engineering (2013). *Water Softening Alternatives Analysis*. Ferndale, WA: Ferndale Publishers.

Xing, D. Y., Peng, N., & Chung, T.-S. (2010). Formation of cellulose acetate membranes via phase inversion using ionic liquid, [BMIM] SCN, as the solvent. *Industrial & Engineering Chemistry Research* 49(18), 8761–8769.

Zulaikha, S., Lau, W. J., Ismail, A. F., & Jaafar, J. (2014). Treatment of restaurant wastewater using ultrafiltration and nanofiltration membranes. *Journal of Water Process Engineering* 2, 58–62.

11 Bio-Nano Filtration as an Abatement Technique Used in the Management and Treatment of Impurities in Industrial Wastewater

Osikemekha Anthony Anani, Maulin P. Shah, Paul Atagamen Aidonojie, and Alex Ajeh Enuneku

CONTENTS

11.1 Introduction 171
11.2 Reports on the Different Abatement Techniques in the Management of Industrial Effluents, Their Beneficial Assets, and Technical Obstacles 172
11.3 Commercialisation of Bio-Nano Filtration for Industrial Water Treatment 177
11.4 Legal Contexts and Processes of Nanotechnology That Are Utilised for Wastewater and Water Abatements across the Globe 177
11.5 Conclusion and Future Recommendations 179
References 179

11.1 INTRODUCTION

It has been estimated that the global human population is billed to reach over nine billion in the year 2050 (Gehrke et al., 2015). Studies have shown that population growth has negatively impacted the global development of water and its associated resources via the activities of humans like mining of minerals, pesticides use in agriculture, heavy-metal contaminants from wastewater effluents, and toxicants from waste sludge (World Water Development, 2014; Gehrke et al., 2015, Anani et al., 2020a, 2020b).

The quality of water depends on the physical, chemical, and biological processes present at the time of exposure. Portable water shortage is a result of extant problems faced by the global industrial sector (Singh et al., 2020). Water pollution is a global crisis that needs global attention in safeguarding its quality for both domestic and industrial uses. The major precursors of water pollution are caused by climate change and anthropogenic activities together with rapid uncontrollable population growth. Polluted water has been known to contain several pathogenic microorganisms that portend the capacity to cause serious health issues such as cholera, vomiting, typhoid fever, and dysentery. It has been established that, for some decades, about a hundred waterborne disease-causing diseases that cut across the protozoa, viruses, and bacteria have been known to influence the quality of water. Most times the contamination takes place during transportation, storage, collection, and withdrawal (Singh et al., 2020).

The use of conventional water treatment techniques like adsorption, filtration, purification, sedimentation, and absorption in the removal or abatement of contaminated residues has been documented by Rashed (2013), Englande et al. (2015), Nicomel et al. (2016), Al-Gheethi et al. (2018) and Oluwole

DOI: 10.1201/9781003165149-11

et al. (2020). However, their efficiency in the decontamination or abatement of waste is very low. This is because of the inherent active behaviour of the pollutants; pathogenic microorganisms, inorganic, and organic contaminants in the traditional system of treating wastewater (Kfir et al., 1995; Upadhyayula et al., 2008). More so, this inefficiency can cause a sudden change in the water parameters, which can alter the evenness between the nonabsolute and absolute elimination of wastes in the treatment plants (Assavasilavasukul et al., 2008; Singh et al., 2020). The utilisation of these conventional methods can also lead to the release of toxins and poisonous strains of cyanobacteria into the water thus resulting in serious health hazards. The chlorination of the water can also lead to serious compromise of certain water features (Nuzzo, 2006; Dong et al., 2019; Mohammed, 2019).

These restrictions of the traditional treatment method by the abatement of waste from water have demanded a clean technique of wastewater decontamination. However, there are several methods used in the abatement of impurities that are available in the water that are responsible for a higher level of water contamination. In recent years, the adoption of new technologies such as nanotechnology has opened a gateway for sustainable abatement of impurity available in the water. This might be linked to their high sorption efficiency when compared to conventional techniques. Novel abatement techniques like bio-nano filtration, photocatalysts, nanomembranes, nanometals, and nanoadsorbents in the purifications of industrial waste effluents have been proven to be more effective, cheap, and sustainable and have been used to validate the efficiency of abatement of waste in water (Gehrke et al., 2015; Butt, 2020; Cheriyamundath and Vavilala, 2020; Yaqoob et al., 2020). This chapter evaluates bio-nano filtration and other effective novel techniques as possible abatement techniques used in the management and treatment of impurities in industrial wastewater.

11.2 REPORTS ON THE DIFFERENT ABATEMENT TECHNIQUES IN THE MANAGEMENT OF INDUSTRIAL EFFLUENTS, THEIR BENEFICIAL ASSETS, AND TECHNICAL OBSTACLES

Recently, the water ecosystem has been seen as a receptor of waste. Freshwater and its associated resources are affected negatively by the discharge of waste effluent from industrial sources. As the rate of human population increases, the domestic, agricultural, and industrial activities in turn increase, thus influencing the biological, chemical, and physical aspects therein. So, there is a need to develop an effective method for the purification and removal of contaminants from wastewater before discharge into water or land. In this context, Ezugbe and Rathilal (2020) evaluated the removal of wastes from wastewater using membrane-modified technologies. The authors reported that wastewater treatment in the past using conventional methods has to an extent aided in the reduction and removal of contaminants in the effluent. However, an improvement on the traditional techniques of waste treatment using membrane advanced technology that has been proven to be effective in the removal of contaminants from domestic, agricultural, and industrial waste has a great utilisation in its application. This emerging technology can also be used to reclaim assorted waste from treatment plants to be recycled and reused using the modules, cleaning, and fouling membrane pathway methods.

Singh et al. (2020) evaluated the decontamination of polluted drinking water from pathogens using nanofiltration. The authors stated that water and its associated resources can be used to support the biological processes of humans. Anthropogenic activities that are the major point of water contamination can disrupt the potable nature of water. This has necessitated that environmental scientists and engineers design and develop an advanced water treatment that will aid in the removal of microscopic pathogens like protozoa, viruses, and bacteria from different drinking water sources (borewells, tap water, and bottled water). The health hazards from drinking water contaminated with pathogenic microbes such as cholera, vomiting, typhoid fever, and dysentery can be life-threatening. Therefore, there is a need to remove these pathogens to portend human health and water security. Several chemical and physical methods have been employed in the removal of pathogens from water, however, they have not proven very effective. Nonetheless, the application of nanofiltration as a membrane method in trapping dangerous microbes has gained global recognition for the treatment of contaminated water. It has an edge over the traditional method because of its efficiency in the removal of pathogen microscopic organisms. More so, its flexibility, robustness, and the absence of disinfection by-products (DBPs) have elevated its utilisation and general performance in the selective membranous process. However, in a large-scale application, the nanofiltration process can be affected by the leaching of contaminants via adsorption, obstruction of the filter beds, and alteration of the properties of the membrane system. Nonetheless, it serves as a first-class advanced system over the conventional techniques and it is highly recommended for biological removal of pathogenic organisms in drinking water.

Human activities like pollutants discharged from the domestic, agricultural, and industrial sectors have globally affected the environment thus causing climate change and global warming. These in turn have affected the quality of water. In contrast to this background, Gehrke et al. (2015) did a review on the application of nanobased materials like nano adsorbents in the removal of waste in water. The application of highly advanced technical know-how to conventional methods has made recent developments in the purification of waste in water. Photocatalysts, nanomembranes, nanometals, and nano adsorbents have made tremendous impacts in recent times in waste removal from treatment plants. This technology has advantages like technical barriers to materials over the traditional techniques. Nanobased materials have highly compatible and adapted specific applications that customers manipulate to integrate different properties in multifunctional ways that could be used to eliminate contaminants from wastewater. However, the utilisation of this method portends some risk potentials since the nanoparticles can emit and release their contents into the environment and amass in the ecosystem for a longer period. However, there has been a stringent approach to use in the regulation of the number of nanobased materials used in water treatment systems. In conclusion, the authors stated that nanobased materials used in the water treatment process serve as potential tools for the next-generation treatment of waste and heavy contaminants in potable water.

Water plays a pivotal role in the life of all living things. Contrary to this, the globe is undergoing a major situation of water pollution from heavy-metal and chemical toxins, microorganisms, and gases generated from the activities of humans. Of late, different types of physical and chemical methods have been used to treat pollutants and contaminants in drinking water; however, they have been proven to be not efficient in their utilisation. In this context, Yaqoob et al. (2020) did a review on the use of nanomaterials in the purification of wastewater. The authors reported that there are different efficient nanomaterials and eco-friendly methods such as nano photocatalysts, nanosorbents, and catalytic membranes that can be deployed in the removal of water impurities like inorganic and organic chemicals, metal, and microorganisms. This is based on their unique properties like rate of reaction, low toxic concentration, and ability to cover a large surface area. However, these methods require a lot of dissipation of energy in the process of water purification from impurities. The authors recommend a novel technique that can be flexible, efficient, and cheap for commercialisation and utilisation purposes. Nonetheless, the utilisation of nanotechniques for water purification can serve as one of the best alternatives to the conventional methods because of their high efficiency disregarding their potential low health and environmental hazards. Stipulated guidelines should also be put in place to regulate excessive use of nanomaterials in water impurities treatment to provide cleaner water for the present and future generations.

Cheriyamundath and Vavilala (2020) evaluated in a review the treatment of wastewater using nanobased technology. The authors recounted that water pollution is a worldwide challenge that influences both human and environmental health with significant impacts on the economy leading to some serious social cost(s). Polluted water resources need an effective treatment technical know-how that is more efficient, safer, and cheaper. Nanobased materials have been used as wastewater and water management tools for the rapid purification of toxins therein. The utilisation of this modern technology that is comprised of nanophotocatalysis, nanofiltration, and nanoadsorption processes is a key potential to improve polluted water sources to meet the needs of the end consumers.

Amin et al. (2014) did a review on the recent advances of using various nanobased materials for the removal of waste from wastewater/water. The authors stated that climate change, reduction of the quality of water resources, floods, droughts, and human population increase via their activities have taken a great toll on the quality of potable water. Recent advances in the development of stable and cost-effective methods and materials for adequate wastewater purification and treatment have made the water industrial sectors of the world meet the global demand for quality and safe water following sustainable development goals (SDGs).

The authors stated that the conventional methods of wastewater/water treatment remain ineffective in the safe treatment and purification of emerging pollutants in water. Nanomaterials like nZVI and TiO_2 have been used as a secondary metal catalyst in the enhancement of the wastewater/water quality when used in purification and treatment purposes thus increasing the reactivity and selectivity of designated materials. This superficial modification could improve the activity of the nanophotocatalytic process and elicit the affinity of nanomaterials in the decontamination of possible contaminants (Amin et al., 2014).

Also, bimetallic nanomaterials have been proven efficient in the removal of pollutants in water. However, the mechanism of action should be fully understood to ascertain the degradation of biological, chemical, and physical pollutants and toxins in the wastewater/water treatment process. The process of electrospinning various nanofibrous and nanomaterial filters has been used successfully as an antifouling method in filtration membranes during the water purification process. This process can provide or ensure good water fitness, low noxiousness with reduced health hazards, active response towards decontaminating waterborne disease-causing organisms, porosity, and excellent volume-to-surface ratio. Electrospinning of nanofibers aids in the selectivity and reactivity of various water contaminants. It is cost-effective, cheap, and effective when used in both small- and large-scale industrial water sectors (Amin et al., 2014; Shah, 2021).

The utilisation of composites of nanofibrous membrane for the removal of waste from wastewater/water is quite a stand-alone process with some limitations in removing complex organic compounds: Heavy metals, viruses, bacteria, and emerging contaminants. Nanostructures composites and nanofibers membranes are used in the degradation of a wide range of inorganic and organic contaminants when applied in water treatment. The authors stated that nanomaterials serve as a perfect alternative to the conventional methods of wastewater/water treatment because of their multifunctional properties in the improvement of water matrices. Thus, in real field deployment, nanocomposite materials are best for the removal of both micro and macro pollutants in potable water.

Human activities like the noncommercial and commercial utilisation of agriculture, personal care products, and pharmaceuticals have introduced novel emerging contaminants that have the potential to cause serious environmental and health hazards. On this note, Fanourakis et al. (2020) evaluated the use of nano photocatalysts and nano adsorbents for the removal of emerging pharmaceutical pollutants at the point of reuse of portable water indirectly. The authors stated that pharmaceutical waste poses a serious challenge to the water ecosystem when discharged into it. In turn, the pollutants can cause unstable ecological problems for humans, animals, and plants. The authors recounted that emergent contaminants have a very unchanging nature that makes their breakdown with traditional wastewater purification methods very hard. More so, Fanourakis et al. (2020) stated that studies have shown that the advanced water treatment method also is unable to decontaminate emergent compounds. However, there is a need to use a sophisticated, advanced technique, like nano photocatalysts and nano adsorbents, for the removal of emerging pharmaceutical pollutants from wastewater and potable water. These methods employ the use of graphene-nanomaterials with high adsorption capacity towards the emergent pollutants when associated with unmodified graphene materials. Fanourakis et al. (2020) stated that adsorbents could be combined in photocatalysts and in a membrane-based material that can be glazed on ocular fibres and magnetic constituents for the enhancement of water treatment purposes. In conclusion, the utilisation of advanced photocatalytic constituents via the visible-light photocatalysts in water treatment research has allowed the expansion of efficiency in the degradation of emergent pollutants from pharmaceutical wastes.

Gautam et al. (2019) evaluated the application and synthesis of biogenic nanomaterials in wastewater and potable water treatment. The authors stated that water pollution is caused by the intrusion of inorganic and organic pollutants, which can result in global water quality issues. The demand for clean water comes with a price of systemic water treatment and purification techniques. Several conventional methods have been used to treat wastes in the water but were proven inefficient and expensive. In light of this, the use of nanotechnology to treat polluted water has been proven to be affordable, clean, and efficient. This technology has certain properties like regeneration for recycling, low energy and cost output, high chemical reactivity, highly power-driven, and volume-to-surface-area ratio. Gautam et al. (2019) reported that even though this method is efficient, there is a possibility of secondary contamination by nanomaterials that comprise volatile and hazardous chemicals. This problem has spurred environmental toxicologists to develop procedures and biogenic alleyways that are inexpensive and environmentally safe. These pathways involve a wide range of pollutant degradation and removal using catalysts and bioinspired adsorbent nanoscale materials. Antioxidants, alkaloids, flavonoids, polymers, proteins, and carbohydrates sourced from algae, fungi, and plants have been shown to have high efficiency as stabilising and capping agents at the point of nanomaterials production. In conclusion, the authors stated that the application of nanomaterials from biogenic sources have been shown to have promising potential in the treatment and purification of waste in potable water.

Liu et al. (2020) evaluated in a study the application of biofilters in the removal of microplastics from treated effluents. Pollution from microplastics in wastewater has generated health and environmental concern hazards and has warranted treatment using various techniques. Of late, the use of conventional methods has proven to be ineffective in the abatement of waste in water. In the study, Liu et al. (2020) used a biofilter, which was sectioned into four zones, permitting the intrusion of the microplastics via the filter with secondary wastewater or effluents. The wastewater treatment plants contained raw effluents of 917 substances m^{-3} with a corresponding concentration on the mass of 24.8 $\mu g\, m^{-3}$. After the biofiltration, the effluent concentration was reduced to a median level of 917 substances m^{-3} and 2.8 $\mu g\, m^{-3}$, with a total degradation efficacy of 79%, and the number of particle sizes was 89%. Liu et al. (2020) noticed a propensity that microplastics of higher particles and larger sizes in mass were possibly retained in the biofilter and microplastics with size 100 μm were mainly removed from the last biofiltration zone. The findings from this study showed that the biofilters were able to reduce the abundance of the microplastics in the treated effluents significantly. However, an overall removal cannot be ascertained because of the presence of microplastics of smaller sizes that have the potentials to be discharged into the receptive water environment. A high-definition bio-nanotechnological process using a combined or integration mechanism is highly recommended for complete filtration of microplastics in possibly potable water.

In a biological experiment, Roudbari and Rezakazemi (2018) investigated the application of ultrasound technology in the removal of hormones in municipal effluent. The authors recounted that estrogens E1 and E2 (beta-estradiol) are two of the environmental micropollutants in municipal effluents that have a serious health influence on living biota. The influence of pH, exposure time, frequency, and power of efficiency was investigated in a cylindrical bioreactor containing E1 and E1 estrogens. The two residual estrogen hormones were measured with an ECL (electro-chemi-luminescence). The outcome from the investigation revealed that the ultrasound removed about 85 to 96% of both hormones at 45 mins with varied changes like the internal parameters like pH, exposure time, frequency, and power of efficiency of the ultrasound. It was noted that the power and frequency of the ultrasound were significantly impacted by the reduced efficacy of the hormones. Meanwhile, there was no significant impact of the exposure of the hormones on the ultrasound. More so, there was a highly significant interaction between the frequency and power on the reduced efficiency of the hormones at 64.3%. The outcome also showed that the waves from the ultrasound can decrease steroid hormones from metropolitan effluents or wastewater. In conclusion, the authors recommended that this method of waste removal is one of the effective methods that has significant techniques in the destruction and reduction of hormones from effluent without generating serious or adverse toxic waste products and with low expiation of energy. The authors recommended this method as a first tier in the class of waste management.

Water plays an important role in most biological organisms. However, globally, many nations of the world have been faced with issues of water pollution. There have been several conventional strategies used in the treatment of contaminants in wastewater and water that have been proven somewhat inefficient. Yaqoob et al. (2020) did a review on the treatment of water effluent using nanomaterials. The authors stated that heavy metals, chemicals, microorganisms, and gases are the major water pollutants that can deface its quality for drinking. The application of nanotechnology in the management of toxins in wastewater and water has gained immense use in recent times because of low concentration, greater volume-to-surface-area ratio, and efficiency. Many nanotechnology techniques like nano photocatalysts, nanosorbents, and nanostructured catalytic membranes have been applied to eliminate contaminants from water effluents. However, to be efficient and eco-friendly they need more energy to purify the contaminated wastewater. More so, Yaqoob et al. (2020) stated that nanoparticles can elicit health and ecological issues when their utilisation exceeds the slated threshold set by defined regulatory bodies. In conclusion, the authors stated that the use of new equipment that can be more efficient, cost-effective, and flexible for commercialisation and should be integrated with nanomaterials for smart water purification without harm to humans and the ecosystem.

TABLE 11.1
Different Methods in the Utilisation of Membrane Nanoparticles for the Removal of Industrial Waste

S/N	Type of Membrane	Method of Production	Efficiency and Target	References
1	Microcrystalline cellulose nanofibers-polyethyleneimine	Electrospinning	99.99% removal of *E. coli*, virus, and bacteriophage (MS2)	Sato et al. (2011)
2	Polystyrene-block-poly(methyl methacrylate)	Ultraviolet radiation and acid washing	100% removal of type 14 human rhinovirus	Yang et al. (2008)
3	Reduced graphene oxide-carbon nanoparticles	Assisted-vacuum filtration	99% removal of humic acid, sugars, dyes, and nanoparticles	Chen et al. (2016)
4	Silver oxide nanowire	Hot-press process and water-thermal synthesis	Removal of *E. coli*, polyethylene oxide, and polyethylene glycol	Zhang et al. (2016)
5	Polysulfone-Nanofibers	Polymerisation of interfacial region and electrospinning	100% removal of soyabean oil	Obaid et al. (2015)
6	Poly (acrylic acid)-grafted	Phase inversion	Greater than 99.99% removal of diesel, toluene, and hexadecane	Zhang et al. (2014)
7	Nanosheets incorporated into sodium alginate matrix	Thermal oxidation engraving	Ethanol	Cao et al. (2015)
8	Ag-polyacrylonitrile	Surface alteration and electroplating	1,2-dibromoethane	Li et al. (2014)
9	Bicontinuous cubic	Self-assemblage	81, 33, 59, and 83% removal of NO^-_3, SO^{2-}_4, Cl^- and Br^-	Henmi et al. (2012)
10	Sigle-layer graphene	Etching of plasma oxygen	−100% removal of Cl^-, Li^+, Na^+, and K^+	Surwade et al. (2015)
11	Aquaporin reconstituted	Adduct formation of amine-catechol and vacuum suction	81.1 and 66.2% removal of $MgCl_2$ and NaCl	Sun et al. (2013)

11.3 COMMERCIALISATION OF BIO-NANO FILTRATION FOR INDUSTRIAL WATER TREATMENT

The commercialisation of bio-nanobased membrane materials for wastewater and water technology hinges strongly on their influence on the water environment. Several studies including dispersion of nanomaterials in water systems, technology pathways and assessment LCA (life cycle analysis), and toxicity examinations have been done to evaluate the possible health hazard nanomaterials portend on their application to water systems (Lovern and Klaper, 2006; Warheit et al., 2007; Griffitt et al., 2008; Hainlaan et al., 2008; Mouchet et al., 2008; Velzeboer et al., 2008; Shah, 2020). The findings from their study showed a better knowledge of the behaviour of nanomaterials like silver, TiO_2, and carbon nanotubes nanoparticles in water systems. However, several works have yielded opposing views on the influences of these nanomaterials since there are no definite conditions and standards for measurements and experimental proceedings to determine such processes (Gehrke et al., 2015). Therefore, there is a need for industries and stakeholders to make novel laws to regulates its utilisation before commercialisation to ensure a clean and better environment for the present and future generations.

11.4 LEGAL CONTEXTS AND PROCESSES OF NANOTECHNOLOGY THAT ARE UTILISED FOR WASTEWATER AND WATER ABATEMENTS ACROSS THE GLOBE

There has been a global demand for the utilisation of nanotechnology for the better management of industrial wastes because of its effectiveness. For example, in the USA in Arlington, VA (National Nanotechnology Initiative), about 1.5 billion dollars have been put into this technological framework to process, promote, and bring it into the industrial market for industrial clean-up use (Nano. gov, 2014). In addition, the European Union spent in seven years the sum of US$110 billion for the novel nanotech, US$92 million for the innovation for water, and US$110 billion for novel research to bring about safe water production and other nanoparticle applications (Horizon, 2020a and b).

The scientific concept of bio-nanotechnology filtration is a scientific method that has been developed in control and prevention of waste generation at the source. Although there is a relevant global legal framework concerning waste generation and waste disposal, the drafter of this legal framework did not contemplate or consider bio-nanotechnology filtration as one of the most potent means or methods in minimising or preventing waste and hazardous waste generation at source. For example, the Basel Convention, which is a global international convention aimed at controlling and curbing the generation and transboundary movement of waste and hazardous waste (Ijaiya et al. (2018) and Sejal (2001)), did not have in any of it provisions that provide for or recognise bio-nanotechnology filtration in controlling waste generation. This concerns the fact that, by Annex III of the Basel convention, it considered waste, which contains harmful substances such as wastewater/oils, arsenic, zinc, cadmium, cyanides, chlorophenols, substances containing thallium, and metal carbonyls as hazardous. Waste that contains substances such as polychlorinated terphenyls, thallium, mercury polychlorinated biphenyls (PCBs), etc. are also considered hazardous. Furthermore, the Basel convention also required that where wastewater or any other waste possesses any characteristics such as being toxic, corrosive, flammable, explosive, ecotoxic, or poisonous, an adequate procedure and sustainable means must be adopted by state parties in preventing, reducing, and controlling such waste in moving from one country to another (Kempel, 1999). In this regard, the Basel convention requires that waste generated at sources should be adequately minimised, treated, and disposed of (Kummer (1992)) without referring to bio-nanotechnology filtration as a relevant scientific method that must be adhered to by member states in controlling waste generation.

However, bio-nanotechnology filtration, which is a recent scientific method of waste treatment and control, was not contemplated by the Basel Convention as one of the relevant sustainable means of reducing and controlling waste generation by members' state. However, it is argued that the reason for not contemplating bio-nanotechnology filtration in the Basel Convention is a result of the fact that it is a current or recent scientific discovery on waste control and generation.

Nonetheless, there have been several steps the researchers and stakeholders have put into place to mitigate any damage this novel technology might pose to human beings, plants, micro-organisms, and the environment when applied to the industrial clean-up processes. Some legal approaches have been set by the EU and USA towards regulating exposure to noxious nanomaterials in the water ecosystem. Such are the Registration, Evaluation, Authorisation, and Restriction of Chemicals (REACH) and the European Water Framework Directive controlled by the EU to protect groundwater, coastal waters, transitional waters, and surface water close to the inlands (European Parliament and Council, 2000; European Parliament and Council, 2006; European Commission, 2011). The United States Environmental Protection Agency is the main regulatory body in the USA, controlling and managing the utilisation of nano chemical materials stipulated in the Toxic Substances Control Act (United State Environmental Protection Agency, 2007).

Also, it suffices to state that the European Union Directives of 2008/98/EC concerning waste is a regional policies framework that provides for waste control and prevention. These EU Directives policies on waste classified waste into hazardous and non-hazardous substances or waste and further provide for the specific method of waste treatment and disposal. No specific reference to bio-nanotechnology filtration as a potent method of controlling and minimising waste was generated at the sources.

Furthermore, another relevant international regional convention that provides for the need to control and minimise waste generation at the source is the Bamako Convention. The Bamako Convention has some similar content like Basel Shearer (1993) and Ovink (1995), which concerns the fact that both conventions share the same objective of controlling and minimising the potentially harmful impacts of hazardous waste (Eguh, 1997; Kaminsky, 1992). However, a perusal of the Bamako Convention, which is an African regional convention in the abatement and control of waste, also reveals the fact that the drafter of the Bamako Convention did not provide for or recognise bio-nanotechnology as a means of controlling and prevention of waste generation at source. Annex III to the Bamako Convention comprehensively provides for various methods of disposing or treatment of solid waste. Some of these methods include but are not limited to the following: It required member states to adopt the method of waste treatment or disposal via solvent regeneration/reclamation, incinerating waste in land or sea, biological (using chemical substance) treatment of waste, releasing waste into water bodies excluding oceans/sea, surface impounding, deep injection of wastewater or any kind of waste into wells or natural repositories, and recycling of such wastes.

Given the aforementioned method of waste treatment and control methods, the scientific method of bio-nanotechnology filtration that has been proven to be a reliable means of waste control and prevention was not contemplated by the Bamako Convention. Although, as earlier stated, it suffices to state that, bio-nanotechnology filtration is a current scientific discovery geared towards minimisation and prevention of waste generated at the source.

However, the drafter of the Bamako Convention contemplated creating amendments and provisions to accommodate discovery of waste control and prevention, given the dynamic nature of law and scientific discoveries. This is concerning the fact that by 10 and 15(3) of the Basel Convention, parties to the convention are required to always consider additional measures that will be relevant in ensuring the maximum implementation of the convention in curbing or preventing waste generation. Furthermore, Article 15(4) and 17(1), further provide for an amendment section, which is to the effect that, where there are available scientific and technical discoveries in minimising and prevention of waste such as 'Bio-nanotechnology filtration', parties who adopt the Bamako Convention should endeavour to consider an amendment to incorporate such scientific and technical discovery. In this regard, it is argued that though the Bamako Convention did not provide for bio-nanotechnology filtration as a potent method of preventing and minimising waste generation, the purport of Articles 15(4) and 17(1) can be interpreted to mean that bio-nanotechnology filtration can be incorporated into the Bamako Convention via an amendment. It can be concluded here that countries that have signed the Bamako Convention as a treaty, should be made to compulsorily adopt Bio-nanotechnology filtration procedures or methods in the reduction and minimising of waste generated at the source during industrial activities for a sustainable ecosystem.

11.5 CONCLUSION AND FUTURE RECOMMENDATIONS

This chapter evaluates bio-nano filtration and other effective novel techniques as possible abatement techniques used in the management and treatment of impurities in industrial wastewater. Several reports on the different abatement techniques in the management of industrial effluents, their beneficial assets, and technical obstacles were highlighted indicating that the emerging technology can also be used to reclaim assorted waste from treatment plants to be recycled and reused using the modules, cleaning, and fouling membrane pathway methods. A high-definition bio-nanotechnological process using a combined or integration mechanism is highly recommended for a complete filtration of toxins in possible potable water. The use of new equipment that can be more efficient, cost-effective, and flexible for commercialisation should be integrated with nanomaterials for smart water purification without harm to humans and the ecosystem. However, there is a need for industries and stakeholders to make novel laws to regulate its utilisation before commercialisation to ensure a clean and better environment. Based on these, extant legal contexts and processes of nanotechnology that are utilised for wastewater and water abatements should be geared towards the subscription and the domestication of certain globally accepted treaties to create a better sustainable environment for the present and future generations.

REFERENCES

Al-Gheethi, A.A., Efaq, A.N., Bala, J.D., et al. (2018). Removal of pathogenic bacteria from sewage-treated effluent and biosolids for agricultural purposes. *Appl Water Sci* 8, 74. https://doi.org/10.1007/s13201-018-0698-6.

Amin, M.T., Alazba, A.A., Manzoor, U. (2014). A review of removal of pollutants from water/wastewater using different types of nanomaterials. *Adv Mat Sc Eng*, 1–12. https://doi.org/10.1155/2014/825910.

Anani, O.A., Mishra, R.R., Mishra, P., Enuneku, A.A., Anani, G.A., Adetunji, C.O. (2020a). Effects of toxicant from pesticides on food security: Current developments. In: Mishra P., Mishra R.R., Adetunji C.O. (eds) *Innovations in Food Technology.* Springer, Singapore, pp. 313–321.

Anani, O.A., Mishra, R.R., Mishra, P., Olomukoro, J.O., Imoobe, T.O.T., Adetunji, C.O. (2020b). Influence of heavy metal on food security: Recent advances. In: Mishra P., Mishra R.R., Adetunji C.O. (eds) *Innovations in Food Technology.* Springer, Singapore. http://doi-org-443.webvpn.fjmu.edu.cn/10.1007/978-981-15-6121-4_18.

Assavasilavasukul, P., Lau, B.L.T., Harrington, G.W., Hoffman, R.M., Borchardt, M.A. (2008). Bruinsma J. By how much do land, water and crop yields need to increase by 2050?. The resource outlook to 2050. *Presented at the Food and Agriculture Organization of the United Nations Expert Meeting entitled "How to Feed the World in 2050".* Rome, Italy.

Butt, B.Z. (2020). Nanotechnology and waste water treatment. In: Javad S. (eds) *Nanoagronomy.* Springer, Cham. https://doi.org/10.1007/978-3-030-41275-3_9.

Cao, K.T., Jiang, Z.Y., Zhang, X.S., Zhang, Y.M., Zhao, J., Xing, R.S., Yang, S., Gao, C.Y., Pan, F.S. (2015). Highly water-selective hybrid membrane by incorporating g-C3N4 nanosheets into polymer matrix. *J Membr Sci* 490: 72–83.

Chen, X.F., Qiu, M.H., Ding, H., Fu, K.Y., Fan, Y.Q. (2016). A reduced graphene oxide nanofiltration membrane intercalated by well-dispersed carbon nanotubes for drinking water purification. *Nanoscale* 8, 5696–5705.

Cheriyamundath, S., Vavilala, S.L. (2020). Nanotechnology-based wastewater treatment. *Water Environ J* 35(1), 123–132. https://doi.org/10.1111/wej.12610.

Dong, H., Qiang, Z., Richardson, S.D. (2019). Formation of iodinated disinfection byproducts (idbps) in drinking water: Emerging concerns and current issues. *Acc Chem Res* 52(4), 896–905.

Eguh, E.C. (1997). Regulations of transboundary movement of hazardous wastes lessons from Koko. *Afr J Inter Comp Law* 9, 130–151.

Englande, A.J., Peter, Jr., Shamas, K.J. (2015). Wastewater treatment & water reclamation. *Ref Mod Earth Syst Environ Sci*, 1–32. doi:10.1016/B978-0-12-409548-9.09508-7

European Commission (2011). *Commission Recommendation of 18 October 2011 on the Definition of Nanomaterial (2011/696/EU).* Available from: http://eur-lex.europa.eu/LexUriServ/LexUriServ.do?uri=OJ:L:2011:27 5:0038:0040:EN:PDF. Accessed July 24, 2014.

European Parliament and Council (2000). *Directive 2000/60/EC of the European Parliament and of the Council of 23 October 2000 Establishing a Framework for Community Action in the Field of Water Policy.* Available from: http://eur-lex.europa.eu/legal-content/EN/TXT/?uri=CELEX:32000L0060. Accessed July 24, 2014.

European Parliament and Council (2006). *Regulation (EC) No 1907/2006 of the European Parliament and of the Council of 18 December 2006 Concerning the Registration, Evaluation, Authorisation and Restriction of Chemicals (REACH), Establishing a European Chemicals Agency.* Available from: http://eur-lex.europa.eu/legal-content/EN/ALL/;ELX_ SESSIONID=GctYTQvNMZSqByVl1X23LRQSM0Jp VbD2gTFpbL NTKKRlfJLB2tym!-1120470413?uri=CELEX:32006R1907. Accessed July 24, 2014.

Ezugbe, E.O., Rathilal, S. (2020). Membrane technologies in wastewater treatment: A review. *Membranes* 10, 89. https://doi:10.3390/membranes10050089.

Fanourakis, S.K., Peña-Bahamonde, J., Bandara, P.C., et al. (2020). Nano-based adsorbent and photocatalyst use for pharmaceutical contaminant removal during indirect potable water reuse. *NPJ Clean Water* 3, 1. https://doi.org/10.1038/s41545-019-0048-8.

Gautam, P.K., Singh, A., Misra, K., Sahoo, A.K., Samanta, S.K. (2019). Synthesis and applications of bio-genic nanomaterials in drinking and wastewater treatment. *J Environ Manage* 1, 231:734–748. https://doi: 10.1016/j.jenvman.2018.10.104.

Gehrke, I., Geiser, A., Somborn-Schulz, A. (2015). Innovations in nanotechnology for water treatment. *Nanotechnol Sci Appl* 8, 1–17. https://doi: 10.2147/NSA.S43773.

Griffitt, R.J., Luo, J., Gao, J., Bonzongo, J.C., Barber, D.S. (2008). Effects of particle composition and spe-cies on toxicity of metallic nanomaterials in aquatic organisms. *Environ Toxicol Chem* 29: 1972–1978.

Hainlaan, M., Ivask, A., Blinova, I., Dubourguier, H.C., Kahru, A. (2008). Toxicity of nanosized and bulk ZnO, CuO and TiO2 to bacteria *Vibrio fischeri* and crustaceans *Daphnia magna* and *Thamnocephalus platyurus*. *Chemosphere* 71, 1308–1316.

Henmi, M., Nakatsuji, K., Ichikawa, T., Tomioka, H., Sakamoto, T., Yoshio, M., Kato, T. (2012). Self-organized liquid-crystalline nanostructured membranes for water treatment: Selective permeation of ions. *Adv Mater* 24, 2238–2241.

Horizon. (2020a). *Work Programme 2014–2015. 5. Leadership in Enabling and Industrial Technologies. II. Nanotechnologies, Advanced Materials, Biotechnology and Advanced Manufacturing and Processing.* Available from: http://ec.europa.eu/research/participants/data/ref/h2020/wp/2014_2015/main/h2020-wp1415-leit-nmp_en.pdf. Accessed July 24, 2014.

Horizon. (2020b). *Work Programme 2014–2015. 12. Climate Action, Environment, Resource Efficiency and Raw Materials.* Available from: http://ec.europa.eu/research/horizon2020/pdf/work-programmes/climate_action_environment_resource_efficiency_and_raw_materials_ draft_work_programme.pdf. Accessed July 24, 2014.

Ijaiya, H., Wardah, I.A., Wuraola, O.T. (2018). Re-examining hazardous waste in Nigeria: Practice possibili-ties within the United Nations system. *Afri J Inter Comp Law* 26(2), 264–282.

Kaminsky, H.S. (1992). Assessment of the bamako convention on the ban of import into Africa and the control of transboundary movement and management of hazardous wastes within Africa. Geo. *Int Env Law Rev* 5: 77–98.

Kempel, W. (1999). The negotiations on the basel convention on the transboundary movement of hazardous wastes and their disposal: A national delegation perspective. *Int Negot* 4(3), 413–434.

Kfir, R., Hilner, C., Preez, M.D., Bateman, B. (1995). Studies evaluating the applicability of utilizing the same concentration techniques for the detection of protozoan parasites and viruses in water. *Water Sci Technol* 31, 417–423.

Kummer, K. (1992). The international regulation of transboundary traffic in hazardous wastes: The 1989 basel convention. *Int Comp Law Q* 41(3), 530–562.

Li, X., Wang, M., Wang, C., Cheng, C., Wang, X.F. (2014). Facile immobilization of Ag nanocluster on nanofibrous membrane for oil/water separation. *ACS Appl Mater Interfaces* 6: 15272–15282.

Bio-Nano Filtration as an Abatement Technique

Liu, F., Nord, N.B., Bester, K., Vollertsen, J. (2020). Microplastics removal from treated wastewater by a biofilter. *Water* 12, 1085. https://doi:10.3390/w12041085.

Lovern, S.B., Klaper, R. (2006). Daphnia magna mortality when exposed to titanium dioxide and fullerene (C60) nanoparticles. *Environ Toxicol Chem* 25, 1132–1137.

Mohammed, A.N. (2019). Resistance of bacterial pathogens to calcium hypochlorite disinfectant and evaluation of the usability of treated filter paper impregnated with nanosilver composite for drinking water purification. *J Global Antimicrob Resist* 16, 28–35.

Mouchet, F., Landois, P., Sarremejean, E., et al. (2008). Characterization and in vivo ecotoxicity evaluation of double-wall carbon nanotubes in larvae of the amphibian *Xenopus laevis*. *Aquat Toxicol* 87: 127–137.

Nano.gov (2014). *What is the NNI? 2014*. Available from: www.nano. gov/about-nni/what. Accessed April 16, 2014.

Nicomel, N.R., Leus, K., Folens, K., Van Der Voort, P., Du Laing, G. (2016). Technologies for arsenic removal from water: Current status and future perspectives. *Int J Environ Res Public Health* 13(1), 62. https://doi: 10.3390/ijerph13010062.

Nuzzo, J.B. (2006). The biological threat to the U.S. water supplies: Toward a national water security policy. *Biosecur Bioterror* 4, 147–159.

Obaid, M., Barakat, N.A.M., Fadali, O.A., Motlak, M., Almajid, A.A., Khalil, K.A. (2015). Effective and reusable oil/water separation membranes based on modified polysulfone electrospun nanofiber mats. *Chem Eng J* 259, 449–456.

Oluwole, A.O., Omotola, E.O., Olatunji, O.S. (2020). Pharmaceuticals and personal care products in water and wastewater: a review of treatment processes and use of photocatalyst immobilized on functionalized carbon in AOP degradation. *BMC Chem* 14, 62. https://doi.org/10.1186/s13065-020-00714-1.

Ovink, B.J. (1995). Transboundary shipments of toxic wastes: The basel and bamako conventions: Do third world countries have a choice? *Penn State Int Law Rev* 13(2), 281–295.

Rashed, N.M. (2013). *Adsorption Technique for the Removal of Organic Pollutants from Water and Wastewater, Organic Pollutants—Monitoring, Risk and Treatment, M. Nageeb Rashed*, IntechOpen. https://doi. org/10.5772/54048. Available from: www.intechopen.com/books/organic-pollutants-monitoring-risk-and-treatment/adsorption-technique-for-the-removal-of-organic-pollutants-from-water-and-wastewater.

Sato, A., Wang, R., Ma, H.Y., Hsiao, B.S., Chu, B. (2011). Novel nanofibrous scaffolds for water filtration with bacteria and virus removal capability. *J Electron Microsc* 60, 201–209.

Sejal, C. (2001). The basel convention on the control of transboundary movement of hazardous waste and their disposal: 1999 protocol on liability and compensation. *Ecology LQ* 28, 509.

Shah, M.P. (2020). *Advanced Oxidation Processes for Effluent Treatment Plants*. Elsevier.

Shah, M.P. (2021). *Removal of Emerging Contaminants through Microbial Processes*. Springer, Singapore.

Shearer, C.R. (1993). Comparative analysis of the Basel and Bamako Conventions on Hazardous Waste. *Env Law* 23, 141–163.

Singh, R., Bhadouria, R., Singh, P., Kumar, A., Pandey, S., Singh, V.K. (2020). Nanofiltration technology for removal of pathogens present in drinking water. *Waterborne Pathogens*, 463–489. https://doi: 10.1016/B978-0-12-818783-8.00021-9.

Sun, G.F., Chung, T.S., Jeyaseelan, K., Armugam, A. (2013). Stabilization and immobilization of aquaporin reconstituted lipid vesicles for water purification. *Colloids Surf* 102, 466–471.

Surwade, S.P., Smirnov, S.N., Vlassiouk, I.V., Unocic, R.R., Veith, G.M., Dai, S., Mahurin, S.M. (2015). Water desalination using nanoporous single-layer graphene. *Nat Nanotechnol* 10: 459–464.

Upadhyayula, V.K.K., Deng, S., Mitchell, M.C., Smith, G.B., Nair, V.S., Ghoshroy, S. (2008). Adsorption kinetics of *Escherichia coli* and *Staphylococcus aureus* on single walled carbonnanotube aggregates. *Water Sci Technol* 58, 179–184.

US Environmental Protection Agency. (2007). *Nanotechnology: An EPA Research Perspective*. Available from: www.epa.gov/ncer/nano/factsheet/nanofactsheetjune07.pdf. Accessed April 16, 2014.

Velzeboer, I., Hendriks, A.J., Ragas, A.M., van de Meent, D. (2008). Aquatic ecotoxicity tests of some nano-materials. *Environ Toxicol Chem* 27, 1942–1947.

Warheit, D.B., Hoke, R.A., Finlay, C., Donner, E.M., Reed, K.L., Sayes, C.M. (2007). Development of a base set of toxicity tests using ultrafine TiO_2 particles as a component of nanoparticle risk management. *Toxicol Lett* 171, 99–110.

World Water Development. (2014). *Report 4 Managing Water Under Uncertainty and Risks*. Available from: www. unesco.org/new/en/natural-sciences/environment/water/wwap/wwdr/wwdr4-2012. Accessed July 24, 2014.

Yang, S.Y., Park, J., Yoon, J., Ree, M., Jang, S.K., Kim, J.K. (2008). Virus filtration membranes prepared from nanoporous block copolymers with good dimensional stability under high pressures and excellent solvent resistance. *Adv. Funct. Mater* 18, 1371–1377 [CrossRef].

Yaqoob, A.A., Parveen, T., Umar, K., Ibrahim, M.N.M. (2020). Role of nanomaterials in the treatment of wastewater: A review. *Water* 12, 495. https://doi:10.3390/w12020495.

Zhang, Q.G., Deng, C., Soyekwo, F., Liu, Q.L., Zhu, A.M. (2016). Sub-10 nm wide cellulose nanofibers for ultrathin nanoporous membranes with high organic permeation. *Adv Funct Mater* 26, 792–800.

Zhang, W.B., Zhu, Y.Z., Liu, X., Wang, D., Li, J.Y., Jiang, L., Jin, J. (2014). Salt-induced fabrication of super-hydrophilic and underwater superoleophobic PAA-g-PVDF membranes for effective separation of oil-in-water emulsions. *Angew Chem Int Ed* 53, 856–860.

12 Removal of VOC and Heavy Metals through Microbial Approach

Osikemekha Anthony Anani, Abel Inobeme, Maulin P. Shah, Kenneth Kennedy Adama, and Ikenna Benedict Onyeachu

CONTENTS

12.1 Introduction..183
12.2 Detailed Methods Involving the Use of Bio-Nano Fillers in the Removal of VOCs and HMs From Industrial Wastes Effluent................................184
12.3 The Mechanism of Bio-Nano Fillers for the Removal of HMs and VOCs 187
12.4 Environmental and Health Impacts of Heavy Metals and VOCs from Wastewater 188
12.5 Derived Benefits from the Utilisation of Bio-Nano Fillers over Other Conventional Methods..189
12.6 Conclusion and the Way Forward 190
References 191

12.1 INTRODUCTION

Water plays a vital role in the biological process of living things. About 0.002% of the earth's water is easily accessible to living things. According to the World Health Organization (WHO), about 1/6th of the current world's population cannot get good access to potable water because of the issues of water contamination and pollution (Meena et al. 2020). Human actions like indiscriminate waste discharge from the household, agriculture, and industrial wastes have been known to cause a negative influence on living things, the environment, and water systems (Alrumman et al. 2016).

Since the inception of industrialisation, water systems have been seen as the recipient of waste and this has elicited environmental worry globally. Owing to the previous facts, pollutants like volatile organic compounds and heavy metals have caused serious damage to water systems and humans for many decades even at low concentrations (Tiwari et al. 2016; Yang et al. 2019; Anani et al. 2020; Anani and Olomukoro 2020; Olatunji and Anani 2020; Barupal et al. 2020).

Various inorganic and organic contaminants are released into the environment through anthropogenic means. Heavy metals and volatile organic matter form a major class of these contaminants. Bio-nanofillers play an indispensable role as a vital tool in nanotechnology during the remediation of these environmental matrices containing these pollutants (Yang et al. 2019). Some of the major volatile organic compounds (VOCs) include polycyclic aromatic hydrocarbons, polychlorinated compounds, polychlorinated biphenyl compounds, pentachloro phenolic derivatives, chlorinated benzene derivatives, dibenzo-p-dioxins and dibenzofurans, and many other by-products of industrial processes and solvents. The physicochemical properties of these compounds vary and also affect the remediation processes. These properties of the contaminants as well as the material used for the remediation affect the implementation of bio-nanotechnology (Yang et al. 2019).

Volatile organic compounds (VOCs) like ketones, paraffin, chloroform, trichloroethylene, methyl tert-butyl ether, chlorohydrocarbons, perfluorocarbons, methane, esters, benzene, and toluene are

DOI: 10.1201/9781003165149-12

used as solvents in refinery and chemical industries (Tiwari et al. 2019; Yang et al. 2019; Liang et al. 2020; Lomonaco et al. 2020). However, most of these compounds are toxic, mutagenic, and carcinogenic to humans and dangerous to the environment (Meena et al. 2020). When light is present, volatile organic compounds can be oxidised and produce chlorinated, aromatic, and some other substances that are more noxious than their primary compounds (Zhao et al. 2014). In addition, volatile organic compounds with nitrogen oxides constituents can produce O_3 with the smog that can be toxic to the environment and humans (Malakar et al. 2017; Meena et al. 2020). Volatile organic compounds also pollute soil, water, and air causing biological alterations in the ecosystems (Mehta and Gaur 2005; Kumar et al. 2011).

On the other hand, HMs are present naturally in the ecosystem. However, it has been established that human activities have been linked to increasing their background level over the years (Tiwari et al. 2016; Anani and Olomukoro 2018a). Heavy metals like stannum, chromium, cadmium, cobalt, nickel, copper, zinc, arsenic, lead, and mercury have been reported to be noxious in nature even at concentrations very low with high relative density (Srivastava and Majumder 2008; Anani and Olomukoro 2017; Anani and Olomukoro 2018b; Enuneku et al. 2018; Qu et al. 2020; Shah 2020, 2021). Sources of heavy metals discharge into soil and water bodies through mining, wastewater from industrial activities, and electroplating (Zhou et al. 2020).

Heavy metals removal from the ecosystem is very tedious due to their persistent nature (Acheampong et al. 2010). The exposure pathways of humans to metals can be via ingestion of food, dermal contact, vaporisation, and inhalation. Heavy metals can result in tissue disorders, diseases, algal blooms, destruction of habitats, and disruption of life in water (Akpor et al. 2014; Anani et al. 2020).

Nonetheless, because of the recalcitrant nature of VOCs and HMs, the use of ion exchange, membrane separation, adsorption, solvent extraction, chemical precipitation Kurniawan et al. (2006), incineration, catalytic oxidation, adsorption, membrane separation, condensation, and ozonation (Kumar et al. 2011) have been widely employed for the removal of heavy metals and VOCs respectively for decades. However, there have been some difficulties in the total removal of these compounds from their environmental matrices. Also some of the methods mentioned earlier are too expensive and not eco-friendly.

There is a need to adopt novel technologies like biologically based methods in the total decontamination VOCs and HMs from the ecosystems. On this note, this chapter intends to discuss in detail several methods for the removal of VOCs and HMs from industrial waste effluent through the use of bio-nano fillers, biosorption, bioscrubbers, biotrickling filters, microorganisms, biofilters, and biofiltration.

12.2 DETAILED METHODS INVOLVING THE USE OF BIO-NANO FILLERS IN THE REMOVAL OF VOCS AND HMS FROM INDUSTRIAL WASTES EFFLUENT

Padalkar and Kumar (2018) evaluated the mechanism of the removal of 83 VOCs from wastewater in different common treatment plants in Mumbai. The mass balance (MB) method was utilised to ascertain the volatile organic compounds removal by biodegradation, adsorption, weir drop, stripping, and volatilisation. The results from the experiment revealed that approximately 17% of volatile organic compounds were eliminated by weir drop and chief clari-flocculator correspondingly. The biodegradation process was the most obvious mechanism in the treatment tank. However, the authors stated that it was not too good for aquaphobic complexes that are more susceptible to the stripping removal methods to remain in the wastewater. The rates of stripping could be decreased by improving the fine pore and active dry mass concentration of the diffusers to decrease the ratio of the effluent/air. Reduction of the concentration of the compound and Henry's constant can alter the major elimination mode of action from stripping breakdown. The outcome also indicates

Removal of VOC and Heavy Metals through Microbial Approach

significant agreement between predicted and measured (67.1 and 71.2%) summed removal mainly in the aerated plants. There were fairly displayed satisfactory differences of the subordinate clarifier components, chief clari-flocculator, and equalisation reservoirs at the actual (29.5, 14.2, and 20.5%) and predicted (16.8, 7.7, and 16.9%) rates respectively. In conclusion, the authors stated that the impact of the mode of action of the removal of the VOCs needs to be explored further for future applications.

Arfè et al. (2020) evaluated in a study the efficiency of certain zeolites in the removal of heavy metals and VOCs from wastewater tanks in Maura; Tito Scalo, Italy. The authors stated that zeolites have been used as efficient materials in the evaluations of the quality of water samples. They employed inductively coupled plasma optical emission spectrometry (ICP-OES) and gas chromatography (GC) in the characterisation of VOCs and heavy metals in effluents correspondingly and x-ray powder diffraction (XRD) and thermal analysis (TG DTA) for the valuation of contaminants present in the adsorbent structural conduits. It was observed that MFI topology (ZSM-5 zeolite) was appropriate for the VOCs with > 87% efficiency removal. Contrarily, FAU topology (13X) was 100% in its efficiency by removing heavy metals selectively in situ. After the heavy metals and VOCs removal, it was noticed that the zeolites were loaded with refined structures of both extra-framework and lattice contents endorsing the immobilisation of the pollutants in the microporosities framework. The presence of these species was also validated by the thermal analysis curves that revealed various occurrences on the level of their numbers and nature of the species of the extra-framework hosted in the microsporic zeolite. In conclusion, the authors suggested that this method can also be applied in the removal of aromatic hydrocarbons and chlorinated solvent contaminants in situ because of their efficiencies.

Salek et al. (2021) in a review evaluated nanotechnology biosensor use in the detection and abatement of cadmium, mercury, lead, and arsenic in the environment. The authors recounted that HMs have been known to cause serious health and environmental issues in humans and the environment respectively. A new nano-bio sensor with aptamers is needed to diagnose the amount of ultra-trace heavy metals in various biological and environmental samples. These composites contain hybridised biofunctional materials that serve as a good transducer that can boost the quality of performance function used. Salek et al. (2021) stated that additional label-free and labelled detection methods with a range of modified nanocomposites and nanoparticles could be used also to detect heavy metals in the environment. The linear dynamism and detection limits can be used to characterise the superiority of the biosensors in the primary removal of toxic metals. More so, the role of nanomaterials as metal absorbers can also be used as a tracking system because of the magnetic, optical, and electrical properties that they use to amplify optical and electrical signals as transducers for heavy-metal detection and abatement. Salek et al. (2021) recounted that aptamers and nanomaterials have been employed in the biorecognition of metal elements with the femtomolar and micromolar range or limit. For cadmium, mercury, lead, and arsenic, the ranges are 0.024×10^{-3}, 0.12×10^{-6}, 2.16×10^{-4}, and 0.003×10^{-3} ppb individually. From the findings, the authors stated that the application of the aptamer and nano biosensors brought about superficial modifications in the improvement of the strategies of heavy-metal detection. More so, the advent of these techniques has brought about the standard of straightforwardness and more sensitive, cheaper, and faster methods of measurements compared to the traditional analysis of metals in wastewater.

Environmental pollution caused by noxious organic chemicals is one of the problems of water quality. Volatile organic compounds are one of the organic pollutants emitted into the environment that are facilitated by anthropogenic factors that can cause indirect or direct risk to humans and the environment resulting in cancers and photochemical smog. However, several methods have been used to control or abate their effects on the environment. On this ground, Tomatis et al. (2016) investigated in a review the catalytic removal of VOCs like xylene, ethylbenzene, toluene, and benzene from combustible flue gas. The authors recounted that the improvement of the catalytic combustor of volatile organic compounds over its likes like dual-functional adsorbent-catalysts, spinel catalysts, perovskite catalysts, non-noble, and noble metal catalysts are tenable because of their efficiency, accuracy, low cost, and rapid results. The authors stated that the nature of the particle sizes and elements aided in the support of their acidic properties and porosity, which is a basis for the performance and catalytic influence. It was noticed that water assisted the oxidation of the VOCs that caused high reactivity and hydrophilia. The authors recommend the catalytic oxidation process as an integral amalgam to boost the mechanisms of the process involved in the abatement procedures.

Meena et al. (2021) did a review of the biological strategies used in the decontamination of HMs and VOCs from wastewater. The authors stated that industrial and population explosions have resulted in the deterioration of the quality of water. In addition, inorganic and organic compounds from point source discharge have also had a share in the pollution of water bodies. The presence of HMs and VOCs in concentration in water and air have been linked to certain negative health and ecological effects. However, there is a need to manage the possible risks HMs and VOCs portend to humans and their immediate environment by the application of biofiltration, which is a modern biotechnology that is cheap, non-expensive, green, and sustainable. This technology can be used to remove HMs, VOCs, bacteria, fungi, algae, and plants from contaminated water or wastewater using the principles of biotrickling and filtration.

Cheng et al. (2018) evaluated in a biological study the biodegradation of Naphthol Green B with strain CF10-13 (*Pseudoalteromonas*) bacterium and the production of Fe-nanoparticles. The authors recounted that dye wastes consist of complex toxic metals that have the potential to cause health and environmental problems. The removal of this contaminant from dye using traditional techniques has become a serious problem in recent times because of the high cost of remediation and high resistance to degradation. The use of strain CF10-13 (*Pseudoalteromonas*), a marine bacterium, in the removal of complex metals in the dye was effective because of the metabolites found in the strains. There was total degradation and decolourisation of the dye via cellular transfer. Naphthalenesulfonate, the main molecular structure of the dye, was degraded to the less noxious compound benzamide. There was the formation of a black Fe-sulfur nanomaterial that is stable. The formation of sludge and metal in the system as a result of H_2S discharge from the black Fe-sulfur nanomaterial was averted using the biological degradation technique to avoid inactiveness and unforeseen damage. It can be concluded from this study that this method of metal removal is a promising vehicle for future decontamination of higher refractory pollutants. So, it is highly recommended because it is cheap and eco-friendly.

On the verge of providing an efficient and sustainable method of pollutant removal from wastewater, Yang et al. (2019) in a review examined the removal of HMs from wastewater using sustainable nano-based products. The authors stated that the decontamination of pollutants by traditional methods has been linked to inefficiencies and non-sustainability. With this, the nanomaterials like nanocomposites, nano metal oxide, zerovalent metals, and carbon-based nanomaterials have been long applied for effective removal of waste from water. Furthermore, this technology serves as a promising tool for other environmental clean-up use. Nanomaterials show a high advantage as adsorbents in the decontamination of metals. Nonetheless, there is still a need to improve on some bottlenecks of their utilisation. First, nanomaterials are well known for their unstable nature. This makes them aggregate, hence decreasing their decontamination ability. Also, it is cumbersome to separate the molecules of nanomaterials from their aqueous form efficiently and swiftly because of the size of their nanoscale. This needs more investigation in solving the issues. Second, the nanomaterials employed commercially for heavy-metal decontaminations are rare and expensive. The cost of production and the synthesis should be improved to meet the demands for their use. Last, the increase in the utilisation of nanomaterials in the management of wastewater can increase their toxicities to humans and the environment. Hence, care should be taken in their applications. There is also the need for the regulation of nanomaterials to avoid and reduce possible environmental and health impacts. More so, the cost, synthesis, separation, reusability, and capability of the deployment of nanomaterials for wastewater treatment should also be looked into for better management of waste and a sustainable water environment.

Huang and Keller (2020) evaluated the decontamination of HMs in soil and sediments using Ligand-DNPs (dense ligand-coated nanoparticles) as a remediation tool. The authors stated that HMs such as lead and cadmium have longed been known as toxic and carcinogenic metals that are recalcitrant in the environment in which they find themselves. The re-suspension of soil and sediment region in water can expose an aquatic organism to heavy metals that can bioconcentrate, bioaccumulate, and biomagnify in them and may cause a serious biological threat to the health of humans who are at the top of the food chain structure. The Ligand-DNPs were able to decontaminate 80 and 60% of lead and cadmium in the soil and sediment matrices through a deeper driven force of the ions of the metals, thus making them sustainably bioavailable. The decontamination happened a few minutes after the application of the dense ligand-coated nanoparticles. It can be concluded that the dense ligand-coated nanoparticles provided efficient, cheap, convenient, and swift removal of the concentration of Pb and Cd in the contaminated soils and sediment effectively.

Shareefdeen and Janjani (2017) analysed the utilisation of ASBF (air stripper–biofilter system) in the removal of VOCs from the waste treatment facility. The VOCs examined were styrene, xylene, toluene, and benzene. Their impacts on the aquatic environment were analysed. Effects of the biofiltration using biodegradation and air stripping processes should be that ASBF have high efficiency in the wastewater clean-up process without leaving any VOCs waste in the decontaminated medium in various operating settings.

12.3 THE MECHANISM OF BIO-NANO FILLERS FOR THE REMOVAL OF HMS AND VOCS

There are various mechanisms involved in the use of bio-nanomaterials during the removal of various volatile organic contaminants and heavy metals from different media within the environment. The different nanoparticles and fillers such as quantum dots, graphenes, and carbon nanotubes can be integrated with other biological processes during the application in remediation. Remediation of contaminants using various categories of nanofillers is dependent on various mechanisms that include absorption, adsorption, filtrations, photocatalysis, and chemical reactions (Anjum et al. 2019).

The mechanism of the removal of these contaminants is affected by the potential of the contaminants to desorb, as well as to ease of recovery of the nanofillers. Sustainable utilisation of a bio nanofiller is also dependent on the aggregation of the nanomaterials and the toxicological effects.

The incessant research in the area of wastewater remediation and treatments has brought about an improvement of the knowledge of the sorption mechanism and its role in enhancing sorbent properties. The use of bio nanofillers in the treatment of conventional water during which volatile matter and inorganic contaminants are removed is enhanced through the sorption mechanisms. One of the vital mechanisms involved in this is the sorption process. Sorption involves absorption and adsorption. In adsorption, the association existing between sorbent and the contaminant is vital, and this occurs at the surface level (Saleem and Zaidi 2020). Several factors affect the association with the nanofillers as well as some microorganisms used in some processes. Some events occur during the interaction between the nanofillers and biota such as absorption, dissolution, and biotransformation. Some of these processes play a vital role in the degradation of pollutants. Adsorption, which is an important mechanism involved in the removal of the contaminants using the nanofiller, is further divided into physisorption and chemisorptions. In chemisorptions, a chemical process is involved while in physisorption physical forces are paramount. Irrespective of this, the pollutants are sequestered, immobilised, and concentrated. Different phenomena come into play due to the chemical, physical, and biochemical associations among the nanoparticles, contaminants, and microorganisms during nano bioremediation (Agboola and Benson 2021). Thus for effective comprehension of the mechanisms involved in the use of bio nanofillers for removal of heavy metals and volatile organic contaminants, kinetic, thermodynamic, and mechanistic investigations are paramount in giving a detailed description of the behaviour of the nanofiller on contact with the pollutants. In some studies, researchers put forward varying models for a description of the behaviour including the biological and chemical phenomena during the process of adsorption. Some of the models that have been used in understanding the mechanisms of the removal processes using the nanoparticles include Temkin isotherms, the Langmuir model, the Freundlich model, and Radushkevich models. Depending on the peculiar nature of the nanofillers, pollutants could be degraded through photocatalytic processes. The resulting product could be further transformed biologically through the biotic systems and the concentration of the contaminants is reduced within the media (Aslam et al. 2021).

Another mechanism involved during the process of adsorption on the surface of the nanofiller is the coordination of the heavy-metal ions and other contaminants above a particular concentration bringing about precipitation of the carbon material thereby reducing the metal content present in the solution (Kumar et al. 2021).

12.4 ENVIRONMENTAL AND HEALTH IMPACTS OF HEAVY METALS AND VOCS FROM WASTEWATER

Yang et al. (2014) evaluated the possible health hazards from the removal of three aromatic congener compounds (xylenes, toluene, and benzene) and four VOCs (tetrachloroethylene, trichloroethylene, carbon tetrachloride, and chloroform) from a wastewater treatment facility in China. The authors stated that the negative health influence of the four VOCs on public health is of great concern because of the attendant issues like cancer and pulmonary complications they portend. The Industrial Source Complex (Gaussian plume model) was employed to excite the aerial activities of the VOCs discharged from the wastewater treatment facility. It was noted that three aromatic compounds, especially benzene, were discharged from the wastewater in the surrounding air. Meanwhile, the VOCs exempting chloroform were found in the water layers during the treatment procedure. However, the basic clarifier was a technique used to release the VOCs into the surrounding air. The degrees of biodegradation or volatilisation of the VOCs waste were to remove by aeration and temperature by seasonal variations (physical) and chemical means to avoid potential risks. It was observed that the human population living about 4 km away was still exposed potentially to cancer risks that were exceeding the standard limits. The findings from this study revealed the recalcitrant nature of the emission of VOCs from the wastewater treatment facility. It was recommended here that an advanced biotechnology method that is green and sustainable should be employed for the effective clean-up of complex VOCs from wastewater to mitigate cancer risks.

Health issues related to cancer and pulmonary complications have been linked to the release of mutagenic and clastogenic pollutants into the breathing zones of humans. In the quest to investigate the possible health risks from potential chemicals, Shuai et al. (2018) examined the health hazard evaluation of VOC exposure released from a dye industry in South Korea. The authors opined that possible noxious chemicals produced by dye manufacturing industries have the potentials to carry emergent toxicants. The passive samplers were used to sample the VOCs from different areas of interest. The deterministic hazard evaluation and lifetime cancer hazard of inhalation pathways were used to assess the exposure of VOCs. The results showed that the concentration of the volatile organic compounds was found more in complex areas per site. Toluene (aromatic compound) and chloroform were found highly significant compared to their corresponding counterparts exempting the controlled sites both at autumn and summer. More so, the outdoor and indoor background concentrations of toluene and chloroform showed positive correlations even when exposed to individuals respectively. The hazard quotient and lifetime cancer hazard were observed to be > 1 and $> 10^{-4}$ with possible diseases like cardiovascular complications, anaphylactic sickness, and respiratory disorderliness. The findings of this study showed serious non-carcinogenic and carcinogenic hazard health influence from VOCs emitted from the dye industrial facility. It is recommended that hazard management procedures should be put in place to curtail the preliminary control of waste from industrial activities.

Tchounwou et al. (2012) evaluated the impact of metals on the environment and human health. The authors recounted that heavy metals are noxious to the environment and humans. Heavy metals can be released by agricultural, domestic, and industrial activities occasioned by humans. The potential toxic effects of heavy metals on humans and the environment depend on the chemical species, route of transfer, exposure, and dosage. In humans, the noxiousness of heavy metals depends on the nutritional position, genetics, gender, and age of the individuals exposed to the metals. Heavy metals like mercury, lead, chromium, cadmium, and arsenic are ranked as the primary metals with public health concern because of their ability to cause organ and system damages even at low concentrations of exposure as well as carcinogenicity, genotoxicity, and noxiousness. In elucidating their mode of action, it is important to evaluate and control metal emission into the environment because of the possible health impact they portend.

Akpor (2014) investigated the sources, remediation, and impact of HMs emitted from wastewater. Heavy metals are one of the recalcitrant chemical pollutants found in the environment. HMs exposure to humans can pass via different routes like inhalation (dust), ingestion (food and water), and vaporisation. Some adverse influences of HMs are long-term noxiousness, increased flow of water, habitation obliteration from sedimentation, algal blooms, and death of water biotas. It can also reduce the quality of the environment they find themselves in in concentrations above threshold standards. Possible health effects from heavy metals to plants and animals are death, organ and nervous system damages, cancer, physiological and metabolic alterations, poor development, and growth. To mitigate this the authors suggested both biological remediation using plants and microbes in treating the toxic metals in the wastewater. This will reduce negative damages the metals portend.

12.5 DERIVED BENEFITS FROM THE UTILISATION OF BIO-NANO FILLERS OVER OTHER CONVENTIONAL METHODS

The use of bio-nanofillers is paramount because the material employed in the process of remediation must not in itself constitute harm when used. The use of biodegradable nanomaterials also provides a safer and greener alternative aside from increasing the confidence of consumers. The use of nanotechnology with nanofillers as substrate can also help in resolving most of the limitations associated with conventional approaches. The development of a method for remediation must take into consideration cost-effectiveness, non-toxicity, reusability, green chemistry, and the possibility of recovery without generation of waste. Nanotechnology using nanofillers is reliable in this regard (Guerra et al. 2018). The use of bio nanomaterial is also advantageous because of the potential toxicity of metal nanoparticles that has been raised by some experts. For metallic nanoparticles, there is also a need for increasing their stability, prevention of their agglomeration, and aiding monodispersity.

The combinations of various nanomaterials with biotechnological-based approaches enhance effective remediation of various volatile compounds and heavy metals due to the inherent advantages in comparison to the conventional techniques. The use of bio-nano materials proffers a cheap cost and can be modified to suit the removal of a range of metals and organic contaminants (Baig et al. 2021).

There are unique benefits associated with the use of nanofillers for remediation purposes that include improvement of the extent and ease of the removal of the targeted contaminant, aiding an extension of the treatable range of the contaminants, the enhancing effect of ease in surface modification, and integration for composite formation, amongst others. Some nanofillers are also used in integration with some microorganisms to give rise to a possible synergistic impact.

The chemical and physical associations between the nanofillers and contaminants are dependent on various environmental factors such as pH, stability of the nanomaterial, the surface morphology of the nanofillers, nature of the nanoparticles, presence of surface coatings, and temperature amongst others. As a result of the various parameters that are capable of affecting the adsorption processes of these contaminants into the nanofiller materials, the mechanisms, therefore, involve different complex processes.

The use of nanofillers as nanomaterials in the remediation of heavy metals and organic contaminants is connected with several advantages due to the properties of the nanomaterials. The nanofillers are easily modified and integrated with other materials to form composites and polymer composites. Most of the nanofillers are either of inorganic or organic origin. Particles such as quantum dots, nanosilica, and titanium dioxide are inorganic in nature while others such as the coir nanofiller and cellulosic nanofiller are organic in nature (Aslam et al. 2021).

Carbon nanotubes and graphenes have been utilised as the most vital and common nanofillers that are carbon-based. The carbon nanofillers have unique characteristics and surface properties that help in improving their adsorption properties for the contaminants. Their properties are also enhanced during the formation of composites such as polystyrene carbon nanotubes and modified graphene for enhancing their properties. Their integration during the formation of composites further broadens their applicability for remediation of various heavy metals and organic contaminants (Mittal et al. 2013).

Apart from the possibility of recycling and reusability, other advantages of using bio nanofillers for remediation are their high adsorption potentials and numerous surface active sites with a significantly larger surface-area-to-volume ratio as well as the high rate of intraparticle diffusion. It is also easy to integrate and incorporate other materials into the nanofillers for more efficient removal, which further gives it the upper hand in comparison with conventional methods. Also, the use of nanofibrous materials that are non-woven that are prepared via the use of exchange concept makes the surfaces act as a resin for ion exchange possessing different functional groups that can collect various undesired volatile organic contaminants and heavy metals through ion exchange mechanism (Yang et al. 2019).

Aji et al. (2018) in their study investigated the use of carbon nanotubes for the removal of nickel ions present in the wastewater. The carbon nanotubes were generated from frying oil through the hydrothermal synthetic method at 300C° for a duration of 2 hs. The findings from the study revealed that the carbon dots have high potential for removing the heavy metal of interest from the wastewater. The electric current within the solution brought about a decrease in the number of carbon nanotubes and the time taken for the removal of the metal.

In a related study, Ciotta et al. (2017) produced a two-dimensional nanofiller through the use of oxidative unfolding of fullerene and revealed the sensitivity selectively of the nanoparticles to the lead and copper ions present.

Jlassi et al. (2020) in a related study prepared a hybrid nanocomposite of chitosan hydrogen as a membrane platform for the removal of cadmium ions from contaminated wastewater. The membrane was shown to possess unique mechanical parameters, which facilitated its ease of handling. Equilibrium in the adsorption process was attained within a period of 5 mins. The removal efficiency of the ion of interest was aided by the UV light illumination. The adsorption process was observed to follow the pseudo form of second-order kinetic and fit effectively into the Langmuir isotherm.

Jaiswal et al. (2012) in their work utilised a biopolymer stabilised quantum dot of zinc sulfide for exchange reactions for a view to eliminating toxic metal ions such as mercury, lead, and silver from polluted water. The zinc oxide quantum dot stabilized using chitosan was prepared in aqueous media and the conversion into their various sulphides was monitored. The quantum dots transformed were characterized by their unique colour formation with mercury showing its unique bright yellow coloration while the other ions possessed a brown colour. The process of cation exchange was observed to be enhanced by their disparity in solubility product.

12.6 CONCLUSION AND THE WAY FORWARD

This chapter discussed the methods in the removal of VOCs and HMs from industrial waste effluent using different methods like bio-nano fillers, biosorption, bioscrubbers, biotrickling filters, microorganisms, biofilters, and biofiltration. Heavy metals and VOCs removal were noticed to have zeolites loaded with refined structures of both extra-framework and lattice contents endorsing the immobilisation of the pollutants in the micro-porosities framework. The presence of chemical species was also validated by the thermal analysis curves that revealed various occurrences on the level of their numbers and nature of the species of the extra-framework hosted in the microsporic zeolite.

The role of nanomaterials as metal absorbers was also used as a tracking system because of the magnetic, optical, and electrical properties, which they use to amplify optical and electrical signals as transducers for heavy metals detection and abatement. Heavy metals can be removed from contaminant like dye using traditional techniques. However, this has become a serious problem in recent times because of the high cost of remediation and high resistance to degradation. But the use of strain CF10-13 (*Pseudoalteromonas*), a marine bacterium, in the removal of complex metals in the dye showed effectiveness because of the metabolites found in the strains. It was noticed that there was total degradation and decolourisation of the dye via cellular transfer. Naphthalenesulfonate, the main molecular structure of the dye, was degraded to the less noxious compound benzamide. There was formation of a black Fe-sulfur nanomaterial that is stable. This compound was noticed to avoid discharging H_2S endogenously; because of sludge formation and metal accumulation in the system it portends to result in system inactiveness. Remediation of contaminants using various categories of nanofillers is dependent on various mechanisms that include absorption, adsorption, filtrations, photocatalysis, and chemical reactions. Sorption involves absorption and adsorption. In adsorption, the association existing between sorbent and the contaminant is vital, and this occurs at surface level.

Thus for effective comprehension of the mechanisms involved in the use of bio nanofillers for removal of heavy metals and volatile organic contaminants, kinetic, thermodynamic, and mechanistic investigations are paramount in giving a detailed description of the behaviour of the nanofiller on contact with the pollutants. The Industrial Source Complex (Gaussian plume model) was employed to excite the aerial activities of the VOCs discharged from the wastewater treatment facility based on records from previous studies. The use of nanotechnology with nanofillers as substrate can also help in resolving most of the limitations associated with conventional approaches. The development of a method for remediation must take into consideration cost-effectiveness, non-toxicity, reusability, green chemistry, and the possibility of recovery without generation of waste.

In humans, the noxiousness of heavy metals depends on the nutritional position, genetics, gender, and age of the individuals exposed to the metals. Meanwhile diseases like cardiovascular complications, anaphylactic sickness, and respiratory disorderliness are associated with VOCs effects. Health issues related to cancer and pulmonary complications have been linked to the release of mutagenic and clastogenic pollutants into the breathing zones of humans when exposed to HMs. The deterministic hazard evaluation and lifetime cancer hazard of inhalation pathways were used to assess the exposure of VOCs and HMs in wastewater.

It is recommended here that an advanced biotechnology method that is green and sustainable should be employed for the effective clean-up of complex VOCs and HMs from wastewater to mitigate possible cancer risks.

REFERENCES

Acheampong M.A., Meulepas R.J. and Lens P.N. (2010). Removal of heavy metals and cyanide from gold mine wastewater. *J Chem Technol Biotechnol*. 85:590–613.

Agboola O.D. and Benson N.U. (2021). Physisorption and chemisorption mechanisms influencing micro (nano) plastics-organic chemical contaminants interactions: A review. *Front Environ Sci*. 9:678574. https://doi.org/10.3389/fenvs.2021.678574.

Aji M.P., Wati A.L., Priyanto A., Karunawan J., Nuryadin B.W., Wibowo, E. and Sulhadi MP. (2018). Polymer carbon dots from plastics waste upcycling. *Environ Nanotechnol Monit Manag*. 9:136–140. https://doi.org/10.1016/j.enmm.2018.01.003.

Akpor O.B. (2014). Heavy metal pollutants in wastewater effluents: Sources, effects and remediation. *Adv Biosci Bioeng.* 2(4):37. https://doi.org/10.11648/j.abb.20140204.11.

Akpor O.B., Ohiobor G.O. and Olaolu D.T. (2014). Heavy metal pollutants in wastewater effluents: Sources, effects and remediation. *Adv Biosci Bioeng.* 2:37–43.

Alrumman S., Keshk S. and El-Kott A. (2016). Water pollution: Source and treatment. *Am J Environ Eng.* 6:88–98.

Anani O.A. and Olomukoro J.O. (2017). The evaluation of heavy metal load in benthic sediment using some pollution indices in Ossiomo River, Benin City, Nigeria. *Funai J Sc Tech.* 3(2):103–119.

Anani O.A. and Olomukoro J.O. (2018a). Trace metal residues in a tropical Watercourse sediment in Nigeria: Health risk implications. *IOP Conference Series: Earth Environ Sc.* 210:012005.

Anani O.A. and Olomukoro J.O. (2018b). Health risk from the consumption of freshwater prawn and crab exposed to heavy metals in a Tropical River, Southern Nigeria. *J Heavy Metal Tox Dis.* 3(2):5. https://doi.org/10.21767/2473-6457.10024.

Anani O.A. and Olomukoro J.O. (2020). Examination and screening of the perceived toxicity influence of produced wastewater using *Allium cepa* assay and health risk model as monitoring tools. *Arch Sci Techn.* 1:33–44.

Anani O.A., Olomukoro J.O. and Ezenwa I.M. (2020). Limnological evaluation in terms of water quality of Ossiomo River, Southern Nigeria. *Int J Cons Sci.* 11(2):571–588.

Anjum R., Afzal M., Baber R., Khan M.A.J., Kanwal W., Sajid W. and Raheel A. (2019). Endophytes: as potential biocontrol agent—review and future prospects. *J Agric Sci.* 11:113.

Arfè M.A., Martucci A., Pasti L., Chenet T., Sarti E., Vergine G. and Belviso C. (2020). Evaluation for the removal efficiency of VOCs and heavy metals by zeolites-based materials in the wastewater: A case study in the tito scalo industrial area maura. *Processes.* 8:1519. https://doi.org/10.3390/pr8111519.

Aslam I., Afridi M.S.K., Siddique W., Afridi M.S.K., Peerzada S., Ishtiaq S., Kamran S.H., Fahham H.H. and Shehzadi N. (2021). Antioxidant, α-amylase inhibition, DNA protection and antidepressant activity of Caralluma edulis extracts. *Biosc Res.* 18(2):1498–1509.

Baig N., Kammakakam I. and Falath W. (2021). Nanomaterials: a review of synthesis methods, properties, recent progress, and challenges. *Mat Adv.* 2(6):1821–1871. https://doi.org/10.1039/d0ma00807a.

Barupal T., Meena M. and Sharma K. (2020). A study on preventive effects of *Lawsonia inermis L.* bioformulations against leaf spot disease of maize. *Biocatal Agric Biotechnol.* 23:101473. https://doi.org/10.1016/j.bcab.2019.101473.

Cheng S., Li N., Jiang L., Li Y., Xu B. and Zhou W. (2018). Biodegradation of metal complex Naphthol Green B and formation of iron-sulfur nanoparticles by marine bacterium *Pseudo alteromonas.* sp CF10-13. *Bioresource Technol.* https://doi.org/10.1016/j.biortech.2018.10.082.

Ciotta E., Paoloni S., Richetta M., Prosposito P., Tagliatesta P., Lorecchio C. and Pizzoferrato R. (2017). Sensitivity to heavy-metal ions of unfolded fullerene quantum dots. *Sensors.* 17(11):2614. https://doi.org/10.3390/s1711261410.3390.

Enuneku A.A., Mohammed O.P., Asemota O.C. and Anani O.A. (2018). Evaluation of health risk concerns of trace metals in borehole water proximal to dumpsites in Benin City, Nigeria. *J Appl Sci Environ Manage.* 22(9):1421–1425. https://doi.org/10.4314/jasem.v22i9.10.

Guerra F.D, Attia M.F. Whitehead D.C. and Alexis F. (2018). Nanotechnology for environmental remediation: Materials and applications. *Molecules.* 23:1760. https://doi.org/10.3390/molecules23071760.

Huang Y. and Keller A.A. (2020). Remediation of heavy metal contamination of sediments and soils using ligand-coated dense nanoparticles. *PLoS ONE* 15(9):e0239137. https://doi.org/10.1371/journal.pone.0239137.

Jaiswal A., Ghsoh S.S. and Chattopadhyay A. (2012). Quantum dot impregnated-chitosan film for heavy metal ion sensing and removal. *Langmuir.* 28:15687–15696. https://doi.org/10.1021/la3027573.

Jlassi K., Eid K., Sliem M.H., Abdullah A.M., Chehimi M.M. and Krupa I. (2020). Rational synthesis, characterization, and application of environmentally friendly (polymer–carbon dot) hybrid composite film for fast and efficient UV-assisted Cd2+ removal from water. *Environ Sci Eur.* 32(1). https://doi.org/10.1186/s12302-020-0292-z.

Kumar A., Howard C.J., Derrick D., Malkina I.L., Mitloehner F.M., Kleeman M.J., Alaimo C.P., Flocchin R.G. and Green, P.G. (2011). Determination of volatile organic compound emissions and ozone formation from spraying solvent-based pesticides. *J Environ Qual.* 40(5):1423. http://doi.org/10.2134/jeq2009.0495.

Kumar R., Rauwel P. and Rauwel E. (2021). Nanoadsorbants for the removal of heavy metals from contaminated water: Current scenario and future directions. *Processes.* 9:1379. https://doi.org/10.3390/pr9081379.

Kurniawan T.A., Chan G.Y.S., Lo W.H. and Babel S. (2006). Physico-chemical treatment techniques for wastewater laden with heavy metals. *Chem Eng J.* 118:83–98.

Liang Y., Liu X., Wu F., Guo Y., Fan X. and Xiao H. (2020). The year-round variations of VOC mixing ratios and their sources in Kuytun City (Northwestern China), near oilfields. *Atmos Pollut Res.* 11(9):1513–1523. https://doi.org/10.1016/j.apr.2020.05.022.

Lomonaco T., Manco E., Corti A., La Nasa J., Ghimenti S., Biagini D. and Castelvetro V. (2020). Release of harmful volatile organic compounds (VOCs) from photo-degraded plastic debris: A neglected source of environmental pollution. *J Hazard Mater.* 394:122596. https://doi.org/10.1016/j.jhazmat.2020.122596.

Malakar S., Saha P.D., Baskaran D. and Rajamanickam R. (2017). Comparative study of biofiltration process for treatment of VOCs emission from petroleum refinery wastewater—a review. *Environ Technol Innov.* 8:441–461.

Meena M., Sonigra P. and Yadav G. (2020). Biological-based methods for the removal of volatile organic compounds (VOCs) and heavy metals. *Environ Sci Poll Res.* https://doi.org/10.1007/s11356-020-11112-4.

Meena M., Sonigra P. and Yadav G. (2021). Biological-based methods for the removal of volatile organic compounds (VOCs) and heavy metals. *Environ Sci Pollut Res Int.* 28(3):2485–2508. http://doi.org/10.1007/s11356-020-11112-4.

Mehta S.K. and Gaur J.P. (2005). Use of algae for removing heavy metal ions from wastewater: Progress and prospects. *Crit Rev Biotechn.* 25:113–152. https://doi.org/10.1080/07388550500248571.

Mittal A.K., Chisti Y. and Banerjee U.C. (2013). Synthesis of metallic nanoparticles using plant extracts. *Biotechnol Adv.* 31:346–356. https://doi.org/10.1016/j.biotechadv.2013.01.003.

Olatunji E.O. and Anani O.A. (2020). Bacteriological and physicochemical evaluation of River Ela, Edo State Nigeria: Water quality and perceived community health concerns. *J Bio Innov.* 9(5):736–749. https://doi.org/10.46344/JBINO.2020.v09i05.09.

Padalkar A.V. and Kumar R. (2018). Removal mechanisms of volatile organic compounds (VOCs) from effluent of common effluent treatment plant (CETP). *Chemosphere.* 199:569–584. http://doi.org/10.1016/j.chemosphere.2018.01.059.

Qu K.C., Li H.Q., Tang K.K., Wang Z.Y. and Fan R.F. (2020). Selenium mitigates cadmium-induced adverse effects on trace elements and amino acids profiles in chicken pectoral muscles. *Biol Trace Elem Res.* 193:234–240. https://doi.org/10.1007/s12011-019-01682-x.

Saleem H. and Zaidi S.J. (2020). Developments in the application of nanomaterials for water treatment and their impact on the environment. *Nanomaterials.* 10(9):1764. doi:10.3390/nano10091764.

Salek M.A., Hassani S., Mirnia K. and Abdollahi M. (2021). Recent advances in nanotechnology-based biosensors development for detection of arsenic, lead, mercury, and cadmium. *Inter J Nanomedicine.* 16:803–832. https://doi.org/10.2147/IJN.S294417.

Shah M.P. (2020). *Advanced Oxidation Processes for Effluent Treatment Plants.* Elsevier.

Shah M.P. (2021). *Removal of Emerging Contaminants through Microbial Processes.* Springer.

Shareefdeen Z. and Janjani S. (2017). A theoretical analysis of an air stripper–biofilter system (ASBF) for industrial wastewater treatment. *Desalination Water Treatment.* 100:268–274. http://doi.org/10.5004/dwt.2017.21343.

Shuai J., Kim S., Ryu H., et al. (2018). Health risk assessment of volatile organic compounds exposure near Daegu dyeing industrial complex in South Korea. *BMC Pub Heal.* 18:528. https://doi.org/10.1186/s12889-018-5454-1.

Srivastava N.K. and Majumder C.B. (2008). Novel biofiltration methods for the treatment of heavy metals from industrial wastewater. *J Hazard Mater.* 151:1–8.

Tchounwou P.B., Yedjou C.G., Patlolla A.K. and Sutton D.J. (2012). Heavy metals toxicity and the environment. *EXS.* 101:133–164. http://doi.org/10.1007/978-3-7643-8340-4_6.

Tiwari A.K., Alam T., Kumar A. and Shukla A.K. (2019). Control of odour, volatile organic compounds (VOCs) and toxic gases through biofiltration—an overview. *Int J Tech Innov Mod Eng Sci.* 5:1–6.

Tiwari A.K., Singh P.K., Singh A.K. and DeMaio M. (2016). Estimation of heavy metal contamination in groundwater and development of a heavy metal pollution index by using GIS technique. *Bull Environ Contam Toxicol.* 96:508–515. https://doi.org/10.1007/s00128-016-1750-6.

Tomatis M., Xu H.-H., He J. and Zhang X.-D. (2016). Recent development of catalysts for removal of volatile organic compounds in flue gas by combustion: A review. *J Chem*. 2016: 8324826,15.

Yang J., Hou B., Wang J., Tian B., Bi J., Wang N., Li X. and Huang X. (2019). Nanomaterials for the removal of heavy metals from wastewater. *Nanomaterials (Basel)*. 9(3):424. http://doi.org/10.3390/nano9030424.

Yang J., Wang K., Zhao Q., Huang L., Yuan C-.S., Chen W-.H. and Yang W-.B. (2014). Underestimated public health risks caused by overestimated VOC removal in wastewater treatment processes. *Environ Sci Process Impacts*. 16(2):271–279. http://doi.org/10.1039/c3em00487b.

Yang K., Wang C., Xue S., Li W., Liu J. and Li L. (2019). The identification, health risks and olfactory effects assessment of VOCs released from the wastewater storage tank in a pesticide plant. *Ecotoxicol Environ Saf.* 184:109665. https://doi.org/10.1016/j.ecoenv.2019.109665.

Zhao L., Huang S. amd Wei Z. (2014). A demonstration of biofiltration for VOC removal in petrochemical industries. *Environ Sci Process Impacts*. 16:1001–1007.

Zhou Q., Yang N., Li Y., Ren B., Ding X., Bian H. and Yao X. (2020). Total concentrations and sources of heavy metal pollution in global river and lake water bodies from 1972 to 2017. *Glob Ecol Conserv*. 22:e00925. https://doi.org/10.1016/j.gecco.2020.e00925.